ATMOSPHERIC PHENOMENA

Readings from

SCIENTIFIC AMERICAN

ATMOSPHERIC PHENOMENA

With Introductions by

David K. Lynch

Hughes Research Laboratories

W. H. Freeman and Company

San Francisco

Some of the *Scientific American* articles in *Atmospheric Phenomena* are available as separate Offprints. For a complete list of articles now available as Offprints, write to W. H. Freeman and Company, 660 Market Street, San Francisco, California 94104.

Library of Congress Cataloging in Publication Data

Main entry under title:

Atmospheric phenomena: readings from Scientific American.

Bibliography: p.
Includes index.
1. Meteorological optics—Addresses, essays, lectures.
I. Lynch, David K., 1946– II. Scientific American.
QC975.6.A85 551.5′6 79-26987
ISBN 0-7167-1165-6
ISBN 0-7167-1166-4 pbk.

Printed in the United States of America

9 8 7 6 5 4 3 2 1

PREFACE

Our deepest source of inspiration is the sky. From the cool detachment of gazing up at the stars to the terrifying involvement of being caught in a lightning storm, we have a natural curiosity about things overhead. Ancient and modern religions speak of wonders from above, and every culture has myths based on the appearance of rainbows, halos, or mirages.

The sky and its treasures were around long before humans were there to see them. Even before our species unwittingly stumbled into its place as the dominant creature on earth, intelligent monkeys were aware of the weather and its importance to their well-being. Gathering clouds and falling snow did not pass unheeded without serious consequences. Survival depended on understanding, interpreting, and reacting to the weather. As often as not, lightning and thunder were regarded with fear; yet the appearance of a ghostly rainbow amid the ragged chaos of a thundercloud must surely have given some rain-soaked Neanderthal, dashing to safety at the break of a storm, reason for pause. And it is not hard to imagine that late at night, weary from the struggle to survive, this early human may have looked up at the heavens and wondered about things too big for words. While gazing at some spectacle in the sky, I have often felt that some primitive cousin, now long entombed in the heavy sediments of time, was somehow looking over my shoulder, whispering, "I've wondered that myself."

The history of atmospheric phenomena and meteorological optics goes back thousands of years, and references to rainbows and halos appear in every ancient literature. By Newton's time, the gross features of many apparitions had been explained. The rainbow remained an isolated curiosity until 1803, when Thomas Young used the supernumerary rainbows to support his theory of interference. This event put meteorological optics before a much wider audience, and by the middle of the nineteenth century, partly as a result of widely expanded farming in the United States, everyone from farmers to politicians was contributing to the swelling volume of observations. Newspapers, professional journals, and trade publications bristled with exciting accounts of lights and colors in the sky. This period was largely empirical, with little in-depth theoretical work. Weather-related occurrences, such as halos and dew formation, and their relevance to weather prediction were also emphasized, as more and more people on farms became aware of the sky. During no other period in human history were so many people so aware of the daily changes overhead. If ever there was a golden era in the study of atmospheric phenomena, especially optical phenomena, this was it.

In the early years after the turn of the century, interest in meteorological optics waned, for two reasons. First, because of corporate farming and the Industrial Revolution, millions of people from the rural countryside moved to the cities. The once vital need for up-to-the-minute weather information faded, and people lost interest in the sky. City lights made it increasingly difficult even to see the twinkling stars, and the delicate auroras danced to ever dwindling audiences.

The second reason was scientific. Physics, a large part of which had been devoted to classical optics, including meteorological optics, was suddenly dominated by two new concepts: quantum mechanics and relativity. Most physicists turned their efforts toward understanding the universe in these new and fascinating contexts. Without a doubt, this work contributed to our knowledge in ways that meteorological optics never could, and, as a result, our understanding of the world is more complete than ever before. Yet the new physics was not without casualties, as fewer and fewer scientists found time to get away from the buzzing laboratory and climb a hill to gaze at the sky.

One man, however, was so charmed and fascinated by the rich variety of optical phenomena around him that he devoted a part of every day to their study. Braving two world wars, Nazi incarceration, and flight to Holland, Marcel Minneart had careers in both biology and astronomy and he contributed to many fields of science. He is best known and loved for a series of books written from 1937 to 1940, one of which was translated into many languages, including English. This book is still in print under the title *The Nature of Light and Colour in the Open Air*, and it remains the best book on the subject. Minnaert defined the subject and inspired us all. In the preface to his book, Minnaert wrote:

> A lover of Nature responds to her phenomena as naturally as he breathes and lives, driven by a deep innate force. Sun and rain, heat and cold are alike welcome to his observation; in towns and woods, on sandy tracts and on the sea he finds new objects of interest. Each moment he is struck by new and interesting occurrences. With buoyant step he wanders over the countryside, eyes and ears alert, sensitive to the subtle influences that surround him, inhaling deeply the scented air, aware of every change of temperature, here and there lightly touching a shrub to feel in closer contact with the things of the earth, a human being supremely conscious of the fullness of life.
>
> It is indeed wrong to think that the poetry of Nature's moods in all their infinite variety is lost on one who observes them scientifically, for the habit of observation refines our sense of beauty and adds a brighter hue to the richly coloured background against which each separate fact is outlined.

I can find no more fitting words to describe the philosophy that we all share.

Meteorological optics is experiencing a long awaited rebirth. Although it is still a field with little practical application (and, consequently, little financial backing), there are enough interested scientists in the world to keep the subject alive and healthy. In August 1978, the scientific community reached a "critical mass." In cooperation with the American Meteorological Society, the Optical Society of America held a topical meeting—the first ever of its kind—devoted to meteorological optics. It was a splendid gathering, and the organizing committee had no trouble in persuading the Optical Society to publish the papers in a special issue of the *Journal of the Optical Society of America,* August 1979.

The topics discussed at this meeting are covered in this reader. Collected here are eighteen articles from *Scientific American* discussing atmospheric phenomena from the bottom of the atmosphere, where precipitation collects, to the ionosphere, where the auroras occur, and beyond.

Since 1976, I have been teaching a course at UCLA entitled "Color and Light in Nature." The students have varied backgrounds, although many are teachers, artists, photographers, and scientists. The most common remarks made by these eager people are "I had no idea that all these things were around me" and "Where can I learn more?" My best answer to this last question is *Atmospheric Phenomena.*

This reader is designed to acquaint one with the field of meteorological optics. The book is divided into two sections, the first providing background material and the second supplying explanations and discussions of atmospheric phenomena. Since it is self-contained, the reader fills two obvious needs. First, it is an ideal reference for anyone who has marveled at atmospheric phenomenona but who has never had a handy reference to the subject. Until now, the source material was scattered haphazardly thoughout the literature and was available only to those who had plenty of time and access to a major library. Second, *Atmospheric Phenomena* is an ideal supplement to many college texts on meteorology and earth science. Most texts do a good job of presenting the structure and dynamics of the earth's atmosphere, but few, if any, touch on the subjects covered here.

One might justify the study of atmospheric phenomena on the grounds that it is relevant to agriculture, communications, or national defense, but I'll use a simpler approach, one to which we can all respond and one that is self-justifying. The sky is absolutely fascinating, and a desire to experience its beauty fully is sufficient reason to explore it. This reader is for everyone who shares my love of the sky.

January 1980 David K. Lynch

Note on cross-references to Scientific American *articles:* Articles included in this book are referred to by title and page number; articles not included in this book but available as Offprints are referred to by title and offprint number; articles not included in this book and not available as Offprints are referred to by title and date of publication.

CONTENTS

The beholding of the light is itself a more excellent
and a fairer thing than all the uses of it.

Francis Bacon, *Novum Organum*

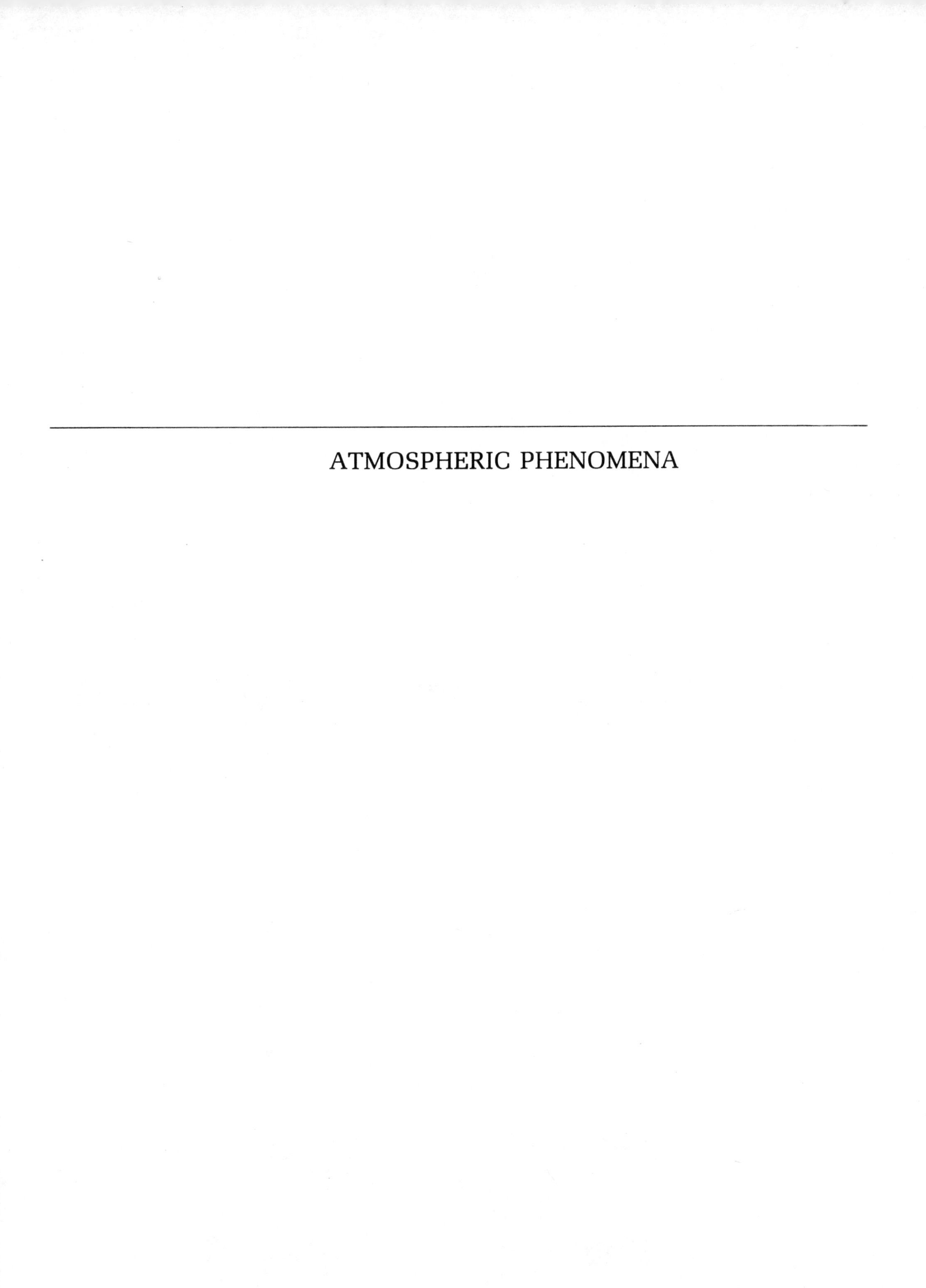

ATMOSPHERIC PHENOMENA

I

BASIC CONSIDERATIONS

BASIC CONSIDERATIONS I

INTRODUCTION

Our atmosphere is like no other in the solar system. Because much of its water vapor hangs in a delicate balance near its saturation point, large areas of the sky can change from totally transparent to nearly opaque in a few minutes, and an entire hemisphere may be cloud-free one day and almost completely obscured the next.

Most of the water vapor in the air is confined to the troposphere, the layer between sea level and the stratosphere. Seventy-five percent of the mass of the atmosphere is found here. Ranging in thickness from 16 kilometers at the equator to 8 kilometers or less at the poles, the troposphere is in constant motion, continually rearranging itself in an attempt to reach equilibrium. The forces that control the physical state of the air are sunlight, surface and air temperatures, water vapor concentration, and cloud cover. Since all four agents influence one another, the feedback interactions are complicated. Cause becomes effect, and vice versa. This never-ending activity is known as *weather*.

Atmospheric water is equal to only 0.1 percent of the total amount of water on the surface of the earth. Yet by acting as a valve to the incoming solar radiation and outgoing infrared radiation, it plays a decisive part in determining the structure of the troposphere and the conditions on the earth's surface. The source and distribution of water is described by the hydrologic cycle, which reveals that evaporation from the oceans accounts for 84 percent of the moisture in the atmosphere, the remainder coming from evapotranspiration (evaporation and transpiration from plants) on the land.

Thermal convection carries moisture from the surface, and the accompanying horizontal winds distribute it around the globe. The initial source of energy is solar radiation, which is stored as the heat of vaporization in water vapor. Air in contact with the warm ground heats and rises. As it ascends, it cools at the mean tropospheric lapse rate of about 6.5°C/km (3.6°F/1000 ft). When the pressure and temperature reach the right stage for the water vapor to condense, clouds form and release the heat of vaporization, which in turn heats the clouds, further promoting the convection. In this way, solar energy is transformed and transported thousands of miles from its origin. The troposphere thus resembles a giant heat engine, drawing power from the sun and driving the weather on the earth.

The violent convection that characterizes the troposphere reaches a maximum near 45,000 ft in the midlatitudes. Here the buoyant forces fade away as the air encounters a natural inversion layer—a stable layer of warm air. This is the *stratosphere*. Little or no moist air penetrates the boundary between the troposphere and the stratosphere (*tropopause*), except in tiny amounts near the equator, an event we will discuss later. Thus, the troposphere is a well-defined thermodynamic environment that does not interact much with the overlying layer of stable air, or stratosphere.

The stratosphere is so named because of its nearly continuous vertical stratification. It was once thought to be stable and nonmoving, but recent work has shown that there are considerable motions in the stratosphere, although not nearly as severe as those below. Between 15 and 50 km, the temperature rises and reaches a comfortable 0°C (32°F), warm compared with the frigid −60°C temperature found at the tropopause. At an altitude of about 22 km, the atmospheric pressure, which does not undergo an inversion, has dropped to about 30 millibars (surface pressure is 1013 millibars). Consequently, the air is too thin to prevent energetic ultraviolet rays from the sun from reaching this level. Here the ozone layer reaches maximum concentration. Curiously, this is the same altitude at which the elusive and inexplicable nacreaous clouds occur, possibly indicating that there may be a physical connection between the two.

The maximum temperature in the stratosphere occurs at the *stratopause*, where the pressure is only about 1 millibar. Here another reversal in the temperature gradient takes place, defining the upper limit to the stratosphere and the lower limit to the *mesosphere*.

In the mesosphere, the temperature drops with increasing altitude but again reaches an isothermal plateau around −90°C. This plateau is called the *mesopause*, which looks down on 99 percent of the atmosphere. It is here, at an altitude of about 80 km (50 miles), that the pearly noctiluscent clouds are found.

Noctiluscent clouds occur in a fairly narrow altitude range of heights near 83 km. Recent work has confirmed that these clouds are made of ice that has condensed onto meteoritic dust. Ordinarily, there is no water vapor at this altitude, but sometimes near the equator water vapor leaks through the tropopause and is raised by stratospheric winds to very high levels. The meteor particles act as sublimation nuclei for this ice.

Beyond the mesosphere is the *thermosphere*, where the temperature lingers near −90°C and then gradually begins to rise. The auroras and airglow are found here. Air densities are low, and what air there is (primarily N_2, N, O_2, O) is ionized as a result of solar ultraviolet radiation. The *ionosphere* is an ill-defined region within the thermosphere, between about 80 and 300 km, where temperatures may be as high as 1000°C. Many satellites orbit in this region, but the air is too thin to greatly affect their thermal balance and they do not heat up.

The *exosphere* ("outersphere") and the *magnetosphere* lie beyond the thermosphere and blend smoothly and imperceptibly with the outer part of the solar atmosphere. This extended corona has two parts, one gaseous and one made of dust. The solar wind, composed of hot, fast-moving electrons and protons streaming away from the sun, is invisible except during a total eclipse of the sun. The dust component, however, is easily observed on a clear night. The so-called zodiacal light is actually a vast cloud of meteoritic dust, each particle of which is in its own orbit around the sun near the ecliptic plane. Of course the zodiacal light is not part of the earth's atmosphere, but its accessibility to the naturalist's eyes makes it an appropriate phenomena with which to close the reader.

Many of the phenomena discussed in Part II involve water. When our planet formed by accretion from interstellar material 4.6 billion years ago, there probably was no water. The interior of the planet grew hot as a result of geostatic pressure and the decay of naturally occurring radioactive elements. Atomic and nuclear chemistry created water, which went into solution with the hot molten rock, or *magma*. When the pressure became too great to contain, the magma forced its way to the surface and erupted as volcanoes. The early days of the earth were violent indeed. As the lava burst from the ground to cover the planet, the catastrophic pressure release also freed water, which escaped as massive, billowing clouds. These clouds covered the earth, and from them came the first rains. For millions of years, torrential storms deposited water into the low-lying areas, and in due course the first oceans developed. Even today, geothermal sources release water into the air, but at a greatly reduced rate.

Since water is so important to our planet and to atmospheric phenomena, we begin with several articles devoted to this remarkable substance.

Water

1

by Arthur M. Buswell and Worth H. Rodebush
April 1956

Although we take its properties for granted, they are most unusual. As an example, it has the rare property of being lighter as a solid than a liquid. If it were not, lakes would freeze from the bottom up!

Water is the only common liquid on our planet. Next to air it is the substance with which we are most intimately acquainted. Because it is so familiar, we are apt to overlook the fact that water is an altogether peculiar substance. Its properties and behavior are quite unlike those of any other liquid. To take just one example, water has the rare property of being denser as a liquid than as a solid, and it is probably the only substance that attains its greatest density at a few degrees above the freezing point (four degrees centigrade). The consequences of this behavior are of great importance to life on our planet. When ice forms on a lake, for example, its lower density (only nine tenths that of liquid water) keeps it on top and it acts as an insulating blanket to retard cooling of the underlying water. As a result lakes in the temperate zones do not freeze solid to the bottom but leave a zone for the winter survival of aquatic life. On the other hand, the same peculiar property of water has fatal consequences for living cells. When water in the cells freezes and becomes less dense, its expansion damages or breaks up the cells.

Even the elements of which water is composed—oxygen and hydrogen—are chemically exceptional. Both are unusually reactive. Oxygen is our chief source of energy, being responsible for the respiration of living organisms and the combustion of fuels. Hydrogen, unique in the fact that it has no enclosing shell but only a single electron, is able to attach itself to other atoms not only by means of its electron (a valence bond) but also by virtue of the attraction of its unoccupied, positively charged side for an electron in a second atom. This attachment is known as the hydrogen bond. In water the two hydrogen atoms attached to each oxygen atom can become linked to other atoms as well by means of these so-called hydrogen bonds. As a consequence the H_2O molecules are joined together, so that water should be considered not a collection of separate molecules but a united association. In effect the whole mass of water in a vessel is a single molecule.

The best method of detecting hydrogen bonds is to study water with the infrared spectrograph. We have found that the hydrogen bond absorbs radiation most strongly at a wavelength of about three microns, which is in the near infrared region of heat radiation—*i.e.*, close to the visible light spectrum. Liquid water absorbs this radiation so powerfully that if our eyes were sensitive to the infrared, water would appear jet black. There is some absorption even in the visible spectrum at the red end. The fact that water absorbs red light accounts for its characteristic blue-green color.

Water's 33 Components

One of the shocks to our familiar notions about water is that its formula is not simply H_2O. Nor is it a single substance. The beginning of this disillusionment came in 1934 when Harold Urey discovered "heavy water." Urey found that the purest water contained besides hydrogen and oxygen another substance like hydrogen but with an atomic weight of two, or twice that of hydrogen. This substance, which is now called deuterium, combines with oxygen to form the compound D_2O. By now, we know, of course, that there is a third isotope of hydrogen, called tritium, and three isotopes of oxygen: 0-16, 0-17 and 0-18. Thus the purest water that can be prepared in the laboratory is made up

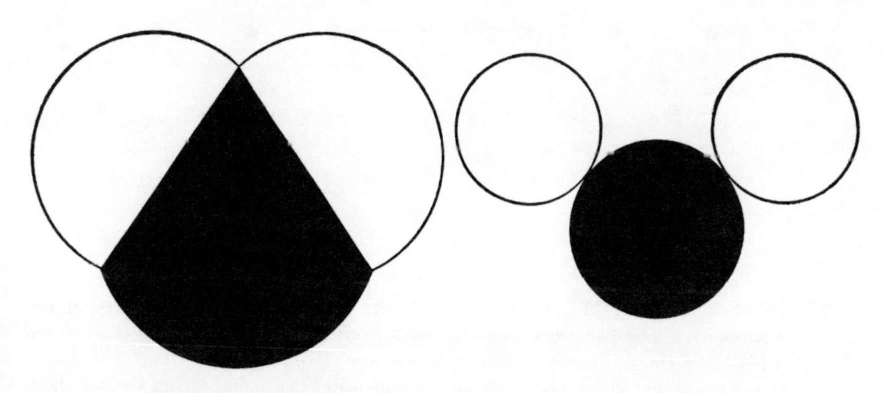

MOLECULE of water consists of one oxygen atom (*black*) and two hydrogen atoms (*white*). The distance between the center of the oxygen atoms and the center of each of the hydrogen atoms is .9 Angstrom unit (one Angstrom unit: .00000001 centimeter). The angle formed by the two hydrogen atoms is 105 degrees. These dimensions are fitted together in the drawing at left. In the more schematic drawing at right the size of the atoms has been reduced. This representation of the molecule is used in the following drawings.

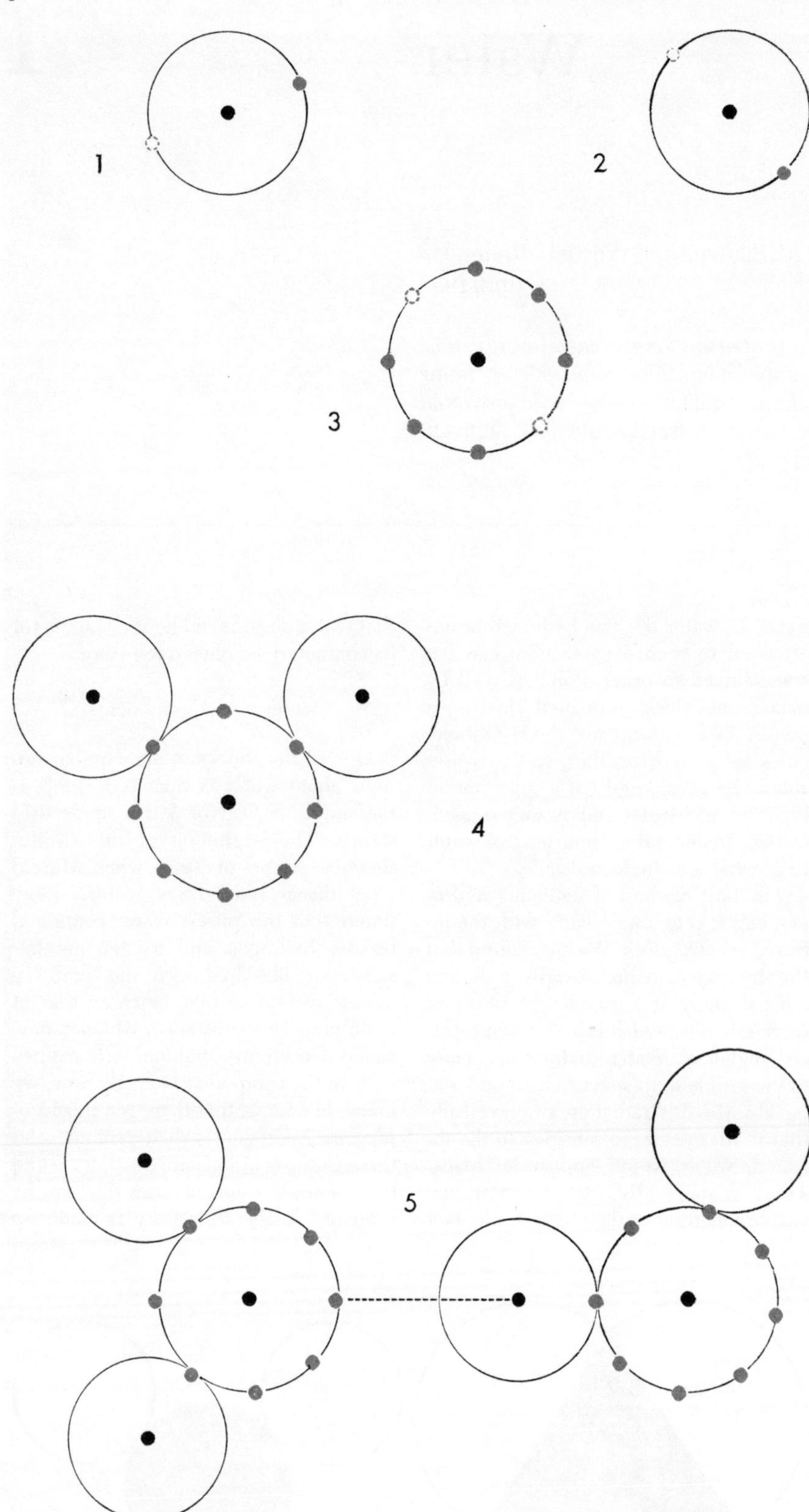

ELECTRONS AND PROTONS of the water molecule account for most of its physical and chemical properties. The hydrogen atom (1 *and* 2 *in this highly schematic picture*) consists of a positively charged proton (*black dot*) and a negatively charged electron (*colored dot*). The oxygen atom (3) has eight electrons, six of which are arranged in an outer shell. Because hydrogen shell has room for one more electron (*broken colored circle*), and the outer shell of oxygen has room for two more electrons, the atoms have an affinity for each other. In the water molecule (4) the electrons of the hydrogen atoms are shared by the oxygen atom. Because the positively charged proton of the hydrogen atom now sticks out from the water molecule, it has an attraction for the negatively charged electrons of a neighboring water molecule (5). This relatively weak force (*broken line*) is called a hydrogen bond.

of six different isotopes, which may be combined in 18 different ways. If we add the various kinds of ions into which the addition or removal of an electron may transform water's atoms, we find that pure water contains no fewer than 33 substances [*see top of page* 9].

Of course the amounts of the isotopes other than common hydrogen and common oxygen (0-16) are tiny. Tritium and oxygen 17 appear only in the minutest traces, and deuterium is present to the extent of about 200 parts per million and oxygen 18 about 1,000 parts per million. However, the properties of heavy water, particularly the D_2O variety, have attracted wide interest and have been extensively studied.

D_2O has a slightly higher boiling point than H_2O (101.4 degrees C.), freezes at a substantially higher temperature (3.8 degrees C.), and is somewhat more viscous than ordinary water. Its physiological properties are surprising. In animals and plants it appears to be entirely inert and useless. Seeds will not sprout in D_2O, and rats given only D_2O to drink will die of thirst.

The largest use of heavy water is as a moderator in nuclear reactors, but it is also widely employed in theoretical research, especially in organic and biological chemistry. If compounds containing active hydrogen are treated with D_2O, deuterium will replace the hydrogen and the compound will show changes in chemical properties resulting from the lesser reactivity of deuterium.

It is interesting to find that the amount of D_2O in natural water appears to be the same whether the water comes from an alpine glacier or the bottom of the ocean, from willow wood or mahogany.

Tritium is more ephemeral and more variably distributed. It is formed in the highest layers of the atmosphere by the bombardment of cosmic rays, and falls in rain and snow [see "Tritium in Nature," by Willard F. Libby; SCIENTIFIC AMERICAN, April, 1954]. Since tritium is radioactive, with a half-life of 12.5 years, it disappears after a time from water which has been out of contact with the atmosphere. Wines, and water in wells, can be dated by their tritium content. An interesting well in the Urbana-Champaign area was found to be devoid of detectable tritium, which means that at least 50 years have elapsed since the water fell as rain.

The functions of water in nature are innumerable. It is the solvent par excellence. It is the medium in which life originated and in which all organisms still exist. The living cell consists largely

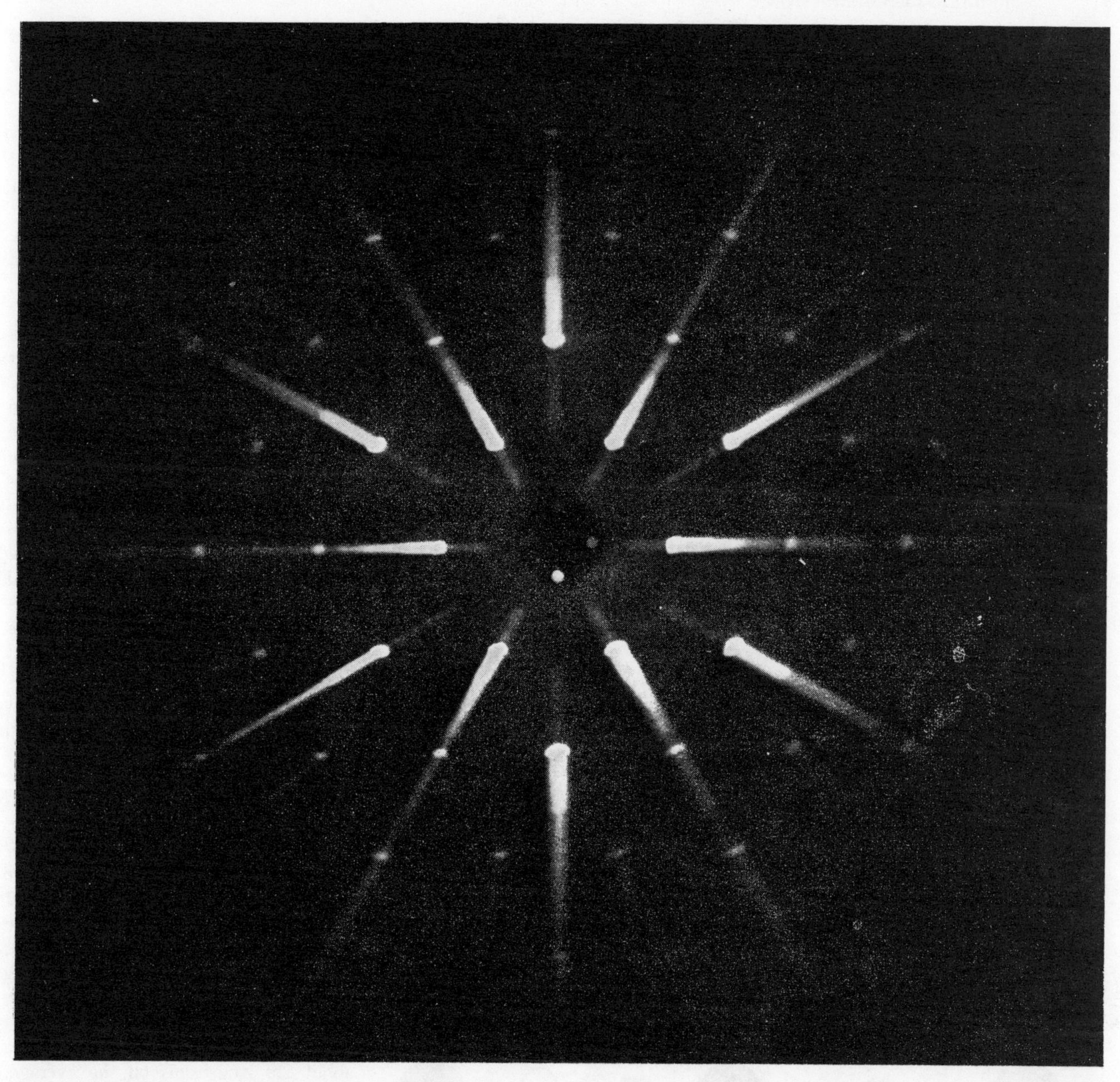

X-RAY DIFFRACTION photograph of ice was made with a precession camera by I. Fankuchen and his colleagues at the Polytechnic Institute of Brooklyn. The position of the spots in the photograph is related to the symmetry of the crystal (*see diagrams on next page*).

of water and literally floats in water. Considering how predominantly living matter is made up of this fluid, the extent to which it takes on a solid shape is surprising indeed.

Water plays a fundamental role in the protein molecule, the basic material of living matter. Proteins have a structure which places them in the class of substances known to chemists as plastics. In order that a plastic may possess flexibility and other desirable physical properties, it must contain a fluid called a plasticizer. Water is a "plasticizer" for proteins. In the chainlike protein molecule the hydrogen bonds of water provide secondary links which fix the pattern of the molecule. Removal of water alters the pattern and "denatures" the protein. Fortunately the process is reversible, so that water in the cells can restore the pattern. On the other hand, the hydrogen bonds of hydrides other than water, such as ammonia or hydrogen cyanide, form a stable denatured configuration which freezes the protein in a dead pattern. This is why all the hydrides except water are extremely toxic. Their action is somewhat like that of a virus, which corrupts the true protein structure into a strange distorted pattern.

Water's Structure

To understand the behavior of water we must understand its structure. It is far from simple. The best approach is a study of the structure of ice. The arrangement of the oxygen and hydrogen atoms in the ice crystal has been determined by X-rays and other means. The two hydrogens are bonded to the oxygen approximately at right angles to each other, more exactly, at an angle of 105 degrees [*see illustrations on page 5*]. If the angle were 109 degrees, the frozen water molecules would form a cubic lattice, as in the diamond crystal. But in this case such a structure would be unstable because of the strain on the distorted bonds. The exact arrangement of the molecules in the ice crystal is not known with certainty; we know that they form a hexagonal structure, which

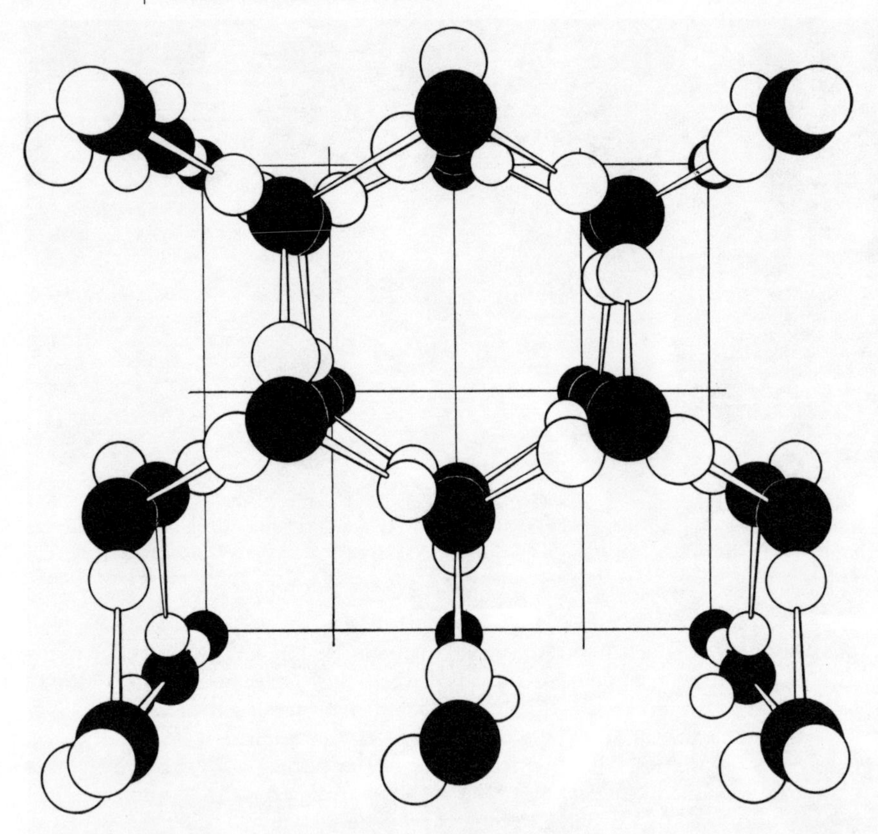

ICE consists of water molecules in this arrangement. The top drawing shows a model of ice seen from one direction. The bottom drawing shows the same model seen as if the reader had turned the top drawing forward on a horizontal axis in the plane of the page. Some hydrogens have been omitted from the molecules which touch the grid. Each hydrogen in each molecule is joined to an oxygen in a neighboring molecule by a hydrogen bond (*rods*). In actuality the molecules of ice are packed more closely together; here they have been pulled apart to show the structure. In a similar model of liquid water the molecules would be much more loosely organized, farther apart and joined by more hydrogen bonds.

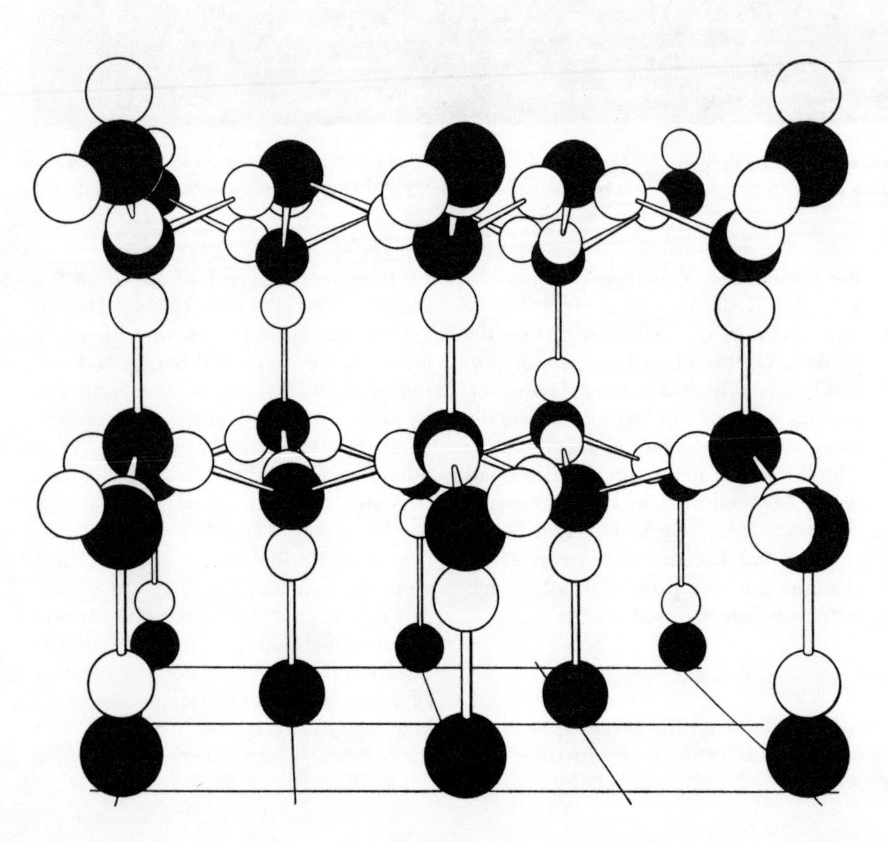

is exhibited on the macroscopic scale in the form of snowflakes. Each molecule is surrounded by four nearest neighbors, so that the group has one molecule at the center and the other four at the corners of a tetrahedron. The molecules and groups of molecules are joined together by hydrogen bonds.

The forces of attraction between the molecules in ice or water produce a strong inward pressure. As we shall see, this accounts for some of water's peculiar properties. In the form of ice, its open structure resembles a bridge arch under heavy downward stress. When the temperature of the ice rises to zero centigrade, the thermal agitation of the molecules is sufficient to cause the ice structure to collapse, and the water becomes fluid. It is well known that the application of pressure from outside will make ice melt at a lower temperature; evidently this reinforces the internal pressure within the ice and assists its collapse. Contrariwise, we can assume that if the internal pressure is reduced in some way, the melting point of ice will rise. Calculations indicate that if this pressure were entirely eliminated, ice would not melt until its temperature reached 15 degrees or more centigrade (59 degrees Fahrenheit).

According to X-ray determinations, the average distance between the center of one oxygen atom and the center of the next in the ice crystal is 2.72 Angstrom units (an Angstrom being one hundred-millionth of a centimeter). When ice melts to liquid water, the hydrogen bonds are stretched and the molecules move farther apart: the distance between oxygens is increased to about 2.9 Angstroms on the average. This stretching would open the structure further and make water less dense were it not for the fact that in the fluid the molecules crowd together in more compact groups. Each molecule is now surrounded by five or more neighbors instead of only four.

The chaotic disorder in which water molecules exist in the liquid state is difficult to picture. Their arrangement shifts continually. The angle between the two hydrogen atoms in the water molecule no longer remains fixed near a right angle but becomes variable, so that the molecule is flexible. Each oxygen atom now attracts by electrical forces not two extra hydrogen atoms as in ice, but three or more. Thus we may find an oxygen atom surrounded by five or six hydrogens and a hydrogen atom surrounded by as many as three oxygens. In the closely knit, flexible structure the hydrogens constantly shift their posi-

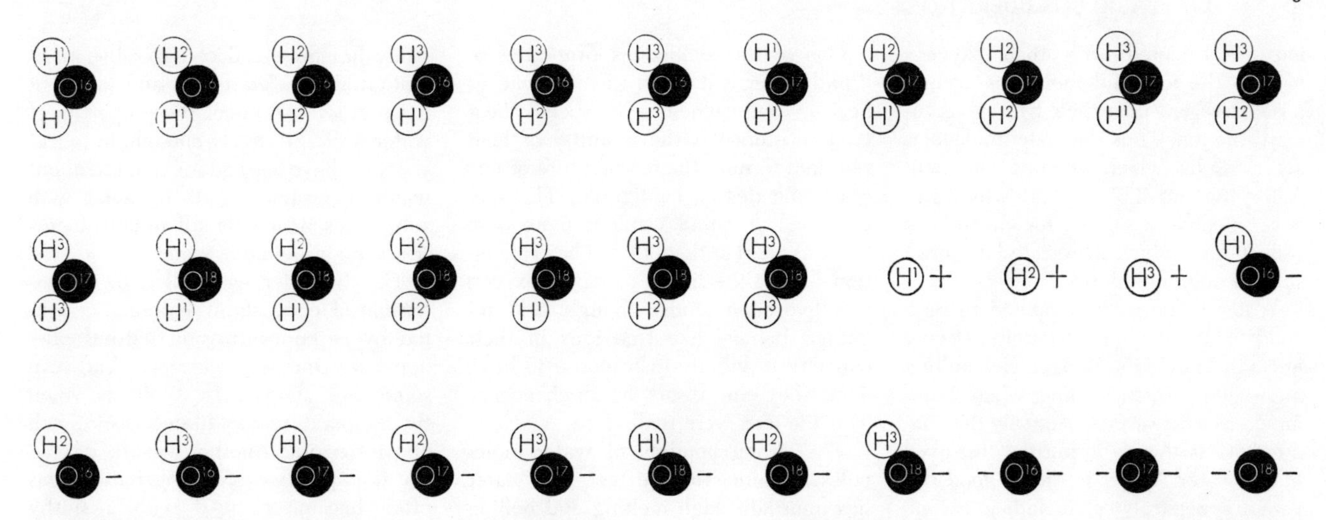

WATER IS NOT H_2O but a mixture of 33 different substances. Eighteen of these are combinations of three isotopes of hydrogen and three of oxygen. The three hydrogen isotopes are ordinary hydrogen (H^1), deuterium (H^2) and tritium (H^3). The three oxygen isotopes are ordinary oxygen (O^{16}), oxygen 17 and oxygen 18. The remaining substances are various ions (*plus or minus signs*).

tions and displace one another. Each such displacement is propagated in a chain or zipper fashion throughout the liquid. This has consequences which affect the viscosity, dielectric constant and electrical conductivity of water.

Water's Properties

In an ordinary unassociated liquid such as benzene the molecules flow by sliding around one another. In water the motion is rolling rather than sliding. Since the molecules are connected by hydrogen bonds, at least one bond must be broken before any flow can occur. From the fact that the molecules are bonded together, it might be expected theoretically that the viscosity of water should be comparatively high. However, each hydrogen bond in water is shared on the average between two other molecules, and one of these weakened bonds is easily broken. The greater viscosity of ice is due to the fact that each hydrogen is bonded to only a single oxygen atom from another molecule, and this firmer bond must be broken before any movement can occur.

The dielectric constant of a liquid is a measure of its capacity to neutralize the attraction between electrical charges. For example, when sodium chloride dissolves in water, the positively charged sodium and negatively charged chlorine ions are separated. They are kept apart because water has a high dielectric constant—the highest of any common liquid. It reduces the force of attraction between the oppositely charged ions in solution to not much more than 1 per cent of the original value. The reason for water's strong neutralizing action lies in the arrangement of its molecules. In an aggregation of water molecules a hydrogen atom does not share its electron equally with the oxygen atom to which it is attached: the electron is closer to the oxygen atom than to the hydrogen. As a result the hydrogen atoms are positively charged and the oxygen negatively charged. Now when a substance is dissolved, separating into ions, the oxygen atoms are attracted to the positive ions and the hydrogens to the negative ones. Consequently water molecules surrounding a positive

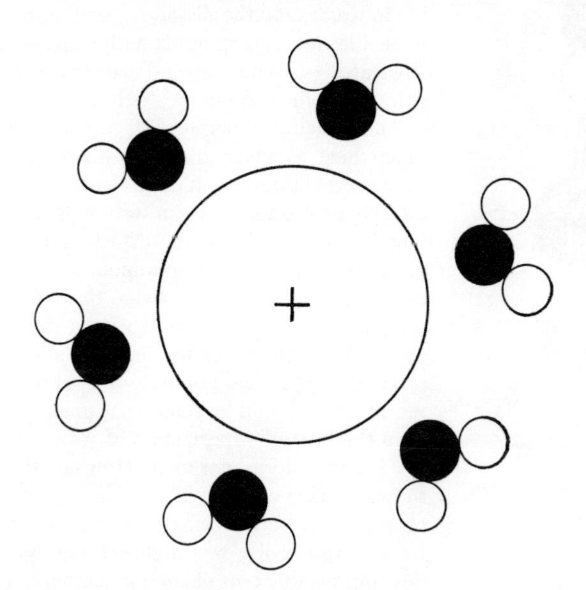

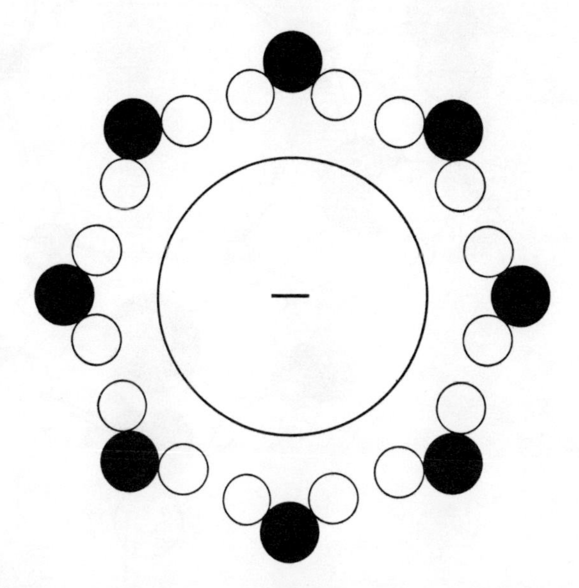

IONS (*circles labeled with plus and minus signs*) in water are kept apart because they polarize the water molecules around them. Because the oxygen atom of the water molecule has more negative charge than the hydrogen atoms, it is attracted to a positive ion (*left*). Because the hydrogen atoms have more positive charge than the oxygen atom, they are attracted to a negative ion (*right*).

ion are oriented with their oxygens next to the ion, and molecules around a negative ion turn their hydrogens toward the ion. Thus the water molecules act as cages which separate and neutralize the ions. This explains why water is so effective a solvent for electrolytes (substances which dissociate into ions) such as sodium chloride.

Water is generally supposed to be a good conductor of electricity. Every lineman knows the danger of handling high-voltage electrical lines when standing on a moist surface. Actually the conductivity is due to impurities dissolved in the water. Water is such a good solvent for electrolytes, including carbon dioxide from the atmosphere, that any moist surface may be assumed to be a good conductor. But pure water (which is difficult to keep pure—it must be kept out of contact with air and in a vessel of an inert material such as quartz) is a very good insulator indeed. The reason is that while the hydrogen and oxygen atoms in a water molecule are in a sense charged, or ionized, they cannot move about separately because they are attached to each other, and hence cannot carry an electric current.

One of the anomalous properties of liquid water is its high specific heat, or heat-holding capacity. The specific heat of a substance is the quantity of heat required to raise the temperature of one gram one degree centigrade. The specific heat of liquid water is more than twice as great as that of ice. The explanation is that the liquid's ionized oxygen and hydrogen atoms, though held together, behave like free ions in their capacity to vibrate in response to heat. Thus they can absorb as much energy as if the ions were really free.

The strong bonding of water molecules accounts for the fact that water has unusually high melting and boiling points. It also explains why it is so difficult to vaporize ice. To do this we must break all the hydrogen bonds holding the molecules together. Calculations indicate that the total energy of the hydrogen bonds in one mole of water (18 grams) is equivalent to 6,000 calories.

Hydrates

For more than 60 years physical chemists have studied water largely in terms of solutions of electrolytes. This study has produced considerable information about electrolytes and ions, but not a great deal about the properties of water itself. Strangely enough, in recent years we have learned much more about water by examining its behavior with substances which for all practical purposes are insoluble in water!

This behavior was called to the attention of chemists in a dramatic fashion by certain surprising natural phenomena. One was the fact that corn sometimes showed frost effects when the temperature was 40 degrees F., well above freezing. Another was the discovery that pipelines carrying natural gas often became clogged with a slushy "snow," containing water, at temperatures as high as 68 degrees F. The plain indication was that these freeze-ups were due to the water. But this raised some startling and interesting questions. What made water freeze at these high temperatures? How could water combine, or become "bound," with substances which were all but insoluble in it? The mystery was not lessened when it was discovered that even the noble gases such as argon and krypton, which refuse all chemical reactions, could join with water to form a quasi compound.

Let us look at these questions in the light of what we have learned about water's structure and properties. Ten years ago in Illinois we began a study of the water-solubility of certain hydrocarbons. Methane gas will serve as an example. The methane molecule does not form ions in water, nor does it accept the hydrogen bonds. There is very little attraction between it and the water molecule. It is, however, slightly soluble in water, and the dissolving methane molecules form compounds with water—"hydrates"—in which several water molecules are joined to one of methane.

The reaction liberates 10 times as much heat as when methane dissolves in hexane, although it is much more soluble in hexane than in water. This fact becomes even more surprising on close examination. The methane molecule occupies more than twice the volume of a water molecule. To form this relatively large cavity for itself on dissolving, a great deal of energy would be required: it should be somewhat greater than the heat of vaporization of water—say 10,000 calories per mole. How could so much energy be provided? The forces of attraction between methane and water are apparently too slight to supply any appreciable part of such an amount.

There is an alternative possibility. The presence of the methane may drastically change the water structure itself.

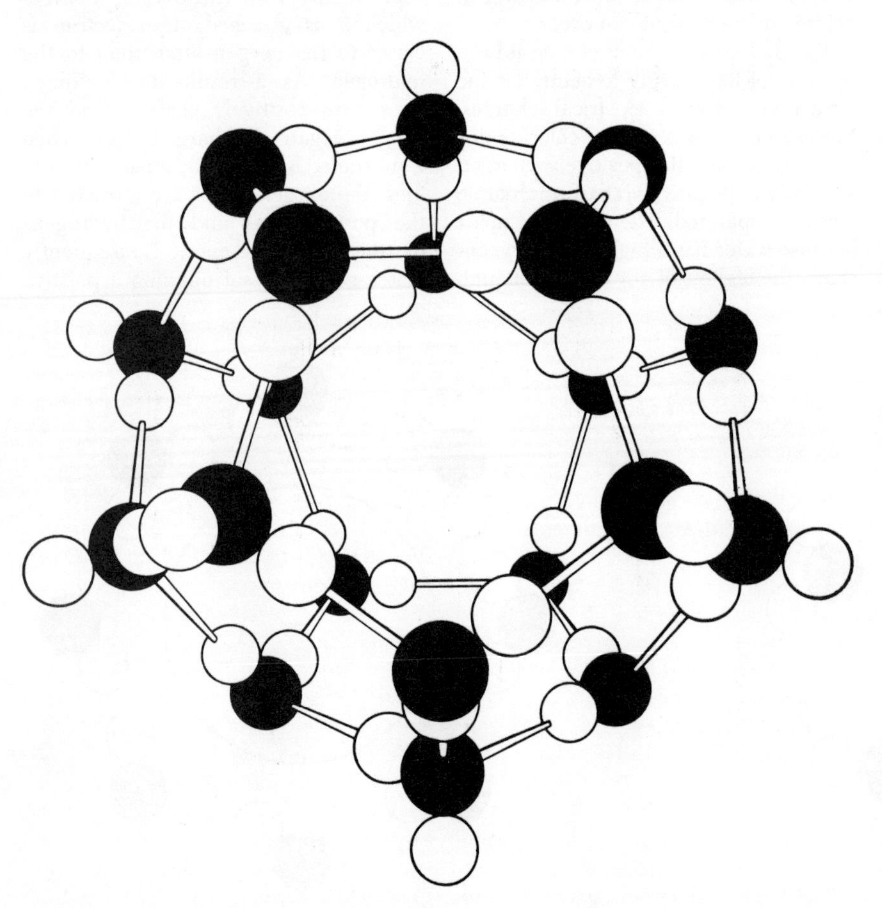

HYDRATE is formed when a foreign molecule in water is electrically neutral and just the right size for the water molecules to collect around it in crystalline cage. This cage can then grow to a much larger crystal. It is part of a repeating unit of 136 molecules.

Let us suppose that the dissolved methane molecule is surrounded by an envelope of 10 or 20 water molecules. The formation of such a structure would account for the heat liberated. In the space occupied by the methane molecule the attractive force on the water molecules, and hence the inward pressure, would disappear. Under these conditions, as we have seen, water will freeze at a higher temperature. Thus the molecules at the interface between the methane and water molecules may crystallize into "ice." The frozen hydrates may accumulate and separate out of the solution.

This hypothesis is known as the "iceberg" theory. It is supported by the fact that practically all the nonelectrolytic substances tested have been found to form solid crystalline hydrates. In contrast, electrolytes show little tendency to form them.

All this leads to an entirely new concept of solubility. Chemists have long supposed that solubility always involves attractive forces. But it now appears that the dissolution of a nonelectrolyte is due not to an attraction between the substance and water but to a lack of attraction. The nonionic substances combine with water because they remove internal pressure and thereby permit formation of a crystalline compound.

In order to understand the formation of these hydrates, it is necessary to consider their molecular structure in detail. They tend to fall into groups according to the number of water molecules they contain.

The ground work for the study of the hydrates structure was laid by M. von Stackelberg in Germany 10 years ago. He showed by X-ray studies that this structure was cubic, in contrast to the hexagonal structure of ice. W. F. Claussen of our laboratory recently attacked the problem of building such cubic structures, each containing a gas molecule, into a repeating lattice. It turns out that there are two possible cubic lattices, one of which was proposed and worked out by Linus Pauling of the California Institute of Technology. This has a spacing of 12 Angstroms between molecules while the other has 17 Angstroms. The smaller lattice contains 46 water molecules and the larger one 136. The holes for gas molecules in the smaller lattice have 12 or 14 walls, while those in the larger one have 12 or 16 sides. These holes are of different sizes and make possible a bewildering array of hydrates. The different sized holes can be filled only with different sized molecules, and not all the holes in a lattice need be filled. The model explains the actual composition of hydrates with remarkable precision.

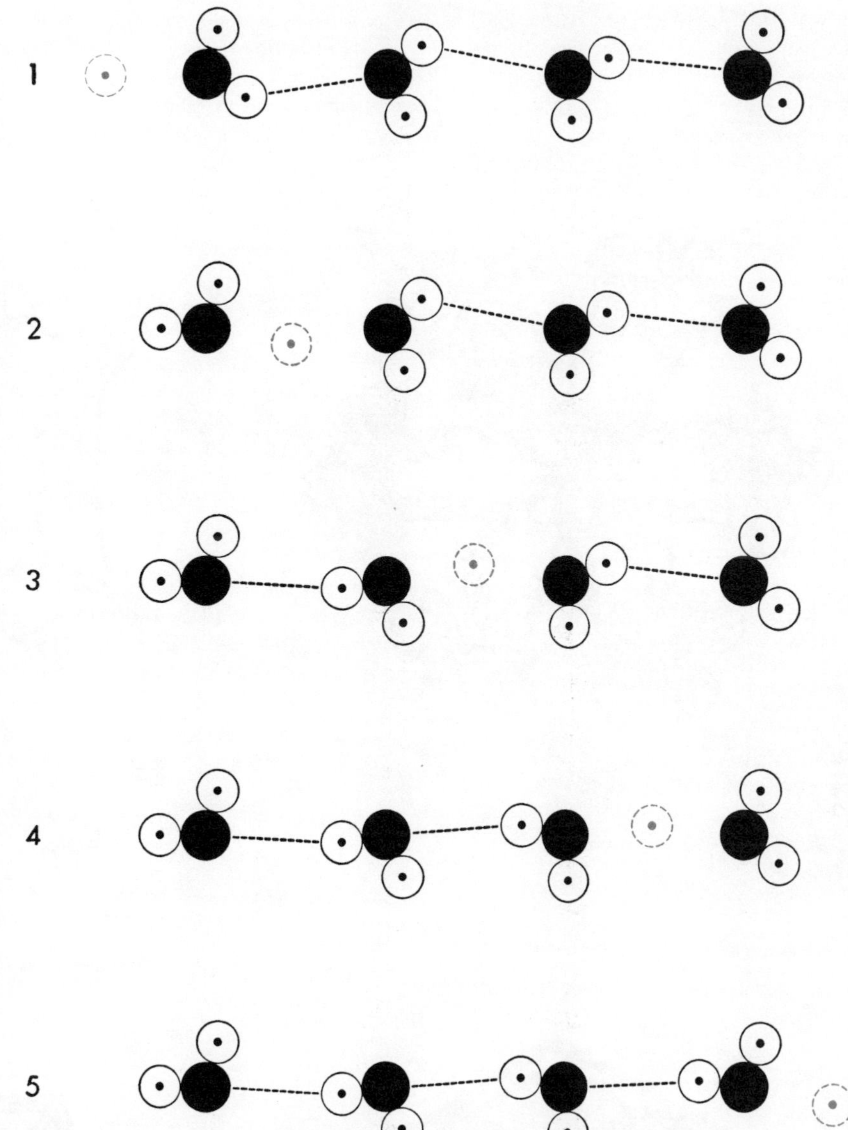

RAPID MIGRATION OF HYDROGEN IONS through water is explained by the assumption that the ions do not actually travel through the water but are passed from one molecule to the next by a process of exchange. Here hydrogen ion is represented by colored dot surrounded by a broken circle. In the first horizontal row the hydrogen ion approaches a water molecule. In the second row the ion has taken the place of one hydrogen atom of the molecule, expelling the atom as a new ion. In the third row the new ion repeats this process.

The importance of this type of hydrate to the processes of life cannot be overemphasized. These processes occur mainly at the interfaces between water and protein molecules. Water has a very strong tendency to crystallize there, for the protein molecule contains large nonionic, or nonpolar groups. Any hydrate so formed has a lower density than ice; consequently its formation can cause a large, destructive expansion.

The freezing of corn at a temperature of 40 degrees F. becomes understandable in terms of the formation of a hydrate. Winter wheat, on the other hand, forms hydrates slowly as temperatures drop through the fall, and under these conditions the hydrate acts as an effective antifreeze protecting the cells from damage.

The frozen food industry uses rapid freezing to avoid the formation of large crystals which would damage the plant cells. But it might be well to explore the possibility of the opposite approach. Very slow cooling of living plant foods might form hydrates which would prevent damage from ice crystals when the plant was frozen.

Let us return now to see how the structure of water may be modified when an electrolyte, say a salt, goes into solu-

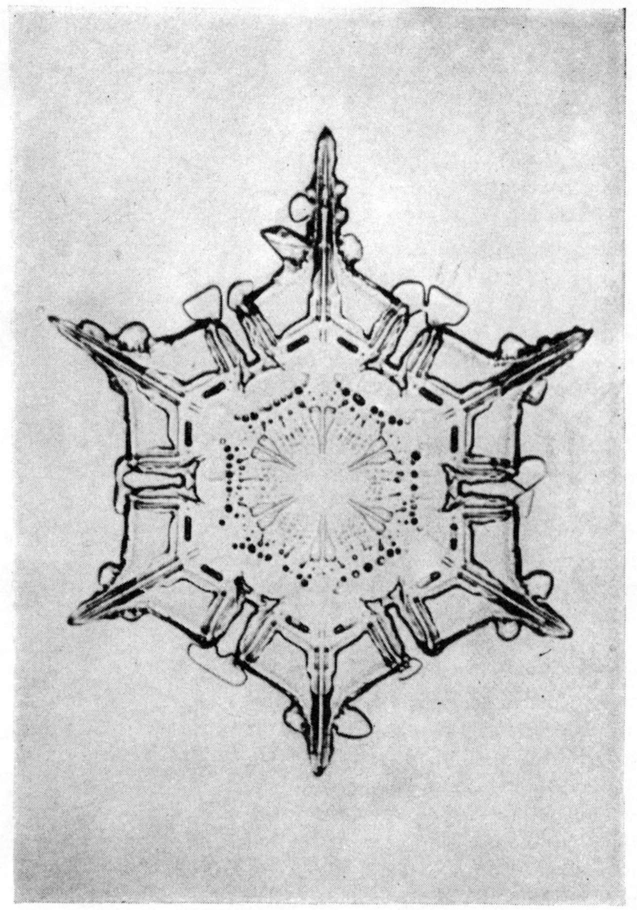

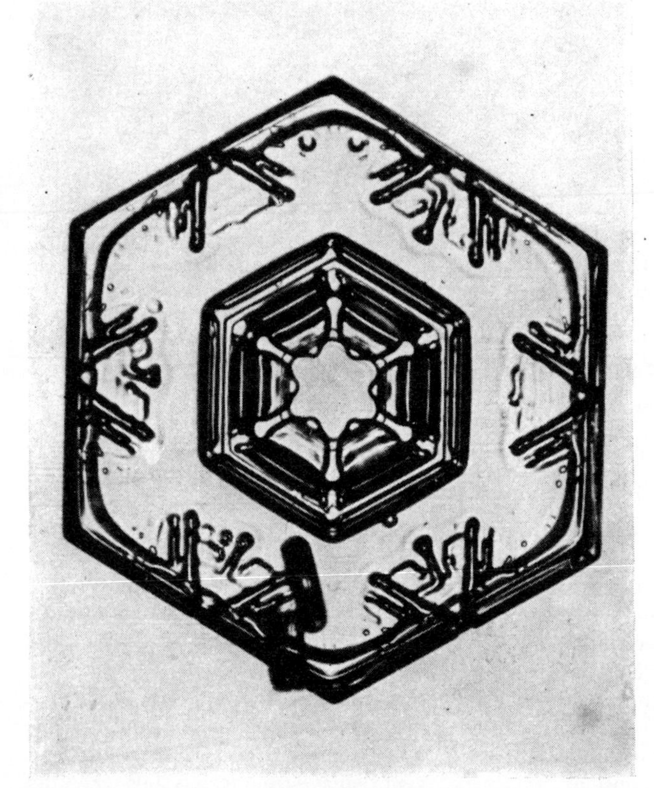

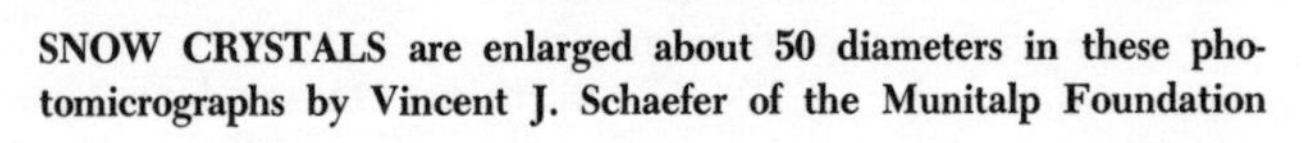

SNOW CRYSTALS are enlarged about 50 diameters in these photomicrographs by Vincent J. Schaefer of the Munitalp Foundation in Schenectady, N.Y. The hexagonal symmetry of the crystals is due to their molecular structure (*see diagram on page 8*).

tion. The only direct physical clue we have lies in the behavior of the salt ions in conducting an electric current. The rate of motion of the ions will depend in part on the resistance they encounter in the liquid, and this in turn will depend on the size of the moving particles. If water molecules are attached to an ion firmly enough to move along with it, they will of course increase the apparent size of the ion. Studies of the mobilities of various ions show that positive ions smaller than potassium carry such a cage of water molecules with them. The positively charged ion attracts the oxygens of two water molecules quite strongly, and if the volume of the ion plus its two water molecules is not greater than that of a methane molecule, a cage of hydrogen-bonded water molecules will form around this group as a nucleus. Positive ions larger than potassium fail to pick up such a cage. The same is true of most, but not all, negative ions.

The positively charged hydrogen ion and the negatively charged hydroxyl ion (OH) are surrounded by cages, and yet they show the highest mobility in carrying a current. We must conclude that they manage in some way to escape from the cage. Actually the mechanism is not hard to picture: they continually form new cages as they travel by a process called proton transfer. Under the influence of an electric field a hydrogen ion may jump from one water molecule to the next. When this has occurred, the hydrogen on the farther side of the water molecule takes up its part in the race like a relay runner and jumps to the next water molecule. Thus a succession of protons, each doing its bit, carries the current. The motion is rapid, because each proton moves only a short step. Transfer of the proton also explains the conduction of electricity by hydroxyl ions. When a proton jumps toward the right, say, and joins a hydroxyl ion, it leaves a hydroxyl ion on its left. The effect is the same as if a hydroxyl itself moved to the left.

Water, then, is not simple H_2O but a unique and complicated material with distinct and varied chemical properties. It has a definite though changing physical structure which depends on the orientation of its molecules with respect to one another, and to the molecules of dissolved substances. Since the behavior of all living nature and much of the inanimate world is inseparably linked to the peculiar characteristics of this liquid, the study of water substance can tell us a great deal about fundamental aspects of the world in which we live.

2 Ice

L. K. Runnels
December 1966

At the molecular level the seemingly rigid perfection of a crystal of ice is disrupted by an astonishingly busy traffic of molecules and migrating lattice faults

Ice is a substance that has fascinated many of the foremost scientists of modern times. Chemists, physicists, mathematicians and even biologists have been drawn to the investigation of its somewhat mysterious crystal structure and its peculiar properties. Their exploration of the problems presented by ice have yielded a better understanding of the hydrogen bond (present in many important molecules, including proteins) and of the activity that goes on within a seemingly rigid crystal. The investigation of ice is a good example of how much can be learned from close study of a most commonplace object.

Let us begin by examining the structure of a single water molecule—say a molecule of water vapor in the air. The easiest way to visualize the molecule is to place it in an imaginary cube with the nucleus of the oxygen atom at the center of the cube and the two hydrogen nuclei at opposite corners of the cube's base [*see illustration on this page*]. Actually the shape of the figure is not exactly cubic—the "cube" should be slightly distorted—because the angle of attachment of the hydrogen atoms to the oxygen is 104.5 degrees, whereas in a perfect cube the lines from the corners to the center form an angle of 109.5 degrees. For our purpose, however, the cube figure is accurate enough. Of the oxygen atom's eight electrons two are held near the oxygen nucleus while another pair joins with two electrons from the hydrogen atoms in binding them to the oxygen; this leaves two "arms" of unshared electrons that point from the oxygen nucleus to other corners of the cube, as the illustration indicates.

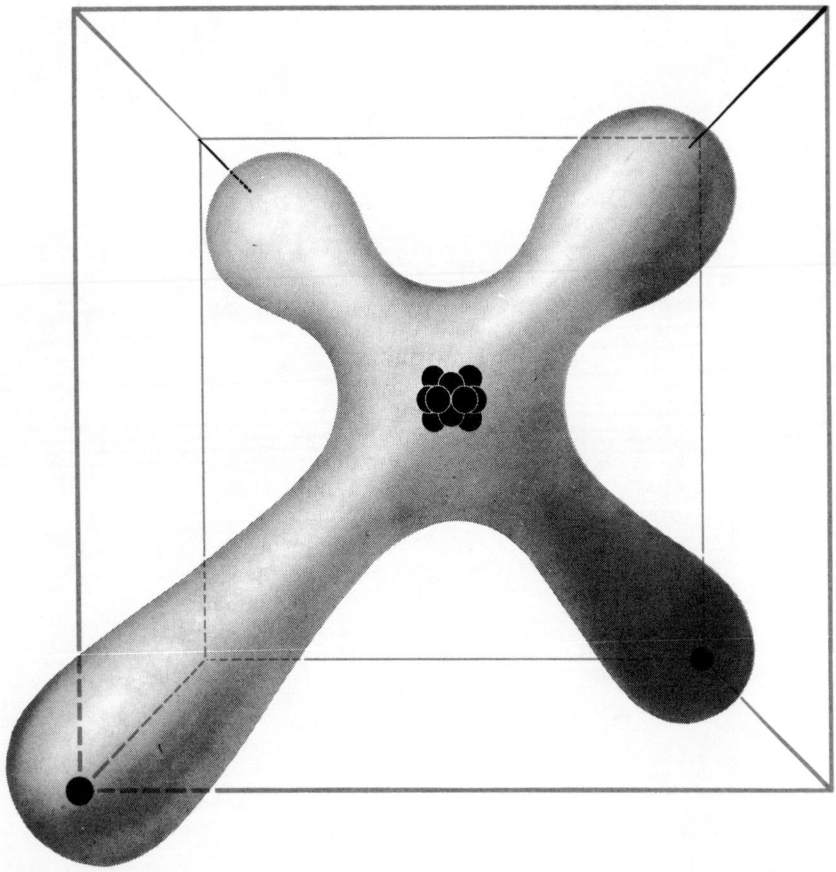

When water molecules are not isolated but packed together, as in the liquid state, these negatively charged arms serve to attach molecules to one another. Each negative arm attracts a hydrogen nucleus in a neighboring water molecule, and the hydrogen atoms thus act to join the molecules with what are called hydrogen bonds. The molecules tend to combine in clusters [see "Water," by Arthur M. Buswell and Worth H. Rodebush on pages 5–13 of this Reader].

In the liquid state the normal thermal motion of the molecules is sufficient to break the bonds as the molecules jostle one another; consequently the clusters are continually being split up and reformed. When water is cooled to the freezing point, however, the thermal motion is so reduced that the molecules form large, stable clusters: the crystals

MOLECULE OF WATER has four "arms" (actually clouds of electrons) extending from the central nucleus of the oxygen atom. Two of the arms contain hydrogen nuclei (protons) and are positively charged. The other two arms contain no protons and hence are negatively charged. In subsequent illustrations the negative arms of molecules will be omitted.

of ice. The first thing that interests us about these crystals is their structure. How are the water molecules arranged in an ice crystal?

The study of ice by the technique of X-ray diffraction showed many years ago that the molecules in the crystal are arrayed in a hexagonal pattern [*see illustration at right*]. This accounts for the six-sided form of snowflakes. More significantly, the analysis revealed that there is a good deal of empty space between the molecules: the crystal has a rather open structure. This finding clearly explained one of the most unusual properties of ice: the fact that, in contrast to almost all other substances, water is less dense in the solid state than in the liquid state, with the result that ice floats on liquid water. In frozen water, it can be seen, the molecules are more loosely packed than in the liquid. When a crystal of ice melts, the breakdown of its structure allows molecules to fill some of the open spaces.

The X-ray studies succeeded in revealing the arrangement of molecules in ice because the oxygen atoms in the molecules serve to deflect the X rays. The technique failed, however, to locate the positions of the hydrogen atoms; their power to deflect X rays is too low. For more than 30 years investigators in Europe and the U.S. have been pursuing the intriguing problem of finding out just where the hydrogen atoms are in the ice crystal, as this information is crucial to understanding the complete structure of ice and some of its most important properties.

The effort took its cue from a suggestion made in 1933 by the British physicists J. D. Bernal and R. H. Fowler. They argued that in all likelihood the orientation of the molecules in the ice crystal is such that each molecule forms good hydrogen bonds with its four nearest neighbors; that is, in each case the molecule's two hydrogen nuclei (protons) are pointed toward its neighbors' negative arms and its own two negative arms are pointed toward its neighbors' protons. This suggestion was accepted as reasonable. Unfortunately, however, it left the problem wide open. It did not specify any particular arrangement of the molecular orientations; indeed, one could picture many different arrangements that would satisfy Bernal and Fowler's rule. Some of these possibilities are shown in the top illustration on the next page, which for the sake of convenience is presented in only two dimensions (instead of the three dimensions of a real

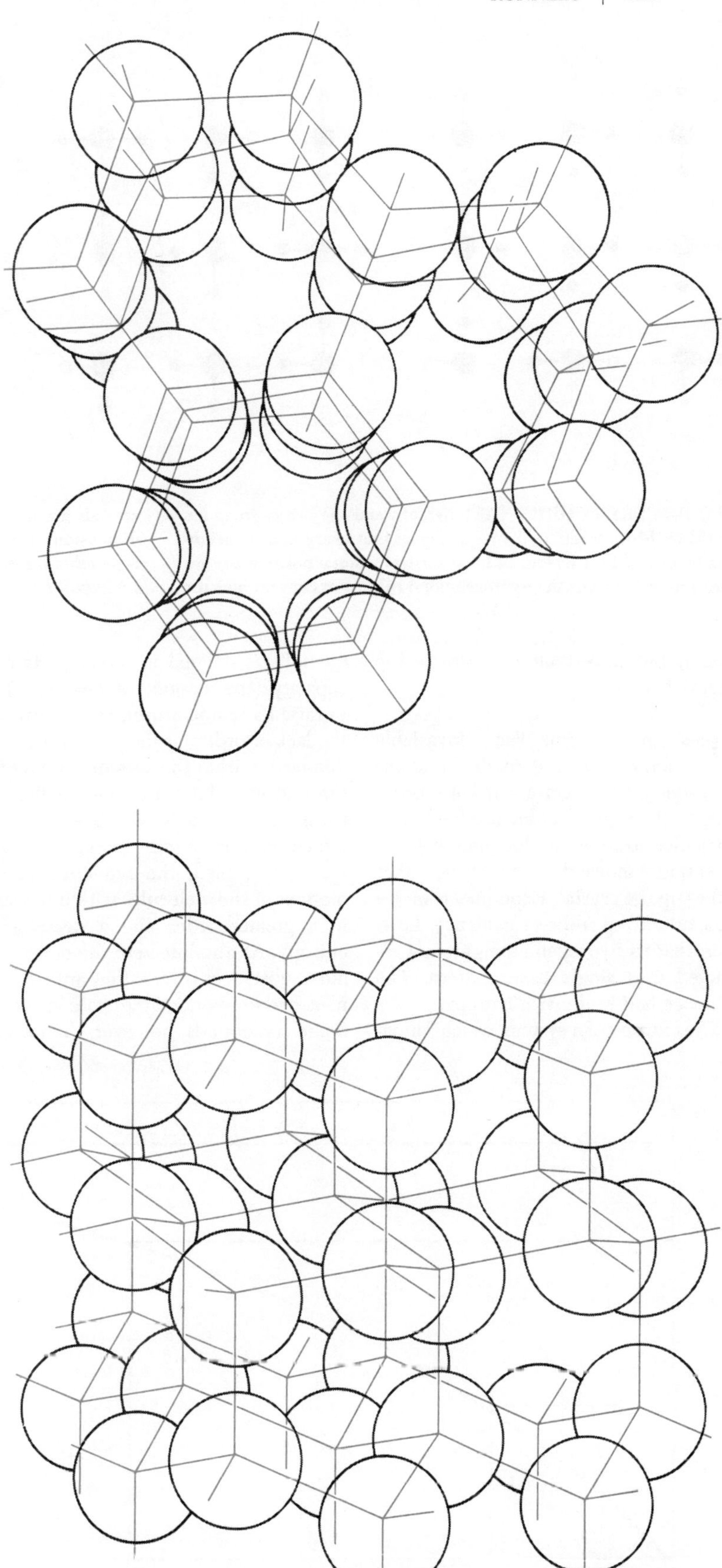

WATER MOLECULES IN ICE CRYSTAL were shown many years ago by the technique of X-ray diffraction to be arrayed in a hexagonal pattern, viewed here from above (*top*) and from the side (*bottom*). The X-ray analysis revealed that there is a good deal of empty space between the molecules, which explains why ice floats on liquid water. The technique failed to locate the positions of hydrogen atoms, because their power to deflect X rays is too low.

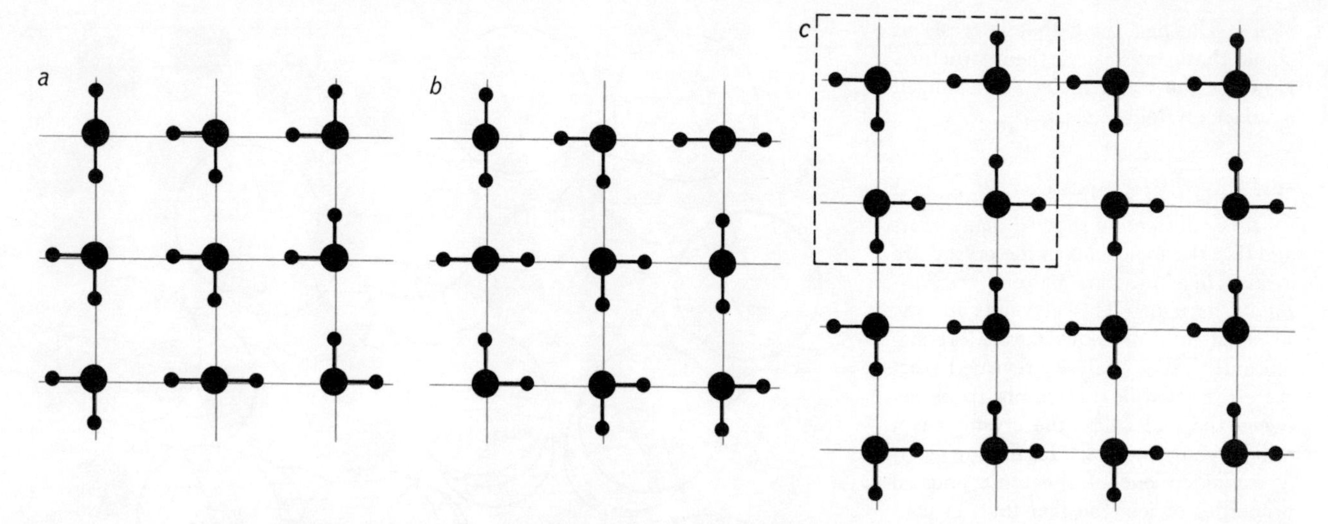

TWO DIFFERENT ORIENTATIONS of water molecules in an ice crystal (a, b) both satisfy the requirement that every pair of neighbors be joined by a hydrogen bond formed when a positive proton faces a negative arm (the hydrogen-bond rule). For convenience the crystals are presented in only two dimensions instead of the three dimensions of a real crystal. A total of 2,604 different arrangements is possible for such an ice crystal containing only nine molecules. In c a hypothetical arrangement is shown with a repeating "unit cell."

crystal) but nevertheless illustrates the essential point.

Does ice have one fixed, invariable structure? In most solids the atoms are arranged in a certain definite order with each type of atom assigned to a particular location in the unit cell, or repeating molecular structure, that makes up the crystal. Experiments on ice soon turned up indirect evidence, however, that its hydrogen atoms are not restricted to a single basic pattern. The evidence had to do with entropy.

The entropy of a system, which fundamentally is defined in terms of its heat capacity (the amount of heat required to raise its temperature), is a measure of the lack of order, or the presence of randomness, within the system. The higher the entropy, the more random the distribution of particles or molecules contained in the system. Now, as a substance is cooled, the reduction of the motions of the molecules within it results in a greater order and a decrease of entropy. At absolute zero the entropy of most substances is zero: they are "frozen" in a certain predictable state of order. Some compounds, however, do not give up all their entropy even when they are cooled to the lowest attainable temperature (almost absolute zero). In one way or another they retain some randomness —a "residual entropy." A good example is carbon monoxide (CO). With a carbon atom at one end and an oxygen atom at the other, this molecule is slightly polarized (the electric charge of one end is slightly more positive than that of the other), but the difference is so small that there is little distinction between the positive and the negative ends of the molecule. Consequently when the substance is cooled to very low tempera-

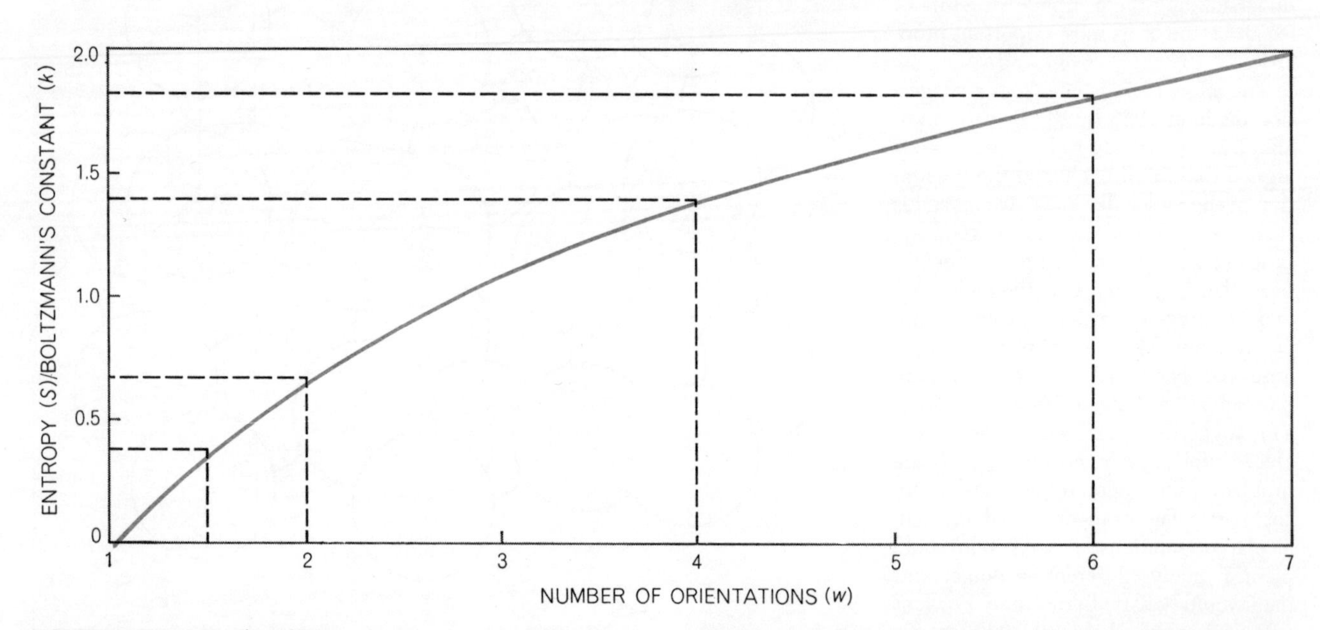

BOLTZMANN RELATIONSHIP, named after the 19th-century Austrian physicist Ludwig Boltzmann, predicts that the entropy (or randomness defined in terms of heat capacity) of any system in which the molecules are independent of one another will be proportional to the natural logarithm of the number of equally likely orientations of the molecules. The values for w of 1.5, 2, 4 and 6 correspond respectively to ice, carbon monoxide, deuterated methane and ice without the hydrogen-bond rule. The experimental value of $S/k = .4$, obtained by measuring the heat capacity of ice to very low temperatures, substantiates hydrogen-bond rule for ice.

tures, the orientation of the molecules in the crystal is ruled largely by chance, some molecules pointing in one direction and others in the opposite direction. For this reason carbon monoxide is said to have a residual entropy at absolute zero.

Bernal and Fowler in their discussion of the structure of ice implied that there might be no energy considerations dictating a particular arrangement of the hydrogen atoms. Amplifying this idea, Linus Pauling at the California Institute of Technology argued that ice (like carbon monoxide and a few other compounds) should have a residual entropy at low temperature. Indeed, this had already been observed experimentally by W. F. Giauque and J. W. Stout of the University of California at Berkeley. By measurement of the heat capacity of ice down to very low temperatures they found that the crystals actually did retain a certain amount of entropy.

It became clear, then, that in the crystals of ice the molecules are oriented in more than one way. Could the measured residual entropy shed any light on just how many different arrangements are actually present? A means of tackling this question was available: it was the famous entropy formula of the 19th-century Austrian physicist Ludwig Boltzmann, one of the founders of statistical mechanics. Boltzmann's formula predicts that the entropy (S) of any system in which the molecules are independent of one another will be proportional to the natural logarithm of the number of equally likely orientations of the molecules. In symbolic terms the formula is $S = k \log w$, with w standing for the number of possible orientations and k for the celebrated Boltzmann constant. In the case of carbon monoxide w is 2 (each molecule has two possible orientations), and the formula predicts that the residual entropy divided by the constant (S/k) should therefore be .69. The measured value of carbon monoxide's residual entropy agrees satisfactorily with this figure.

Suppose now we apply the formula to ice. In the ice crystal a molecule has six possible orientations: its hydrogen atoms can be oriented in six different ways. According to the Boltzmann formula, with $w = 6$ the residual entropy (more precisely, S/k) should have the value 1.8 [*see bottom illustration on preceding page*]. Actually in the experimental measurements the value turns out to be only .4. The discrepancy is not surprising; in the Boltzmann formula, as we have noted, w is the number of *independent* alternatives. The orientations of the

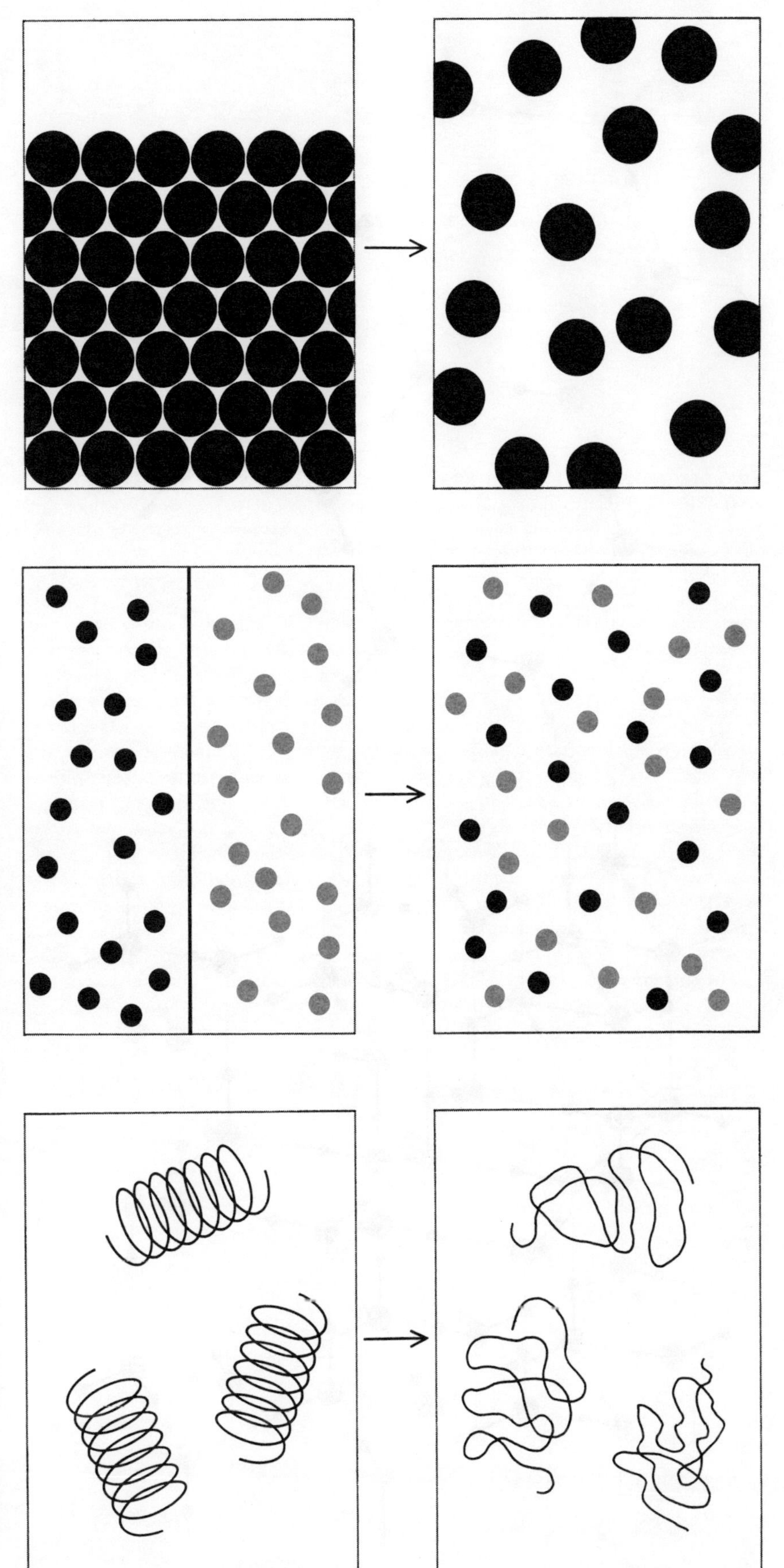

ENTROPY AND DISORDER are closely related, the more random arrangement of a system (*right*) having a higher entropy than the more orderly one (*left*). The three typical entropy-increasing processes shown here are the vaporization of a solid (*top*), the mixing of two gases (*middle*) and the helix-to-random-coil transition of protein molecules (*bottom*).

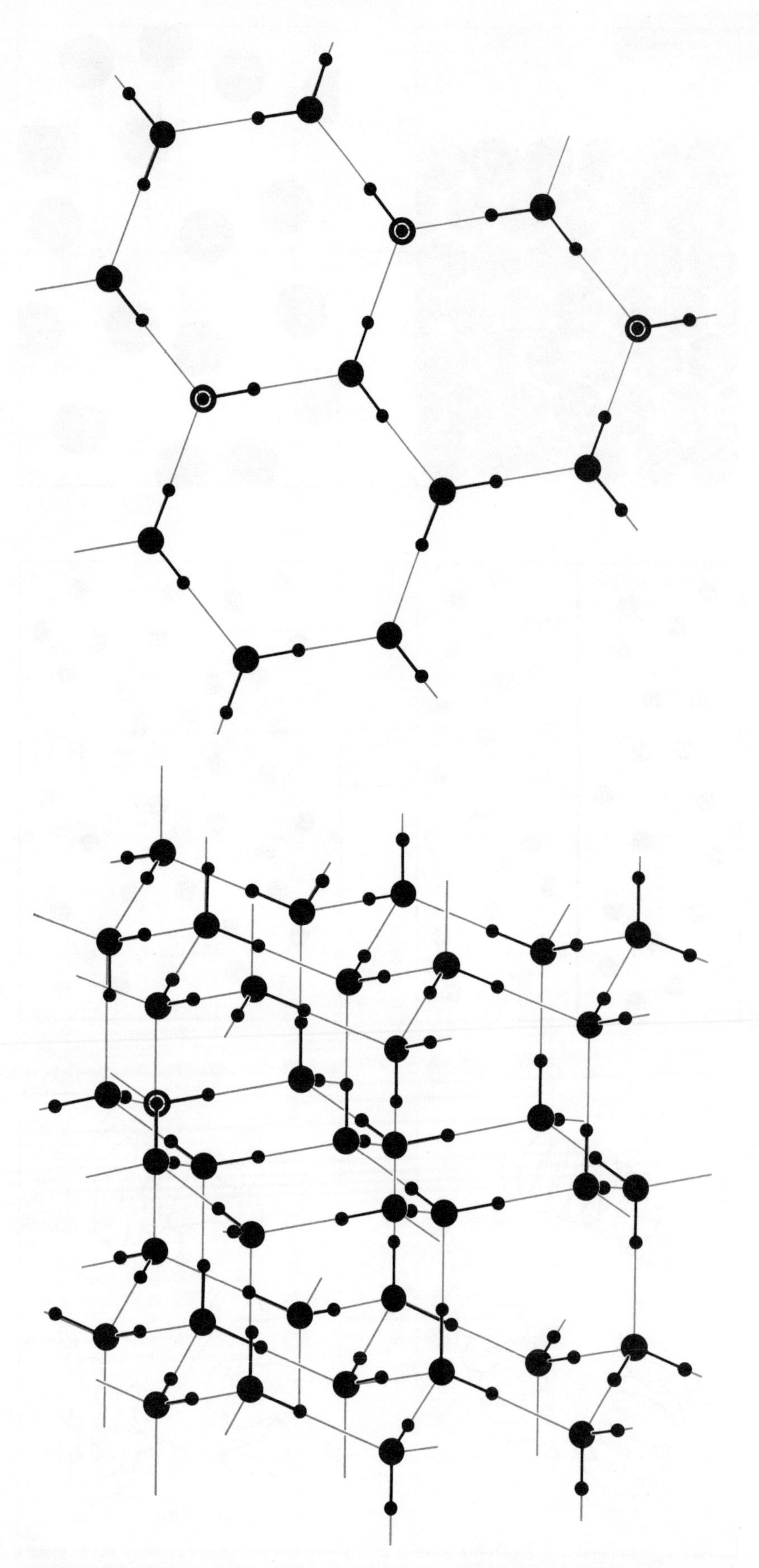

HYDROGEN ATOMS IN ICE CRYSTAL appear to be oriented at random, even at very low temperatures. There is no unit cell in the crystal for hydrogen nuclei. This conclusion is based on measurements of the residual entropy of ice. Crystal is viewed from above (*top*) and from the side (*bottom*). The molecules are somewhat reduced in size for clarity.

water molecules in ice are not independent of one another. Once we have selected a particular orientation for one molecule there are only three possible ways a second molecule can orient itself to form a good hydrogen bond with the first, and the molecules that join up with the first have other neighbors that in turn restrict the orientation choices still further. Pauling calculated that the average number of orientation choices for the system (that is, the value of w) is about 1.5 per molecule.

This estimate agreed well with the measurements Giauque and Stout had obtained for the residual entropy of ice; with $w = 1.5$, the value of S/k would be almost exactly .4. Lars Onsager of Yale University suggested in 1939 that it was important to compute as precise a value as possible for the average number of possible orientations. Pauling's estimate of 1.5 was the lower limit for this figure; Onsager noted that the actual value must be somewhat higher. The reason it was important to calculate the number exactly was that, if w turned out to be significantly higher than 1.5, it would indicate that there was some ordering influence within the ice crystals; in other words, that the possible orientations of the molecules were not all equally likely.

For more than 30 years chemists and mathematicians have struggled with this interesting mathematical puzzle, attempting to determine just how many geometric orientations are open to a collection of ice molecules. So far the problem has defied the most sophisticated assaults. Two years ago, however, John Nagle, a student of Onsager's at Yale, pursued some ideas suggested by Edmund A. DiMarzio and Frank H. Stillinger, Jr., of the Bell Telephone Laboratories and arrived at a reliable estimate that narrows the area of uncertainty to very close limits. Nagle found that the exact value must lie between 1.5065 and 1.5068. This result, bringing the theoretical prediction of ice's residual entropy into close agreement with the experimental measurements, shows that if there are any ordering forces in the crystals favoring one arrangement over another, they must be extremely weak.

We have been considering an ideal picture in which the structure of ice is governed strictly by the known rules—in particular the hydrogen-bond rule of Bernal and Fowler, which would lock the molecules in a fixed pattern. Now we must look into certain complications that lead to a surprising new view of

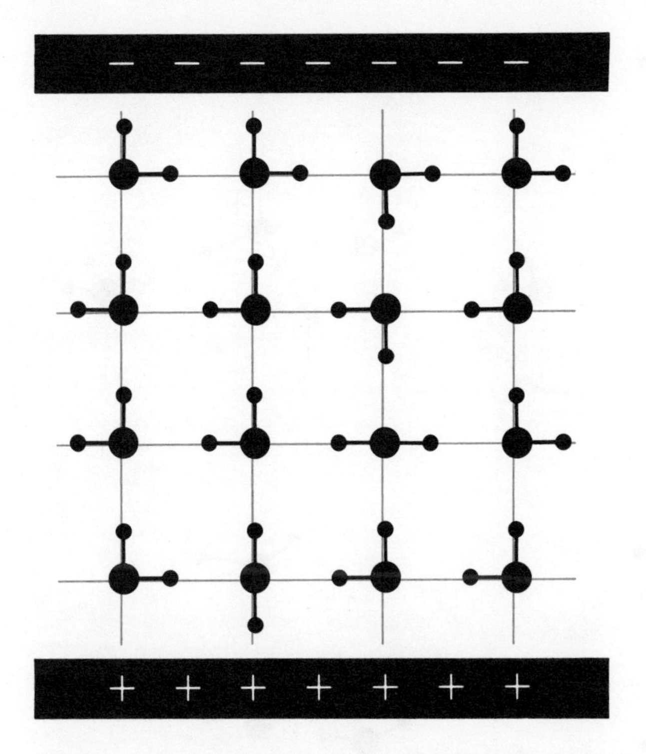

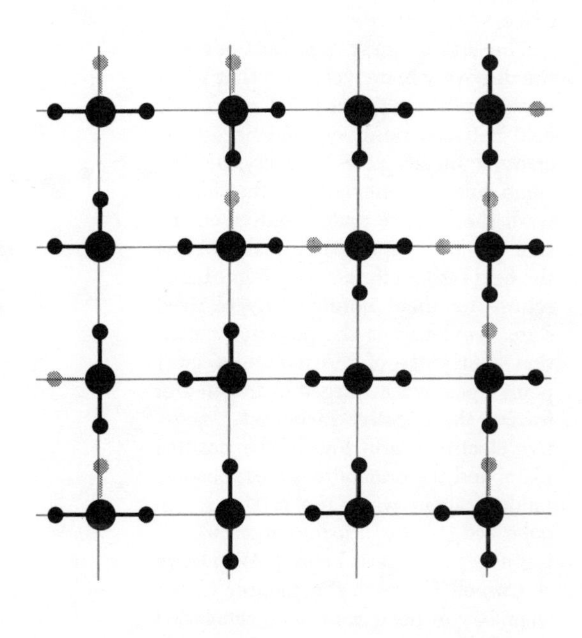

ICE CRYSTALS BECOME POLARIZED when they are placed in an electric field between positively and negatively charged metal plates (*left*). When the field is shut off, the crystals gradually return to their normally "relaxed," or depolarized, state (*right*).

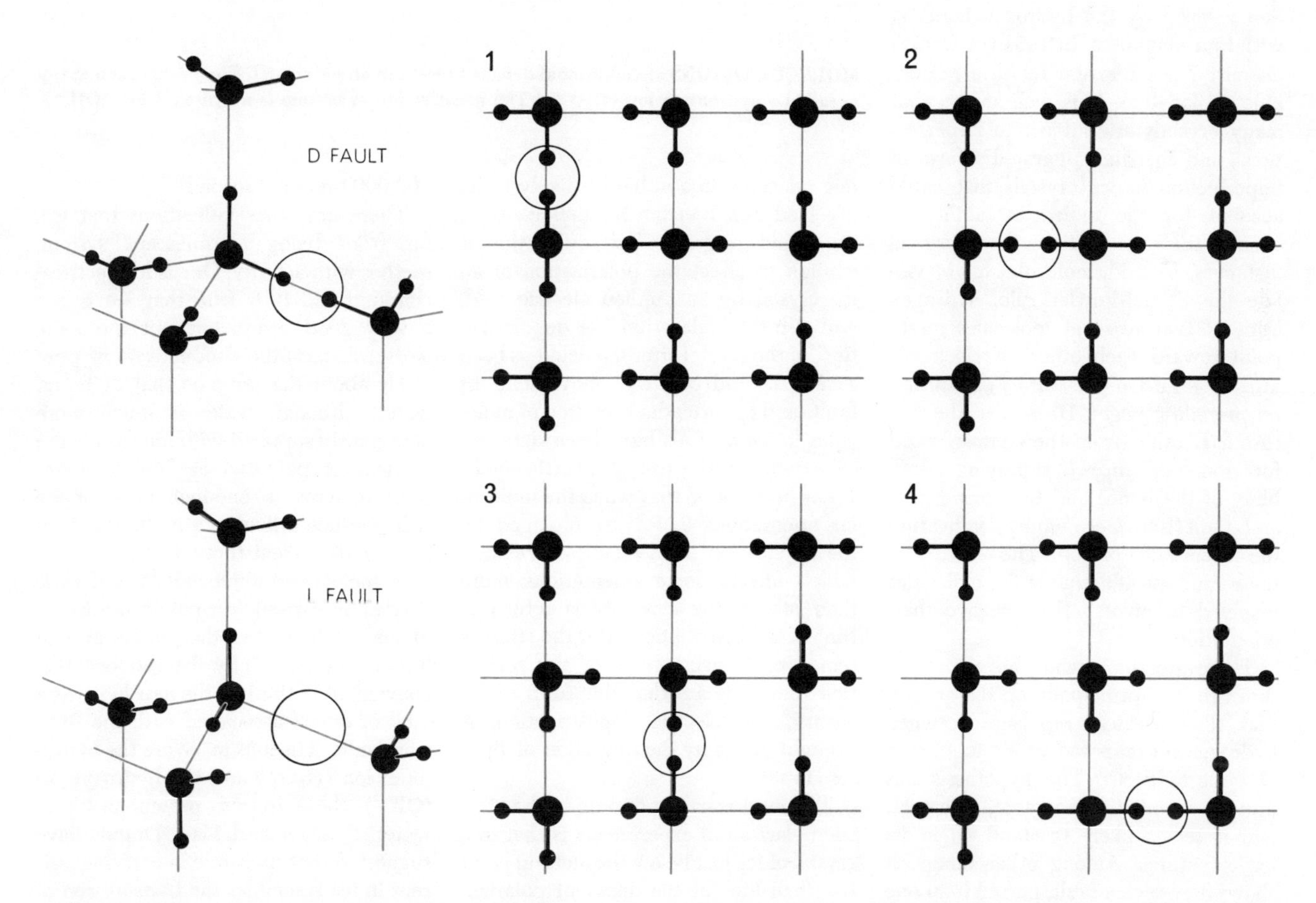

IMPERFECTIONS in the crystal lattice of ice explain how the molecules rotate. The two defects at left, proposed by the Danish chemist Niels Bjerrum in 1951, represent occasional violations of the hydrogen-bond rule. In a *D* fault the positively charged hydrogen arms of two adjacent molecules point toward each other. In an *L* fault the two negatively charged electronic arms do so. At right a *D* fault is shown wandering through a hypothetical two-dimensional crystal, leaving behind a chain of reoriented molecules. The decay of polarization (or dielectric relaxation) illustrated at the top of the page results from the migrations of such Bjerrum faults.

what actually goes on inside the crystals of ice.

The first of these complications was the discovery many years ago that when ice crystals are placed in an electric field between positively and negatively charged metal plates, the crystals become polarized—negative on the side toward the positive plate, positive on the side toward the negative plate. When the field is shut off, the crystals gradually return to their normally depolarized state. Obviously in the polarized condition a majority of the molecules must point a positively charged hydrogen arm toward the negative plate and a negative electronic arm toward the positive plate, and the orientations must become random again when the crystal is depolarized [*see top illustration on preceding page*]. The late Peter J. W. Debye of Cornell University, a pioneer in the chemistry of polar molecules, concluded that the molecules in ice must be able somehow to shift their orientations!

How could they do so? It was difficult to see how the individual molecules could rotate in the interlocked crystal system, where each molecule's orientation is fixed by the hydrogen bonding with four neighbors. In 1951 the Danish chemist Niels Bjerrum found a reasonable explanation. It is well known that many crystals are subject to imperfections, and Bjerrum suggested a form of imperfection in ice crystals that could account for the molecules' ability to change their orientation. In occasional instances, he said, molecules might violate the Bernal-Fowler rule: hydrogen arms of two adjacent molecules might point toward each other, or electronic arms might do so [*see bottom illustration on preceding page*]. He named the first case a *D* fault (from the German word for "doubled," *doppelt*, signifying a doubling of the bond) and the second case an *L* fault (from *leer*, "empty," indicating the absence of a proton). The presence of these faults would enable the molecules involved to pivot and so change their orientation.

There are some who object to Bjerrum's theory, principally on the ground that the electrical repulsion between hydrogen nuclei is too strong to allow a *D* fault to occur. The hypothesis has proved eminently workable, however, and it seems likely to stand up in its basic features. Among other things, it shows how such a fault, passed from one molecule to the next, can cause many molecules in a crystal to shift their orientation.

Experiments suggest that in ice at a temperature just below the melting point one molecule in a million is likely to be involved in a Bjerrum fault. Because of the rapid migration of faults this is enough to effect the polarization of an ice crystal by an applied electric field, and also the "relaxation," or depolarization, of the crystal after the field has been removed. Indeed, the movement of faults and the attendant rotation of molecules in ice crystals have been detected even without the use of electric fields. It has been found that when the molecular orientations in ice are changed by the application of mechanical pressure, which affects some orientations more than others, the orientations return to the normal distribution after the stress is removed. Measurements of the relaxation time indicate that this form of recovery, like electric depolarization, is brought about by the migration of Bjerrum faults.

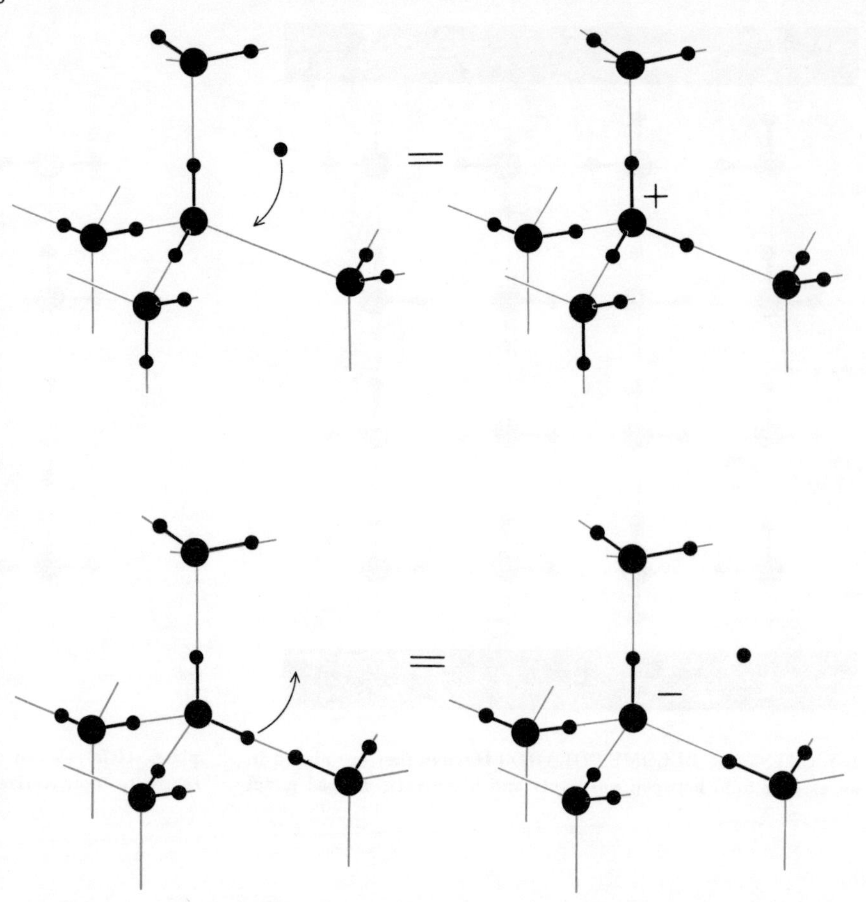

MOLECULAR IONS are additional defects present in an ice crystal. The positive ion at top is called a hydronium ion (H_3O^+). The negative ion at bottom is a hydroxyl ion (OH^-).

The most startling finding in the electric-polarization experiments is that in a crystal of ice just below the melting point the "half-life" of the decay of polarization after the shutoff of the electric field is about a hundred-thousandth of a second. We are forced to the conclusion that every molecule in such a crystal normally rotates at the rate of about 100,000 times per second!

There are other indications that ice, far from being a quiescent system, seethes with activity. One is its electrical conductivity. It is true that ice is not a very good conductor, but there are worse; in fact, the conductivity of pure ice is about the same as that of liquid water, although water is much more abundantly supplied with ions as charge carriers. A potential of 200 volts impressed across a one-inch cube of ice will produce a flow of a millionth of an ampere of current through the cube.

In metals and semiconductors electric current is carried by mobile electrons; in ice, as in some other solids and in liquids, it is carried by charged ions. It is convenient to think of the ions in ice as a kind of defect associated with the Bjerrum faults. The ions in ice are the hydronium ion (H_3O^+) and the hydroxyl ion (OH^-); these are also present in liquid water. Onsager and Marc Dupuis have suggested that their role in carrying current in ice resembles the transmission of Bjerrum faults. The hydronium ion, for instance, can transmit current by transferring its extra proton to the next ice molecule, which in turn will pass a proton to the next and so on [*see illus-*

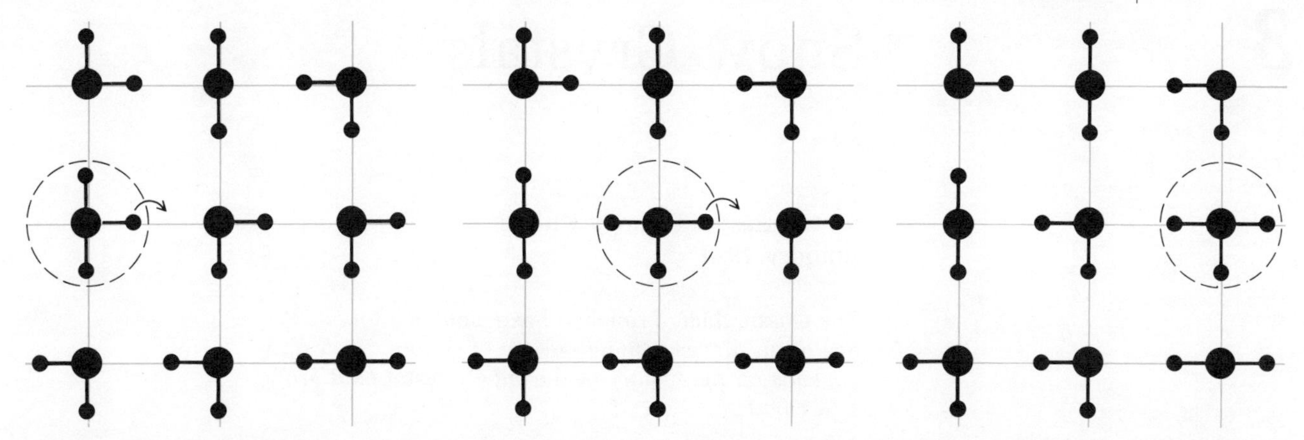

HYDRONIUM ION MOVES at an extremely high speed through the crystal lattice of ice because the entire ion need not jump, just the extra proton (*arrows*). Because of the great mobility of its protons, ice shares many properties with electronic semiconductors.

tration above]. The ions are extremely rare in ice; according to Manfred Eigen and L. De Maeyer of the University of Göttingen, there are only one hydronium ion and one hydroxyl ion for about every million million molecules. The ions move so fast, however, that they convey an appreciable amount of current. Furthermore, in ice a "catcher" is always in position to catch the proton "pitched" by its neighbor, whereas in liquid, amorphous water there may be a wait for a molecule to arrive in a position to receive the pitch. This compensating factor enables ice to conduct electricity about as well as water even though it contains fewer ions.

Ice can be called a semiconductor. Since its current-carriers are protons, whereas in the better-known semiconductors the carriers are electrons, Eigen describes ice as a "protonic" semiconductor. Indeed, Eigen has shown that ice can act as a transistor. The usual practice in making a transistor is to "dope" the semiconducting element with traces of impurities to add extra electrons. Eigen doped ice with small amounts of acids or bases (such as hydrogen fluoride, ammonia or lithium hydroxide) that increased the number of mobile protons, and he was able to construct a *p-n* junction device that rectified current.

Several years ago two Swiss chemists, W. Kuhn and M. Thürkauf of the University of Basel, reported a curious finding that has led to the discovery of still another kind of movement in the ice crystal. They used labeled molecules of water, tagged with a heavy isotope of hydrogen (deuterium) or an uncommon heavy isotope of oxygen (oxygen 18), as tracers whose travels in the crystals could be followed. Rather surprisingly it turned out that the rate of diffusion through the crystal was the same in both cases: the spread of deuterium was as rapid as that of oxygen 18. This indicated that the tagged atoms moved about in the crystal primarily as members of intact molecules, not as separate hydrogen or hydroxide ions.

How can complete molecules move in the crystal lattice? The most satisfactory explanation seems to be that they are able to make their way through the open spaces, or channels, within the lattice. The normal thermal vibrations of the molecules may occasionally cause a molecule to jump out of its lattice site into an interstitial space between other molecules. From there it may either jump back or begin to wander about in the interstices of the lattice, perhaps to wind up kicking another molecule out of position and taking its place.

From various experiments, including measurements of the "relaxation" of nuclear magnetic resonance, which is a very sensitive indicator of molecular motion, we can derive an estimate of how often these jumps occur. The calculations lead to the conclusion that in ice at a temperature just below the melting point the average molecule jumps out of its lattice position about once every millionth of a second, and it travels an average distance of about eight molecules before regaining a normal lattice site. Thus the process takes place some 10 times faster than the in-place rotations described earlier!

The placid and symmetrical appearance of a crystal of ice is certainly deceiving. Inside, at the molecular level, the perfection is punctuated by imperfections, the order is disrupted by fascinating varieties of disorder, and an astonishingly busy traffic of molecules and defects is continuously charging about through the lattice. In short, there is a great deal more to ice than meets the eye.

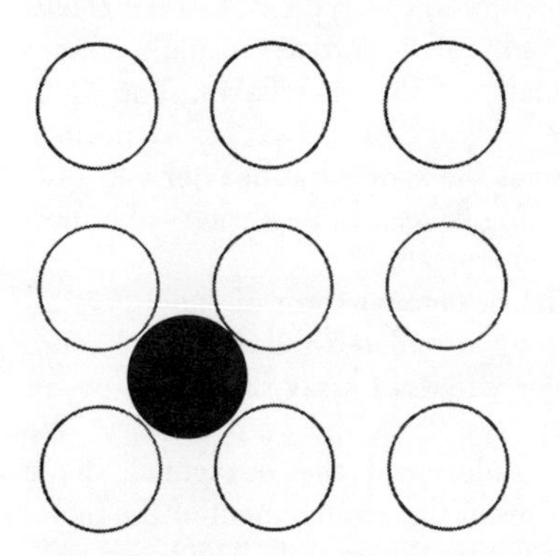

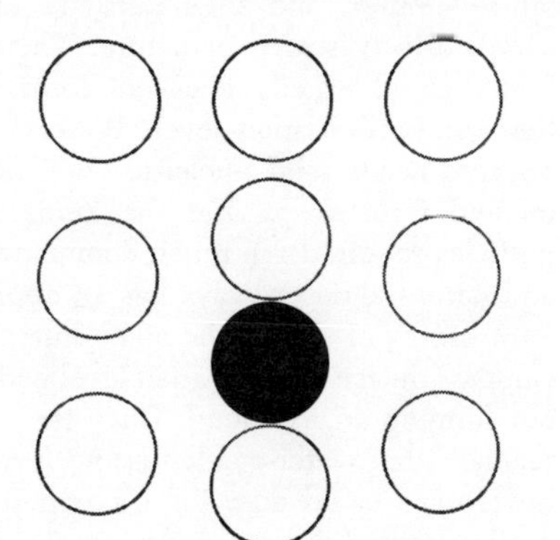

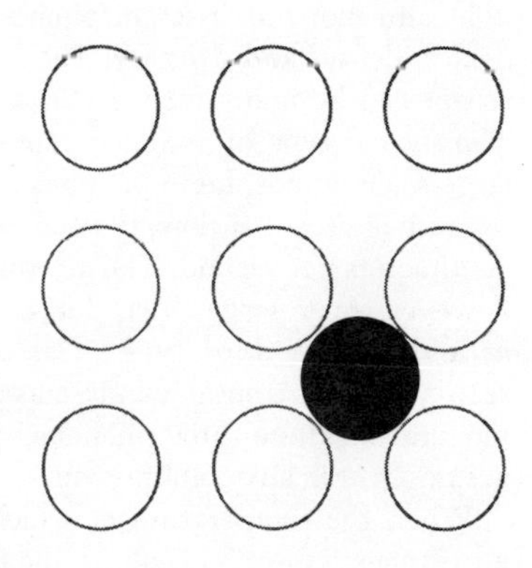

INTERSTITIAL MOLECULE is another type of imperfection in ice crystals. Normal thermal vibrations may occasionally cause a molecule to jump out of its lattice site and begin to wander about through the channels, or open spaces, within the crystal lattice.

3

Snow Crystals

by Charles and Nancy Knight
January 1973

The classic finely branched hexagonal crystal is only one of an almost infinite variety of shapes. Each form depends on the history of the snow crystal as it grows in a cloud

The familiar six-sided figure of a snow crystal reflects the arrangement of water molecules in the structure of ice. A close look at natural snow crystals reveals, however, that beyond the basic hexagonal shape the crystals are far from uniform. If fresh snow crystals are collected and examined under the microscope, there is a good chance that many of the shapes will not be at all familiar. Some are flat plates; others are long needles; others are so complex and apparently irregular that they nearly defy description. The habit, or specific appearance, of the crystal depends on the conditions under which it was formed. By catching snow crystals in their natural environment and examining them it is possible to gain knowledge of these conditions.

Since snow is formed in clouds, let us first consider some of the general properties of clouds. Clouds form when air cools below the temperature at which it is saturated with water vapor. The cooling is usually caused by the air's expanding as it ascends from a region of higher pressure to one of lower pressure. There are three common types of ascent. The first is convection: relatively warm, light and moist air rises in plumes into colder, denser and drier air. The rate of ascent can be many meters per second. The second type of ascent is caused by large-scale atmospheric motions. Here the rate of ascent is slow, typically a few centimeters per second. The air can rise, however, over large areas for as long as a day. The third type of ascent is seen when horizontal winds encounter mountains. These three phenomena can act singly or in any combination.

When the temperature at which the air reaches saturation is above the freezing point of zero degrees Celsius, a cloud of water droplets forms. The base of the cloud is flat, and the height of the base is the water-condensation level. When the temperature at which the air becomes saturated is below zero degrees C., as is typical in clouds that produce snow, there are two condensation levels, one for ice and one for supercooled water: water that remains liquid below zero degrees C. This situation arises because supercooled water always has a higher vapor pressure than ice at the same temperature. (Vapor pressure is the equilibrium pressure between a substance's vapor phase and its liquid or solid phase.) Thus when the two condensation levels coexist, the ice-condensation level is always the lowest: the first level that the air reaches as it ascends.

Since the rising and cooling air first reaches saturation with respect to ice, one might expect to find flat-bottomed ice clouds with their bases at that level. This, however, does not normally happen. Cold air can remain supersaturated with respect to ice without crystals forming for the same reason that supercooled water need not freeze. In both cases foreign particles are needed as nuclei for the ice to crystallize around, and such particles are comparatively scarce in nature. Therefore a cloud usually does not form at the first condensation level. Water vapor also needs solid nuclei to condense on and form drops, but the kinds of particles required are much commoner, and nature almost always has an abundant supply of them. The ascending air can pass the ice-condensation level without forming an ice cloud. Once the air reaches the water-condensation level, however, a cloud does form, composed of tiny drops of supercooled water.

Although nature is generally deficient in freezing nuclei, those nuclei that are present seem to be more effective in freezing supercooled water drops than in forming ice directly from a vapor. Some of the supercooled drops in the cloud contain or collide with freezing nuclei, and these drops freeze into ice. At this point the difference in the vapor pressures of ice and supercooled water becomes important. When supercooled water drops and ice particles coexist, vapor molecules around the water diffuse toward the ice. The water drop starts to evaporate, and the ice grows by direct condensation from the vapor.

The crystals grow as they fall, but since they are falling in updrafts they may actually be rising with respect to the ground, particularly in convective clouds. The clouds change with time, and the growth history of a single snow crystal can follow almost any course. Probably the commonest single habit is the planar dendrite, a flat crystal with delicate branches that is often regarded as the typical snowflake. Some of the prettiest of these snow crystals come from gentle updrafts where most of the crystal growth occurs as the crystals fall below the water-condensation level. Light snows on calm, cold nights are frequently of this type. In heavier snows the individual particles usually clump into large, fluffy snowflakes. The light, large aggregates consist of dendrites, whereas the somewhat heavier and faster-falling flakes can be clusters of bullet-shaped crystals.

The hexagonal form of snow crystals has long stimulated reflection. Johannes Kepler wrote an essay on the shapes of snow crystals in the 16th century. We now understand the hexagonal shape in terms of the arrangement of the molecules within the crystals. The speed with which water molecules find bonding sites on the crystal's surface depends on

PLANAR, OR FLAT, SNOW CRYSTALS show the familiar hexagonal form. The two crystals at the top are plates; the two at the bottom, dendrites. These forms are not replicas but original flakes. They were caught by Teisaku Kobayashi of the Institute of Low Temperature Science at Sapporo in Japan and were photographed on a microscope slide in both transmitted and reflected light.

the structure of the surface. The structure of the surface in turn depends on the surface's orientation with respect to the hexagonal lattice of ice molecules of which it consists [*see bottom illustration on opposite page*]. Although the details of the process of crystal growth are difficult to study because the scale is so small, it is clear that some orientations of the crystal's surface can take molecules into the crystal lattice faster than others. The process gives rise to crystal shapes whose external hexagonal symmetry reflects the internal symmetry of the molecular bonding.

The simplest method of studying snow crystals is to replicate them by letting them fall into a thin layer of a dilute solution of plastic and solvent. The solvent evaporates rapidly, leaving a thin plastic cast of the snow crystal. (The ice itself evaporates through the cast.) The technique works beautifully with planar crystals, although it fails badly when the crystals are complex and three-dimensional. It nonetheless makes inspecting snow crystals quite convenient. One needs a low-power microscope to study the crystals, and working with a microscope is much easier at room temperature than in the cold.

Studying the more complicated snowflakes requires that the student suffer some discomfort. Snow can be captured and examined under a microscope in an unheated room or outdoors. Since snow evaporates rather rapidly even below zero degrees C., such examination must be fast in order to ensure that the details are faithfully recorded. One way to preserve the crystals without replicating them is to collect them in a cold liquid that does not dissolve ice. They can be studied and photographed at leisure, although the environment must remain cold. That is the method we have chosen in our work at the National Center for Atmospheric Research. Hexane is the liquid we use, but kerosene or unleaded gasoline would be equally satisfactory. Our procedure has the added advantage that it reduces the contrast between the crystal and the background, because the index of refraction of the liquid is not very different from that of ice. Thus the details of the more complicated snow particles are much clearer.

The traditional method of collecting snow crystals is to catch them on a piece of cardboard covered with black velvet, and then to pick them up with a dampened toothpick. This procedure can lead

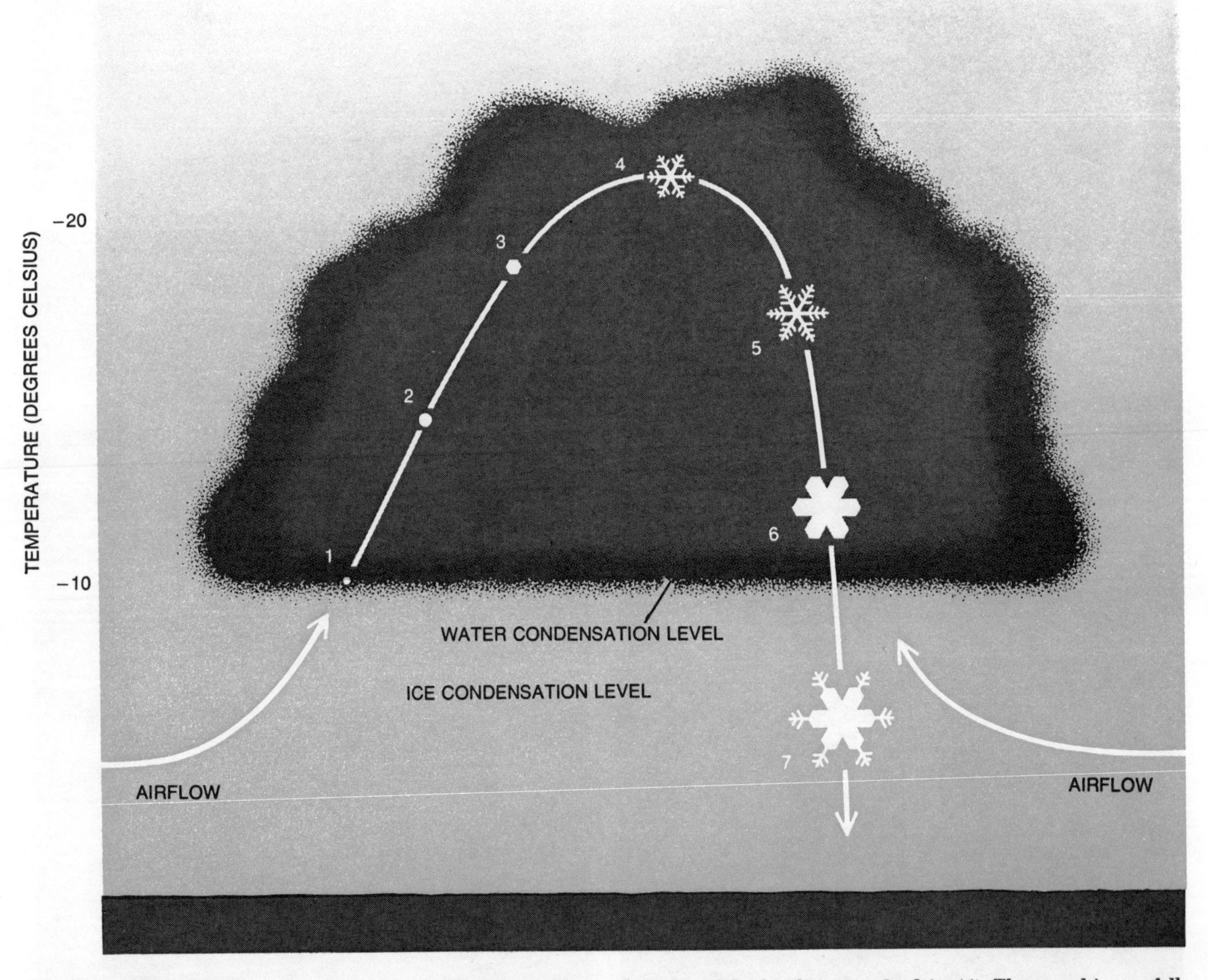

SNOW CRYSTAL GROWS within a cloud that has been formed by convection: warm, moist air rising through a layer of cold, dry air. A droplet of water condenses (*1*) at the base of the cloud at the water-condensation level. It grows (*2*) as it rises in an updraft and eventually freezes (*3*) into an ice crystal. Water-vapor molecules in the cloud attach themselves to the lattice of the crystal, creating the branches of the familiar snow dendrite (*4*). The crystal is now falling (*5*) and starts to rime (*6*), or collide with relatively large water droplets. It falls out of the cloud altogether and continues to grow from vapor (*7*) until it drops below the ice-condensation level on its way to the ground. The growth of the snow crystal can follow almost any course of events. Temperatures shown are arbitrary.

to a very distorted view of snow crystals, since one is inclined to select particularly large or symmetrical crystals for replication or examination. It is a good idea to occasionally look at everything that falls in a certain small area in 10 or 20 seconds to appreciate exactly what the storm cloud is producing.

The crystal lattice of ice guarantees that a snow crystal will be a hexagonal prism, but it does not determine whether the prism will be a thin hexagonal plate or a long hexagonal needle. Both crystal habits are found in nature. Experiments have shown that the habit of a snow crystal is almost entirely a function of temperature. As the temperature descends below the freezing point, the habit changes from planar to needlelike, back to planar and again to needlelike in four distinct regimes.

The amount of moisture in the air also affects the shape of the snow crystal. As the air becomes progressively more supersaturated the rate of growth increases and the crystal habits are more exaggerated: needles grow longer and plates become larger and thinner. The plates also become dendritic: adorned with fine branches. This branching is a result of the way the water molecules in the air diffuse to the plate. At high supersaturation the corners of the hexagonal plates are able to grow much faster than the sides because they can collect water molecules much more effectively. They develop into branches, which then split into more branches, giving rise to the familiar snow dendrites.

If the diffusion field of the moisture around growing snow crystals were perfectly symmetrical, one might expect perfectly symmetrical crystals. Very good symmetry of dendritic snow crystals is actually quite rare. This seems to be partly due to the way these crystals start and partly due to the fact that they are falling as they grow, giving a one-sidedness to the supply of material.

When a supercooled water droplet freezes, it can be transformed into a single crystal or into several different crystals, depending on the size of the drop and on its temperature when it is nucleated. A small drop with a diameter of from .01 to .05 millimeter becomes a single crystal if it freezes at a temperature above −20 degrees C.; it becomes two or more crystals if it freezes at a much lower temperature. Evidence of this fact is that clusters of bullet-shaped crystals are quite common. The crystals grow from a vapor in the lowest temperature regime, giving rise to the needle-like habit. The frozen droplet is com-

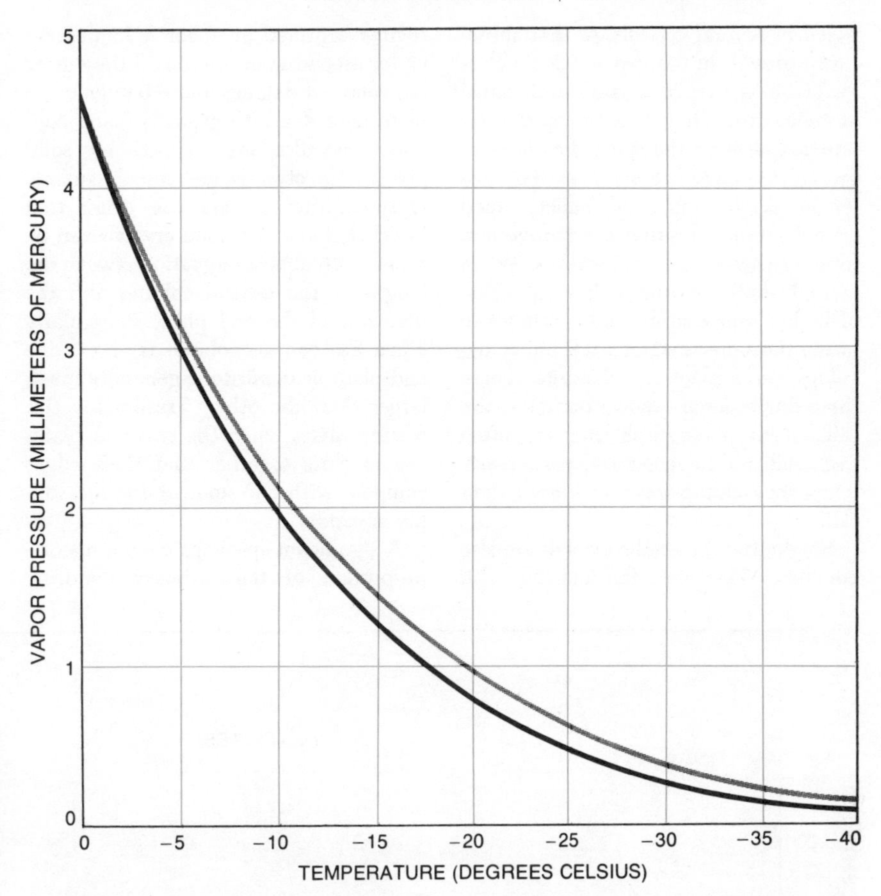

VAPOR PRESSURE of liquid water (*gray*) at any temperature below zero degrees Celsius is always higher than that of ice at the same temperature (*black*). Whenever a water droplet and an ice crystal are near each other, droplet evaporates and the ice crystal grows. Crystals grow most rapidly at −13 degrees C., where difference in vapor pressure is greatest.

FACES OF A SNOW CRYSTAL (*white shapes*) can be oriented only in particular ways with respect to the hexagonal lattice of ice molecules (*dots*) of which it is composed. Some areas of the crystal's surface grow faster than others, giving rise to crystal shapes whose hexagonal symmetry reflects the internal symmetry of the way the molecules are bound.

posed of several crystals, so that subsequent growth in the vapor yields a cluster of bullet-shaped crystals and joined at their tips. They taper toward their junction because the space they have to grow in is limited. Sometimes the clusters break apart into single bullet-shaped crystals; sometimes they clump together, forming the complex snowflakes. When a single bullet cluster falls to a region of higher temperature and continues to grow, the outer end of each bullet develops into a plate or a dendrite. These three-dimensional snow particles are difficult to photograph and are often even difficult to recognize, particularly when they clump together during their fall.

Single straight-needle crystals are also common. When they fall into the habit regime centered at about −15 degrees C. (or ascend in an updraft if they grew between −4 degrees and −8 degrees), a plate or a dendrite grows at each end. When needles are capped by solid plates, the crystals are called tsuzumi crystals, after a Japanese drum that has that shape. Tsuzumi crystals can be found with almost any ratio between the length of the central column and the diameter of the end plate. Particularly when the central column is short, one end plate or dendrite is generally much larger than the other. Presumably this feature arises when the two end plates are so close together that their edges compete with one another for the supply of vapor.

A significant—perhaps even a major—proportion of the ordinary dendritic snow crystals are actually in a sense tsuzumi crystals, although they have an extremely short central column. In fact, the central "column" can be just the original water droplet frozen into a single crystal with a plate or a dendrite developed at each side along the axis of hexagonal symmetry. When the two end plates are less than .1 millimeter apart, the competition between them becomes very strong, leading to dendritic snow crystals with fewer than six branches. One end plate develops one, two or three of the six branches; the opposite end plate develops the other five, four or three branches. Then the two end plates may break apart, as they often do when they strike the velvet of the collecting pad. The result is a dendrite with fewer than six arms, although the angle be-

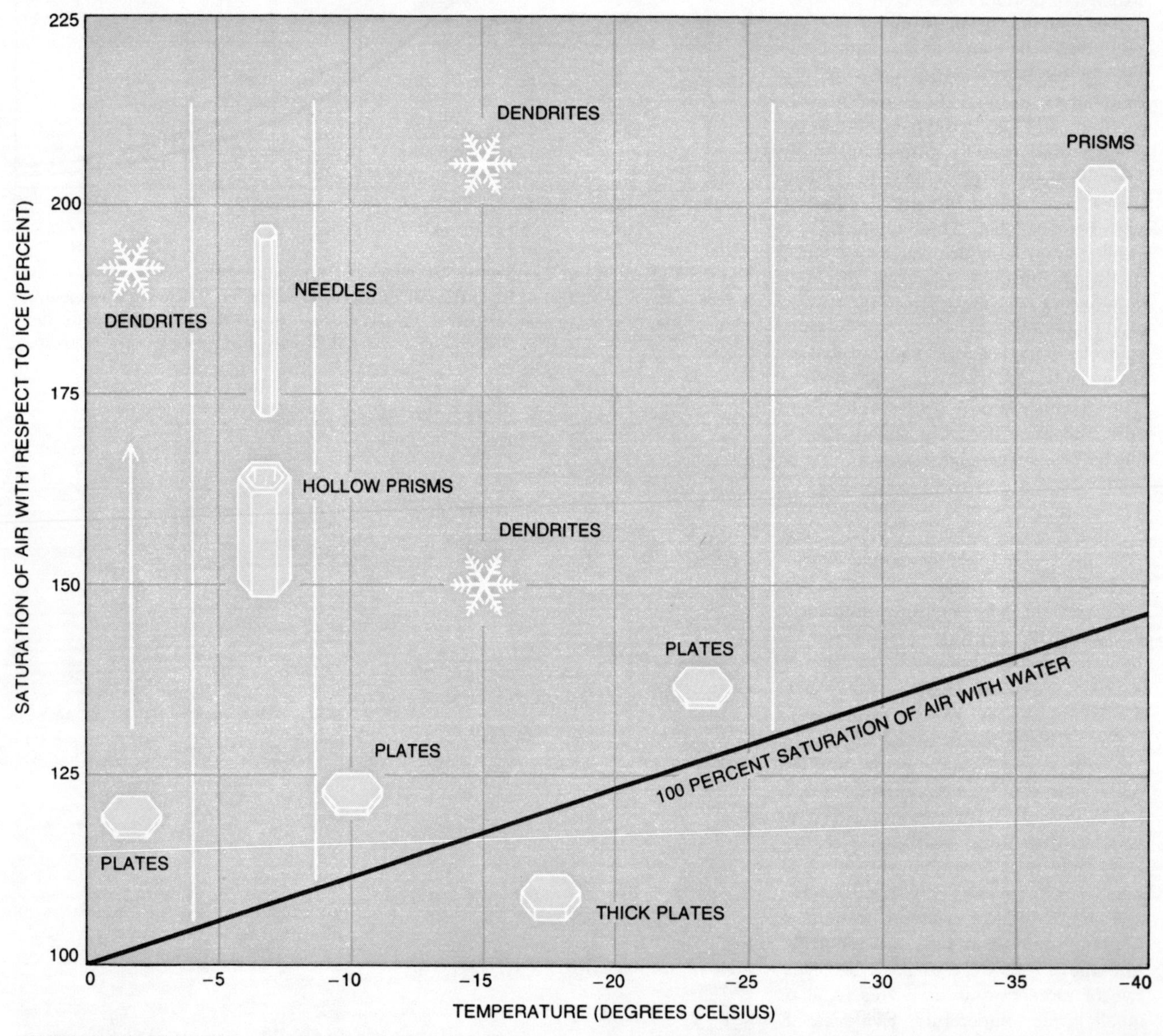

SNOW CRYSTAL HABIT, or form of growth, depends on the temperature and the amount of water vapor in the air from which the crystal grows. The habits are divided into four regimes; in two of the regimes the crystals grow as flat plates, and in the other two they grow as prisms and needles. High water-vapor content yields more exaggerated habits: dendrites instead of plates and long, thin needles instead of short, thick ones. Sometimes crystals are combinations when they fall through several regimes as they form.

tween each pair of arms is always a multiple of 60 degrees. When such crystals fail to break apart, it takes careful scrutiny to determine that they actually have two layers.

When a snow crystal evolves within a cloud of supercooled water drops, it can grow not only by stealing vapor from around the drops but also by actually colliding with individual drops. The process is called riming, and it is an important complication in all types of snowflakes and snow crystals. As a crystal grows by diffusion from the droplets, its velocity increases as it falls through the cloud. At first the small droplets in the clouds are swept around the growing crystal along with the air, some of the drops evaporating away completely as they pass close to the crystal. If the crystal grows big enough, however, the larger drops can neither evaporate away nor bypass the crystal, and they collide with it.

In any particular cloud a snow crystal must reach a certain size and attain a certain speed as it falls before it can be rimed appreciably. Growth by riming increases the crystal's speed of fall much faster than growth from vapor. When a crystal rimes, material is added mostly on its underside, thus increasing its weight without greatly increasing its air resistance. On the other hand, when the crystal grows from a vapor, material tends to be added at its sides, increasing its air resistance along with its weight. Therefore riming is an accelerating process: the more a crystal is rimed, the faster it rimes. It is not uncommon for rime deposits to have many times the mass of the original crystal.

A snow crystal may rime and then continue to grow from a vapor if it falls into the very lowest regions of the cloud, where the water drops are too small to collide with it, or if it falls into the region below the cloud but above the ice-condensation level. When the riming creates new orientations of the crystal lattice, the result is quite picturesque. A crystal type known as the spatial dendrite originates in this fashion. The new crystals are not oriented at random with respect to the original crystal. They assume certain new orientations related to the old one in such a way that the lattice structures tend to fit one another across the interface. The production of new crystal orientations in riming is probably the same in principle as the freezing of water drops at low temperatures into several crystals.

When riming continues to such an extent that the original crystal is unrecognizable or nearly so, the resulting snow is called graupel. Clouds must be thick or updrafts fairly strong before the crystals remain in them long enough to grow into graupel. Since graupel particles have the largest mass and the highest speed of fall of any of the particles in a snowfall, they can be a stage in the formation of rain. Much rain is melted graupel. When the bases of the clouds are fairly high and there is a thick layer of warm, undersaturated air below them, the individual elements of the precipitation in the cloud must be large before they can fall to the ground as rain without evaporating on the way. The mechanism that produces graupel enables many clouds to rain, and it is this mechanism that rainmakers try to encourage by adding artificial freezing nuclei to clouds. Graupel particles are also the embryos for hail [see "Hailstones," by Charles and Nancy Knight on pages

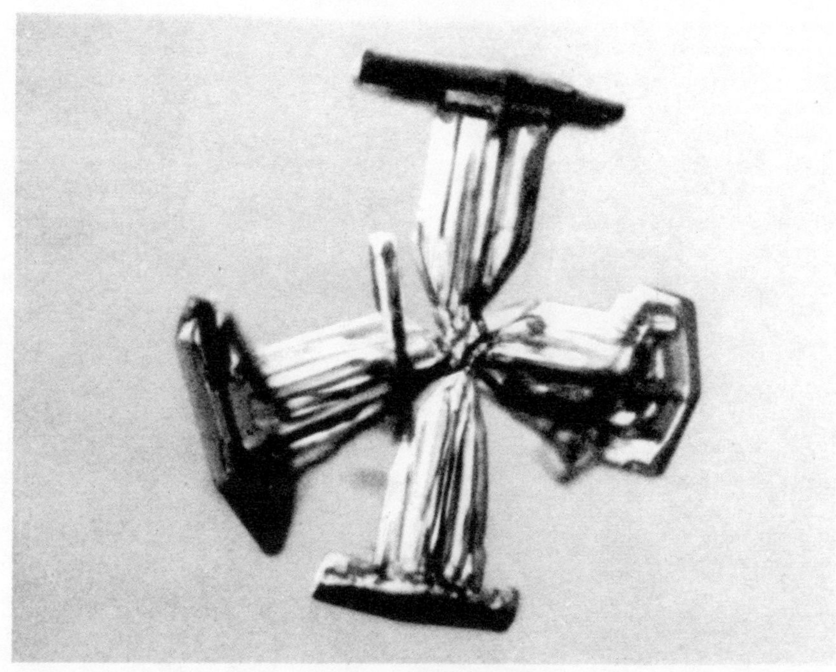

BULLET CLUSTERS originate when a tiny droplet freezes at very low temperature, forming several crystals shaped like bullets. The crystals are all joined at their tips, and the orientation of the ice lattice in each bullet is different from that of every other. If such crystals fall into higher temperature regimes, their ends grow plates (*top*) or dendrites (*bottom*).

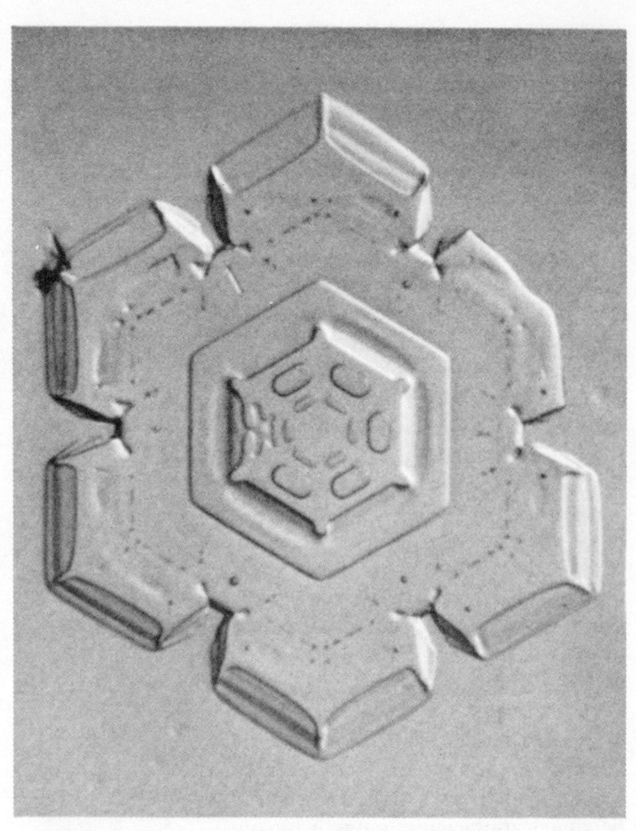

TSUZUMI CRYSTALS, named after a Japanese drum of the same shape, start out as a central hexagonal column that grows a plate at each end. Sometimes the crystals are clearly of the tsuzumi shape (*left*); at other times they look almost like an ordinary planar crystal (*right*). Here the crystal has a very short, thick column with one end plate much larger than the other. The perfect inner hexagon is the smaller end plate, and the markings inside it are the details of the central column. If a central column grows two dendrites instead of two hexagonal plates, one dendrite may develop a few of the six branches and the opposite dendrite the remaining ones.

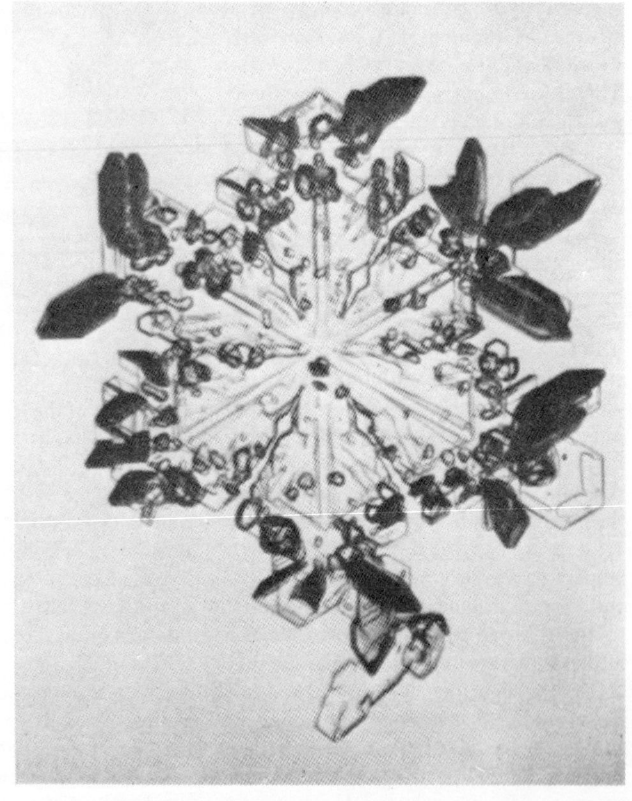

RIMED SNOW CRYSTALS have grown by capturing water droplets as well as vapor molecules. A lightly rimed dendrite (*left*) has a peculiar rounded appearance. New orientations of crystals are produced in many cases when a rimed crystal grows further from a vapor (*right*). The original crystal was a flat dendrite; the end result is a three-dimensional crystal known as a spatial dendrite.

40–47 of this Reader]. Clouds have been seeded to suppress the formation of hail by furnishing nuclei for more graupel to freeze from supercooled water drops. The purpose is to make the resulting hailstones more numerous and smaller, so that they melt before they reach the ground.

A number of the phenomena related to snow crystals are imperfectly understood. Explaining why the growth habits of the crystals vary as they do is the classical problem. Other aspects, such as the formation of new crystal orientations in riming and the origin of certain types of graupel, are not yet completely explained. By far the most important problem, however, is one that has arisen rather recently. Over the past several years a number of workers have measured the concentration of ice nuclei with respect to the concentration of ice crystals in clouds that form in the same air. Particularly at the warmer subfreezing temperatures around −10 degrees C., they have found that there are literally thousands of times more ice crystals than their measurements of the concentration of ice nuclei had led them to predict. As a result of this discrepancy they have been led to postulate a process by which one snow crystal or ice particle, perhaps during riming, could give rise to many more ice crystals. Intensive laboratory study has so far failed to reveal such a process. The problem is unsolved: either the measurements of the concentration of ice nuclei are wrong or the crystal-multiplication process exists and up to now has simply eluded discovery. Once such processes are understood it may be possible to predict and to alter them.

Let us conclude with a calculation bearing on the proverbial problem of whether or not there have ever been two identical snow crystals. A typical snow crystal weighs about 10^{-6} gram. If the average amount of snow formed on the earth each year (including snow that melts or evaporates before it reaches the ground) is equivalent to a layer of liquid water three centimeters deep over the entire surface of the earth (probably an underestimate), and if the earth is three billion years old, then some 10^{35} snow crystals have formed in that time. This comes to some 10^{29} grams—about 50 times the mass of the earth. Each snow crystal, however, consists of some 10^{18} molecules of water. Considering the huge variety of ways that number of molecules can be arranged, it may very well be that there have never been two identical snow crystals.

in them. Findeisen proposed as early as 1938 that for ice crystals to appear in supercooled clouds a nucleating or seeding agent might be required. He suggested that the agent might be dust particles of the proper configuration to start the nucleation of snow crystals, but he was never able to demonstrate its existence.

The subsequent history of cloud-seeding is well known. In 1946 Vincent J. Schaefer, then working at the General Electric Research Laboratory, discovered that ice crystals could be nucleated in a supercooled cloud by dropping dry ice into it. He discerned correctly that dry ice, at 78.5 degrees below zero C., causes water droplets to freeze spontaneously. Within months Bernard Vonnegut, then also with General Electric, conceived the use of silver iodide as an ice-nucleating, or seeding, agent. These two discoveries provided the impetus for rain-making experiments that have been conducted in many parts of the world. After a dozen years the success of these experiments is still debated. Although it has been convincingly demonstrated that the behavior of individual clouds may be modified by seeding them from aircraft, the outcome of operations aimed at producing economically significant increases in rain over large areas is much less conclusive. Evidence is accumulating, however, that modest increases of 10 to 15 per cent may be produced in favorable circumstances.

In our laboratory we have been studying the precise conditions under which supercooled water freezes and how the freezing point may be influenced by nucleating agents of various sorts. We have found that the freezing point of water varies over a range of more than 40 degrees C., depending upon the volume of the sample, the rate of cooling and the presence of impurities that may function as nucleating agents. We have frozen many thousands of water droplets, varying in diameter from one centimeter down to a thousandth of a centimeter and all containing small foreign nuclei. The water droplets are held between layers of two liquids that are practically immiscible with water and with each other. The system is cooled at a constant rate in a refrigerator, and we record the temperature at which each drop freezes. We have found that there is a linear relationship between the freezing temperature and the logarithm of the drop diameter. Thus if one-centimeter drops of a certain sample of water freeze at 18 degrees below zero C., one-millimeter drops will freeze at 24 degrees below zero, and one-tenth-millimeter drops at 31 degrees below zero [*see illustration on page 34*]. This relationship characterizes the nucleation of water droplets by foreign particles and indicates that a decrease in temperature makes a logarithmically increasing number of atmospheric particles capable of acting as nuclei.

We have recently been successful in purifying water to such an extent that we can produce large numbers of drops entirely free of foreign particles. We accomplish this by repeatedly filtering and distilling water in a closed apparatus from which atmospheric air is rigidly excluded. One-millimeter droplets of such very pure water may be supercooled to 33 degrees below zero C. before freezing, and droplets one-thousandth of a millimeter in diameter may be cooled to 41 degrees below zero. These droplets of pure water freeze spontaneously.

Presumably small groups of water molecules, undergoing random fluctuations in position and velocity, become locked by chance into an icelike arrangement and thereby serve as nuclei to initiate the freezing process. One can calculate the rate at which such aggregates form and hence the probability that a drop of a given size will freeze. The lower curve in the illustration on page 34 indicates the computed temperatures at which droplets of various sizes should freeze within one second. The curve coincides rather well with experimental observations, and is distinctly different from the freezing curve where foreign particles play a nucleating role.

Except at very low temperatures—lower than 40 degrees below zero C.—the ice-nucleus content of the air is of fundamental importance for snow formation. It is not easily measured. A favorite method requires a cloud chamber in which a sample of atmospheric air is saturated with water vapor and rapidly

than 30 degrees below zero centigrade. (These columns happen to be formed from heavy water.) The thin hexagonal plate (*center*)

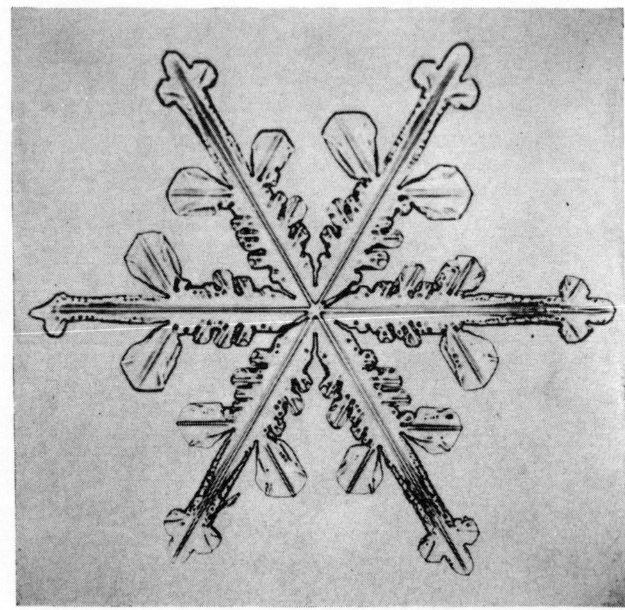

is one millimeter in diameter and has petal-like extensions. Star-shaped crystal (*right*) forms the basis of the typical snowflake.

4

The Growth of Snow Crystals

by B. J. Mason
January 1961

Much of the world's precipitation is triggered by natural dusts that act as nuclei in causing water droplets in clouds to freeze. Some artificial nuclei work more effectively than natural ones.

The remarkable beauty of snow crystals, revealed in the classic elegance of their simple geometrical shapes and the delicate tracery of their more intricate forms, has long been recognized and recorded by the scientist, the artist and the industrial designer. It is only in recent years that a serious scientific study has been made of their structure, germination and growth. These studies have been largely motivated by the increasing interest in the physics of clouds and the formation of rain, and in the possibility of modifying these processes artificially. It appears that over large portions of the earth raindrops first begin their lives as snow crystals; then they melt before they reach the ground.

My colleagues and I at the Imperial College of Science and Technology in London have spent a number of years studying the birth and growth of snow crystals in the laboratory, hoping to learn something about the way the crystals develop in clouds. Except for the very cold, high-altitude cirrus types, which are thin veils of small ice crystals, clouds consist mainly of water droplets so tiny and so dispersed that they stay suspended in the air like smoke particles. For years meteorologists puzzled over this stability of clouds, and were hard pressed to explain how the tiny water droplets ever grow large enough to fall as rain. It was equally puzzling that the water droplets often refuse to freeze even though the cloud may be many degrees below the nominal freezing point of water: zero degrees centigrade, or 32 degrees Fahrenheit. Even on a hot summer day the temperature of the air above 15,000 feet is usually below freezing.

During the 1930's Tor Bergeron of Sweden and Walter Findeisen of Germany provided a theory of cloud behavior that seems to account satisfactorily for much of the world's precipitation. They proposed that clouds remain stable until a small percentage of the cloud droplets finally freeze, spontaneously or otherwise. When water molecules are locked into place in an ice crystal, they evaporate much less readily than they do from a drop of water. Thus if a cloud contains both water droplets and ice crystals, the water molecules that diffuse from the vapor state onto the ice crystals tend to be bound fast, and those that condense on the water droplets are relatively free to evaporate again. As a result the crystals grow more rapidly than the droplets; finally, as the air is denuded of moisture by the ice crystals, the water droplets evaporate and disappear. The ice crystals meanwhile grow large enough to fall toward the earth. After growing for about an hour in a deep layer of cloud, a snow crystal will reach the size of a drop in a drizzle, or perhaps the size of a small raindrop. Such crystals fall at the rate of about one foot per second. Several of them may become joined together, as they settle through the air in a fluttering or tumbling motion, to form a snowflake which, in falling into the warmer regions of the cloud, may melt and reach the ground as a raindrop.

Bergeron and Findeisen originally believed that virtually all the world's precipitation—snow or rain—originated with this ice-crystal mechanism, but it is now known that, especially in the tropics, rain sometimes falls from clouds so warm that ice could never have formed

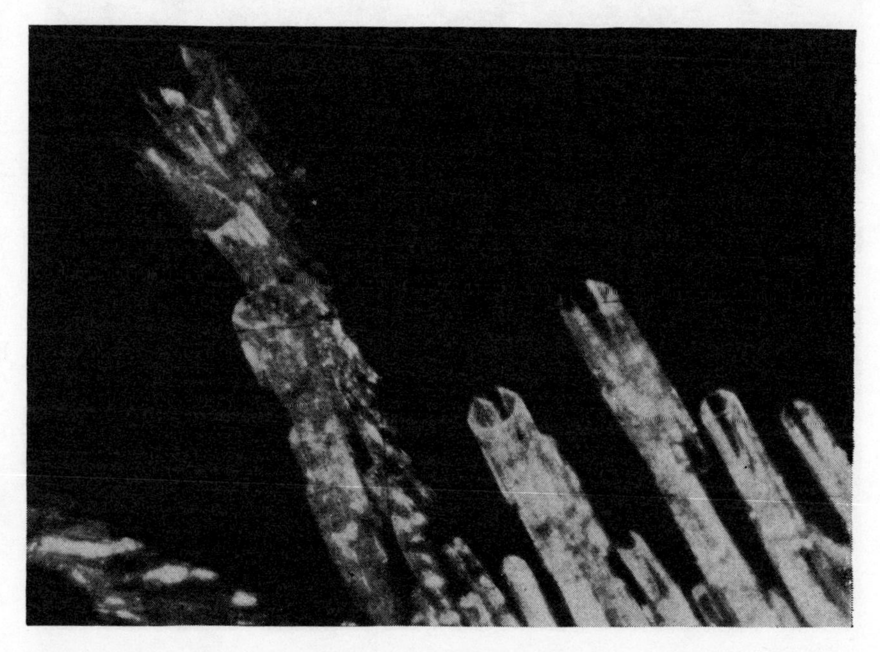

THREE BASIC FORMS OF SNOW CRYSTALS provide the basis for an infinite variety of shapes. Hollow prismatic columns (*left*) populate cirrus clouds, which are usually colder

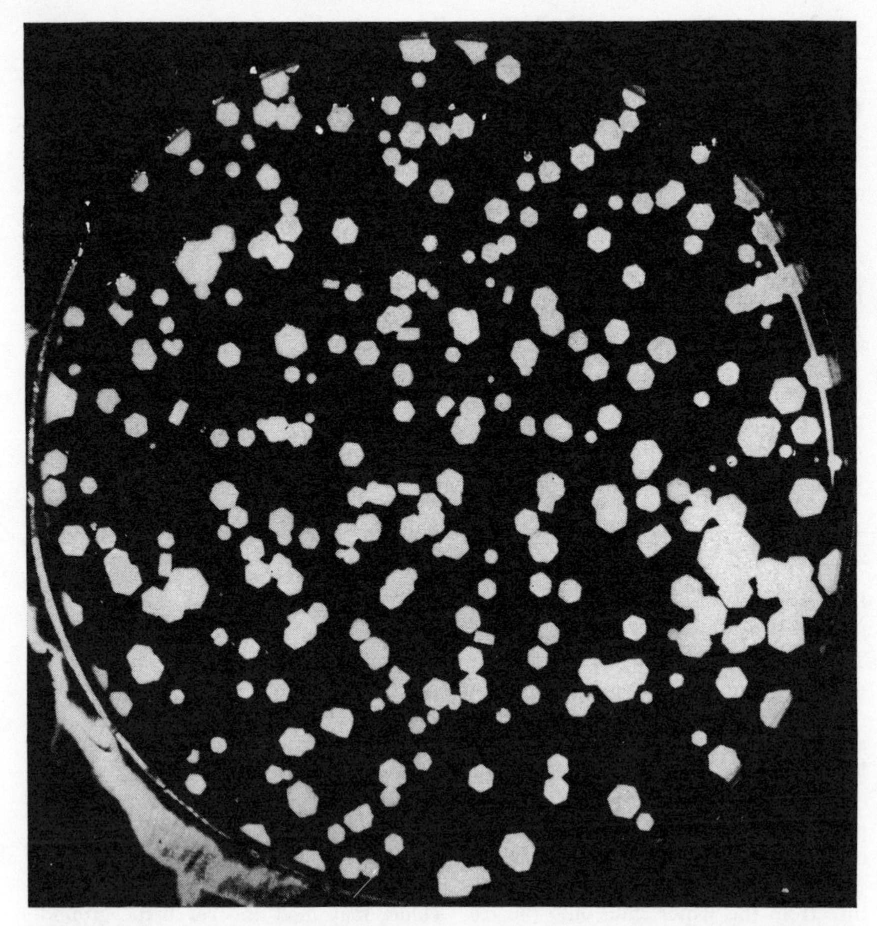

SUPERCOOLED SUGAR SOLUTION provides a way to count tiny ice crystals created in a cloud chamber. After falling on the sugar solution, crystals grow to appreciable size.

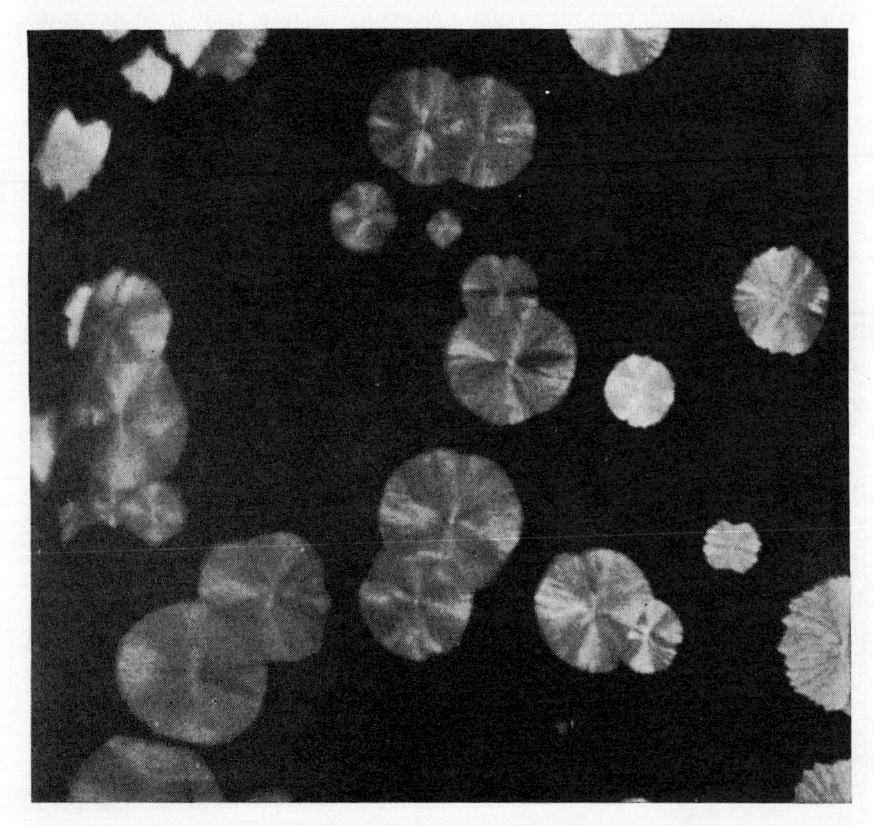

SUPERCOOLED SOAP FILM is a simple detector for determining nuclei-content of the atmosphere. Tiny ice crystals enclosing nuclei are counted after they land on film and grow.

cooled by sudden expansion. During the rapid cooling, water vapor condenses on some of the airborne particles to produce a cloud of tiny supercooled droplets. Some of these contain ice nuclei; they freeze and grow into ice crystals. The technique is then to count the number of crystals glittering in an illuminated volume of the cloud, successive measurements being made at lower and lower temperatures achieved by larger and larger expansions. Because it is not easy to discern small numbers of crystals swirling about in a thick fog, direct visual counts are not very accurate. This led my former colleague Keith Bigg to devise an ingenious technique in which the ice crystals fall into a tray of sugar solution placed at the bottom of the cloud chamber. The water in the solution supercools, and when the tiny ice crystals fall into it, they quickly grow to visible size and may be easily counted [*see illustration at left*].

Measurements made from aircraft over both land and sea show that the ice-nucleus population of the atmosphere varies considerably from day to day and from place to place. On some occasions it appears to fall below the minimum value required for the efficient release of precipitation from clouds. This is the justification for rain-making experiments.

The nature and origin of the nuclei necessary to initiate the formation of ice crystals are subjects of considerable interest and controversy. While I believe that they originate mainly from the earth's surface as dust particles carried aloft by the wind, E. G. Bowen of Australia's Commonwealth Scientific and Industrial Research Organization has suggested that the debris of meteorites may be an important source. He has made analyses of world rainfall patterns which seem to show some correlation with the annual meteor showers. In an attempt to test these rival hypotheses John Maybank and I have recently examined, in the laboratory, the ice-nucleating ability of various types of soil particles and mineral dust and also of meteoritic dust.

Of the 30 terrestrial dusts we have tested, 16 (mainly silicate minerals of the clay and mica variety) produced ice crystals in supercooled clouds at temperatures between 10 and 15 degrees below zero C. [*see illustration on pages 38 and 39*]. These substances are all minor constituents of the earth's crust. It is significant that common materials such as sea sand were not active. (Since the quartz of ordinary sand has a hexagonal

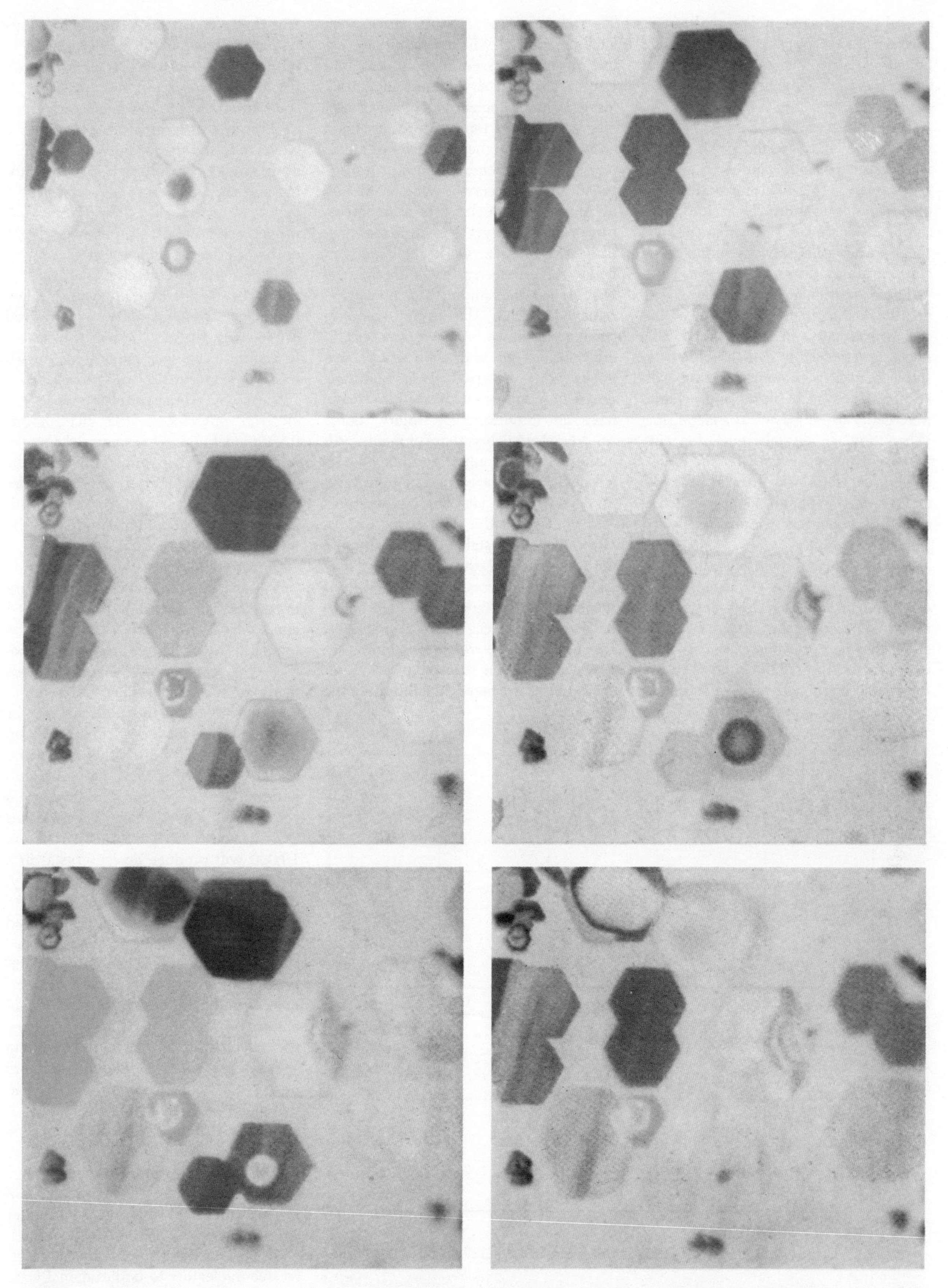

GROWTH OF SNOW CRYSTALS is revealed in this series of photomicrographs made with reflected light. The hexagonal ice crystals, growing on a single crystal of natural cupric sulfide, appear to change color (due to canceling of certain wavelengths by interference) as they become thicker. Time interval between first picture (*top left*) and second (*top right*) was 45 seconds. The rest of the series followed at 15-second intervals. Ice crystals tend to grow in diameter until they meet another crystal, then they thicken. Crystals that are of differing thickness when separate tend to acquire the same thickness after coming in contact.

crystal-structure resembling that of ice, Findeisen had thought that quartz might be an effective nucleating agent. But a superficial resemblance in structure is not enough.)

The most abundant of the active substances we have tested is kaolinite, which initiates ice formation at nine degrees below zero C. This mineral is common enough to provide an important source of ice nuclei, but not so common that the atmosphere always contains high concentrations of its particles.

These particular experiments were greatly facilitated by the use of a simple, convenient and readily renewable nucleus detector. It consists of a very stable soap film (obtained from a half-and-half mixture of water and a liquid detergent) stretched across a metal ring. When tiny ice crystals, enclosing submicroscopic nuclei, land on the supercooled soap film, they grow rapidly into crystals large enough to be easily detected and counted [*see bottom illustration on page 32*].

In the course of our nucleation experiments we made a surprising discovery: Ten of the terrestrial dusts were found to become more effective ice nuclei if they had previously been involved in ice-crystal formation. In other words, they could be preactivated, or "trained." Thus when ice crystals grown on kaolinite nuclei, which are initially active at nine degrees below zero, are evaporated in a dry atmosphere, they leave behind nuclei which are thereafter effective at temperatures as much as five degrees higher. Particles of montmorillonite (another important constituent of some clays), which initially become active nuclei only at temperatures some 25 degrees below zero, may be preactivated to work at 10 degrees below zero. It seems that, although the bulk of ice surrounding the nucleus is removed during the drying process, small germs of ice, retained in pores and crevices, survive and serve as effective nuclei when the particle is again exposed to a supercooled cloud. We now have an interesting possibility. Some soil particles, such as those of montmorillonite, which are initially rather poor as ice nuclei may be carried aloft to form ice crystals at the very low temperatures associated with the high cirrus clouds. Later, if the crystals should evaporate without reaching the earth, they may leave behind trained nuclei capable of nucleating lower clouds at temperatures only a few degrees below freezing. If we accept this possibility of training initially unpromising material in the upper atmosphere, we need not interpret the fact that efficient nuclei are occasionally more abundant at higher levels as implying that they must have entered the atmosphere from outer space.

In an attempt to provide a direct test of Bowen's meteoritic-dust hypothesis, we have tested the ice-nucleating ability of the fine dust resulting from the grinding and vaporization of several different types of stony meteorite. None has proved effective at temperatures higher than 17 degrees below zero C.

The evidence therefore appears to favor the theory that atmospheric ice-nuclei are predominantly of terrestrial origin, with the clay minerals, especially kaolinite, being a major source. Additional confirmation is provided by Japanese workers who have used the electron microscope and electron-diffraction techniques to examine the nuclei at the centers of natural snow crystals. More than three quarters of the particles were identified as soil particles, with kaolinite and montmorillonite as the most likely constituents.

Since the discovery by Vonnegut that tiny particles of silver iodide, introduced as a smoke into a supercooled cloud, cause ice crystals to appear at temperatures as high as four degrees below zero C., an intensive search has been made for other substances that might be even more effective and cheaper for cloud-seeding purposes. The table on pages 38 and 39 lists those artificial nuclei that have proved active at temperatures between four and 14 degrees below zero. The temperature shown is that at which at least one particle in 10,000 will produce an ice crystal in a supercooled cloud in the laboratory. Greater numbers of effective nuclei are obtained as the temperature is lowered below the threshold value. The first seven substances in the table are active at temperatures between four and 11 degrees, where only a very small proportion of natural ice nuclei are effective; hence the seven are all potential seeding agents. But silver iodide, being more potent and more easily dispersed than its rivals, retains first place. Nuclei of ammonium fluoride, cadmium iodide and iodine, being soluble substances, would dissolve within a minute or two of entering a water cloud, but they can be made to act as ice nuclei under special laboratory conditions.

There is a tendency for the more effective nucleators to be hexagonal crystals in which the atomic arrangement is reasonably similar to that of ice, but there are exceptions. Nevertheless for all those substances that are active above

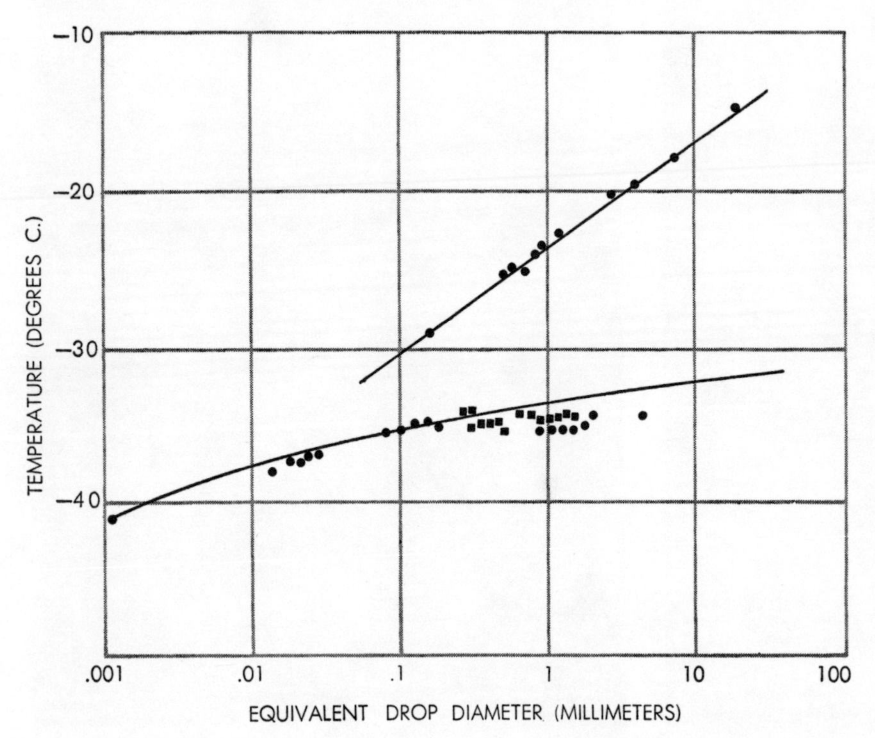

FREEZING POINT OF WATER varies with drop size. Upper curve shows freezing point of water containing impurities. Curve falls linearly with logarithmic decrease in drop size. Lower curve is theoretical freezing point for pure water and agrees well with author's values (*circles*) and those (*squares*) of Stanley Mossop of the University of Oxford.

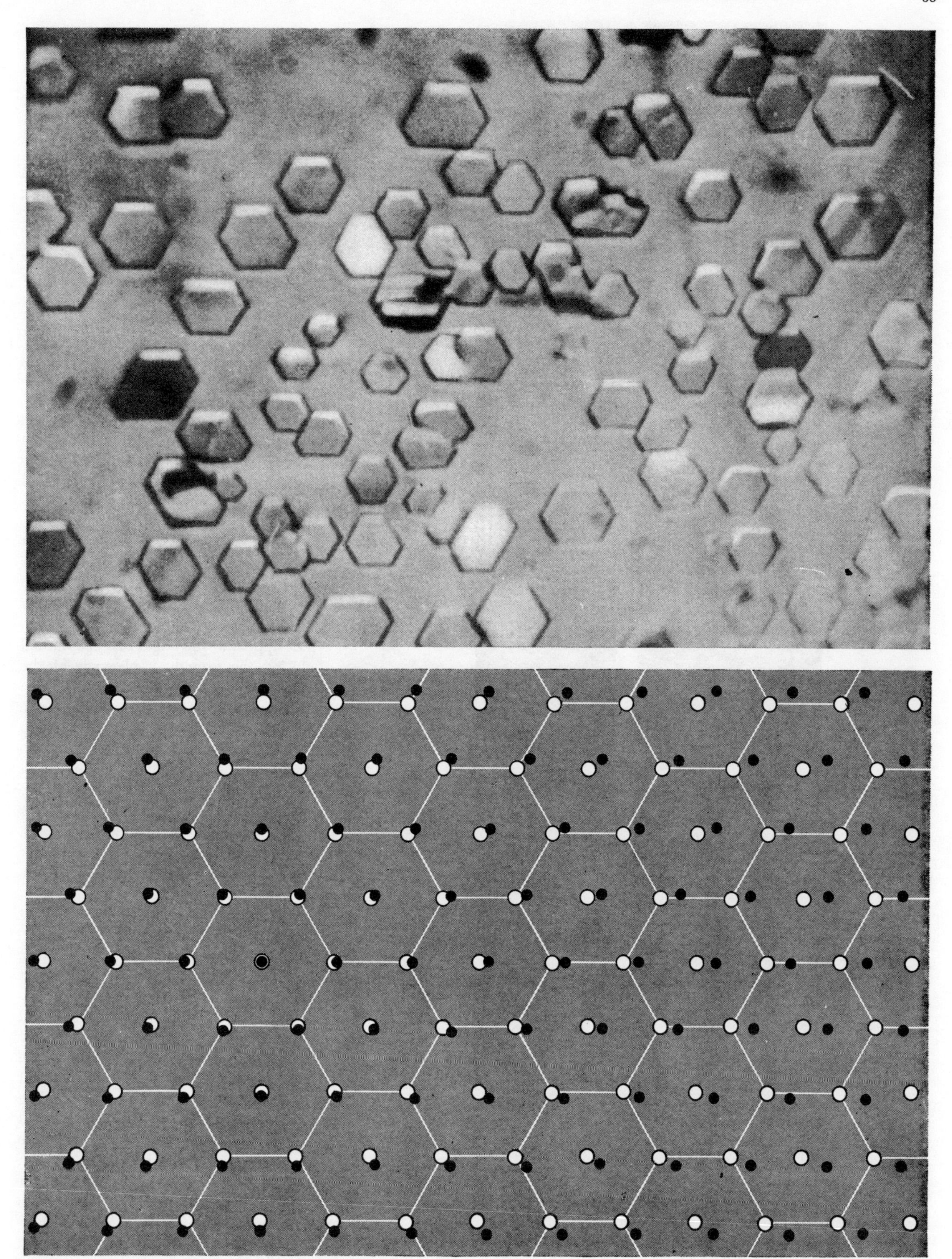

ICE GROWING ON SILVER IODIDE (*top*) shows how underlying symmetry of a large single crystal forces ice crystals to assume a parallel orientation. The relationship between the lattices of the two crystals is diagrammed below. Oxygen ions (*open circles*) define the corners of the hexagonal ice lattice; silver ions (*black dots*) lie at the corners of the silver iodide lattice. Assuming the two crystals are in perfect superposition at left of center, the match between them becomes progressively poorer in all directions, here exaggerated about threefold. Silver iodide provides best lattice fit in three dimensions of all artificial nuclei.

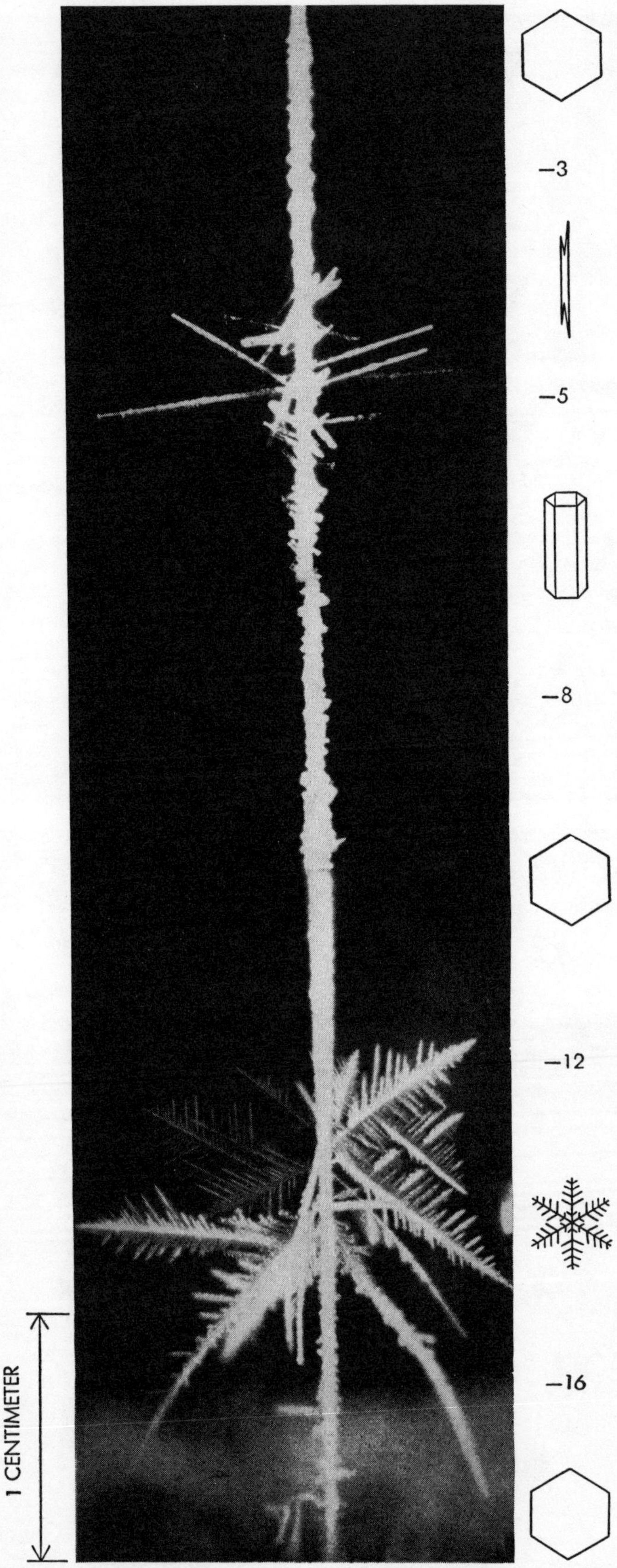

GAMUT OF ICE CRYSTAL SHAPES grows on a filament suspended in a diffusion chamber with controlled temperature gradient. Crystals take characteristic forms at various temperatures as indicated along the right edge of the photograph. Reading from the top, the symbols represent: thin hexagonal plates, needles (which are defective prisms), hollow prismatic columns, hexagonal plates, branched star-shaped crystals (or dendrites), and hexagonal plates. At temperatures lower than 25 degrees below zero C., prisms appear again.

15 degrees below zero C. it is possible to find a crystal face on which the atomic spacings differ from those of ice by only a few per cent. On the other hand, we now know that the nucleating ability of a particle is not determined solely by the degree to which its atomic structure matches that of ice, and that other factors, not yet fully understood, play a role.

To investigate such factors in more detail my colleagues and I have studied, under carefully controlled conditions of temperature and humidity, the growth of ice on individual faces of single crystals of various nucleating agents. We have observed that ice crystals always assume a parallel orientation when they are grown on the hexagonal crystals of silver iodide, lead iodide, cupric sulfide, cadmium iodide and brucite, and also on crystals of calcite, mercuric iodide, iodine, vanadium pentoxide and freshly cleaved mica.

We also find that ice crystals start growing preferentially around local imperfections on the crystal surface, which show up as small dark spots in our photomicrographs. Often the ice crystals appear at the edges of steps formed during the growth or cleavage of the nucleating crystal. They also show a definite preference for the deeper steps. On lead iodide, for example, steps which are about a tenth of a micron (one ten-thousandth of a millimeter) in height provide effective nucleation sites if the air is supersaturated by about 10 per cent, but much higher supersaturations, exceeding 100 per cent, are required for nucleation on the very flat areas of the host crystal.

A number of striking color effects that appear during the growth of ice crystals on a blue crystal of natural cupric sulfide (covellite) reveal much about the growth mechanism. Being only a few hundred millimicrons high, and thus comparable to the wavelength of visible light, the growing hexagonal ice plates produce interference colors that give a very accurate measure of plate thickness. The series of six photographs on page 33 are typical of many we have made. They show that in the first stages some crystals grow in diameter with no change in color, indicating no appreciable change in thickness. Evidently the water molecules arriving at the upper surface of the crystal are not captured but migrate over the surface to be built in at the crystal edges. When two crystals touch, however, they begin to thicken as colored growth-layers spread, with a speed of a few microns per second, from the point of contact across the crystal surface. Crystals that are of differing thickness when separate tend to acquire the same thickness after coming into contact. We now know that perfect crystals grow only very slowly under normal supersaturations. It seems that to allow a crystal to grow at an observable rate, imperfections must be set up, as when a crystal accidentally hits a step or another crystal. Thus crystals sitting astride a step are often of different thickness on either side, as indicated by their two-toned appearance.

Although snow crystals occur in almost infinite variety, they can all be classified into three basic forms: the hexagonal prismatic column, the thin hexagonal plate and the branching star-shaped form, sometimes called dendritic [*see illustrations on pages 30 and 31*]. Until recently experts disagreed on the reasons for the differences in form. Some argued that the form depended less on temperature of formation than on the degree of supersaturation of the surrounding air.

It is almost impossible to assign reasons for the different crystal forms by collecting snowflakes on the ground. Accordingly much effort has gone into collecting ice crystals from aircraft and noting the temperature of the surrounding air. Such studies have revealed that at a given temperature a particular crystal type tends to be dominant. The high cirrus clouds, which are usually colder than 30 degrees below zero, consist of prismatic columns, typically half a millimeter long and containing pronounced funnel-shaped cavities at each end. The medium-altitude clouds, whose temperatures range between 15 and 30 degrees below zero, contain both plates and prisms. The greatest variety of crystals are found in the lower portions of supercooled clouds, where temperatures run from about five degrees below zero up to zero. Here occur hexagonal plates, short prisms, long thin needles and, most striking of all, beautiful, intricate stars up to several millimeters in diameter. Snowflakes, which are composed of from two to several hundred individual crystals, form at temperatures only a few degrees below zero.

The observation of clouds thus suggests that the shape of a snow crystal is largely controlled by the temperature of the air in which it grows. This has been confirmed in a striking manner by growing crystals under carefully controlled conditions in the laboratory. We grow crystals on a thin fiber running vertically through the center of a diffusion cloud chamber in which the vertical gradients of temperature and vapor density can be accurately controlled and measured. The results of many experiments, covering a temperature range from zero down to 50 degrees below zero and vapor supersaturations varying from a few per cent to 300 per cent, consistently show that the crystal shape varies with temperature along the length of the fiber. The cycle of shapes —from the hexagonal plates at zero degrees through needles at three degrees below zero—is always precisely reproducible. The boundaries between one form and another are very sharp [*see illustration on opposite page*]. For example, the transitions between plates and needles at three degrees below zero and those between hollow prisms and plates at eight degrees below zero occur within temperature intervals of less than one degree. Crystals grown from heavy water are almost identical with ordinary ice crystals except that the transition temperature between forms is shifted upward by about four degrees. This conforms with the melting point of heavy ice, 3.8 degrees above zero.

Our experiments have shown conclusively that it is temperature alone, and not supersaturation of the surrounding vapor, that governs the crystal form. The effect of supersaturation is simply to alter the growth rate: the greater the supersaturation, the faster the growth. The dependence of crystal growth upon temperature may be demonstrated in dramatic fashion merely by raising or lowering the fiber in the chamber. Whenever a crystal is thus transferred to a new environment, its further growth takes the form characteristic of the new temperature regime. By

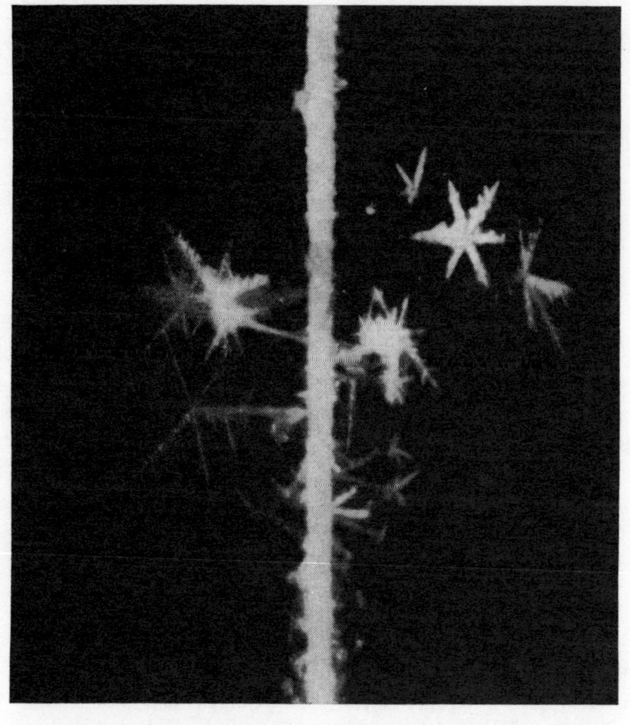

CRYSTAL HYBRIDS show how form is dictated by temperature. Needles grown at five degrees below zero C. developed plates on their ends when shifted to a temperature of 10 degrees below zero (*left*); stars when shifted to 14 degrees below zero (*right*).

thus altering the temperature we have been able to produce hybrid combinations of all the basic crystal types [*see illustrations on preceding page*].

The exact nature of the growth mechanism that can completely change the crystal shape in the space of a degree or two, and that produces five complete changes of habit in the space of only 25 degrees, is still something of a mystery. Our studies indicate, however, that the rate of migration of molecules from one crystal face to another, which appears to be very sensitive to the temperature, will be an important ingredient of the final explanation.

The work I have described does not tell us whether we can hope eventually to modify the weather, but it is aimed at establishing some of the basic physical processes that are involved in the natural formation of rain. We know that water droplets of the size found in clouds will rarely freeze spontaneously except at temperatures of 30 degrees or more below zero. We also know that there are terrestrial dusts capable of acting as nucleating agents at temperatures as high as five degrees below zero. Some

NATURAL NUCLEI					ARTIFICIAL	
SUBSTANCE	CRYSTAL SYMMETRY	CRYSTAL FORM	LATTICE MISFIT WITH ICE (PER CENT)	THRESHOLD TEMPERATURE (DEGREES C.)	SUBSTANCE	CRYSTAL SYMMETRY
ICE CRYSTAL	HEXAGONAL		0	0	SILVER IODIDE	HEXAGONAL
COVELLITE	HEXAGONAL		−2.8	−5	LEAD IODIDE	HEXAGONAL
BETA TRIDYMITE	HEXAGONAL		−3.5	−7	CUPRIC SULFIDE	HEXAGONAL
MAGNETITE	CUBIC		−7.1	−8	MERCURIC IODIDE	TETRAGONAL
KAOLINITE	TRICLINIC		−1.1	−9	SILVER SULFIDE	MONOCLINIC
GLACIAL DEBRIS				−10	AMMONIUM FLUORIDE	HEXAGONAL
HEMATITE	HEXAGONAL		−3.5	−10	SILVER OXIDE	CUBIC
GIBBSITE	MONOCLINIC		+12	−11	CADMIUM IODIDE	HEXAGONAL
VOLCANIC ASH				−13	VANADIUM PENTOXIDE	ORTHORHOMBIC
VERMICULITE	MONOCLINIC			−15	IODINE	ORTHORHOMBIC

NATURAL AND ARTIFICIAL ICE-NUCLEATING AGENTS tend to have hexagonal crystal habits, though there are notable exceptions. At left are nine of the 16 atmospheric dusts found to be effective nuclei at temperatures of 15 degrees below zero C. or higher. At right are 10 of the most effective artificial nuclei. The dimensional agreement between the crystal lattices of effective nuclei and the lattice of ice is usually a good one, but it does not necessarily correlate with threshold temperatures. The best fit may

of the dusts can even be trained to be more effective than they are normally. But every stable, supercooled cloud is visible evidence that effective nucleating agents are frequently lacking. Whether providing them artificially, under favorable conditions, will significantly increase snowfall or rainfall is a matter still to be resolved to the general satisfaction of meteorologists.

NUCLEI

CRYSTAL FORM	LATTICE MISFIT WITH ICE (PER CENT)	THRESHOLD TEMPERATURE (DEGREES C.)
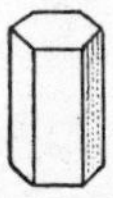	+1.3	−4
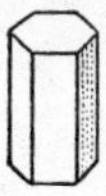	+0.4	−6
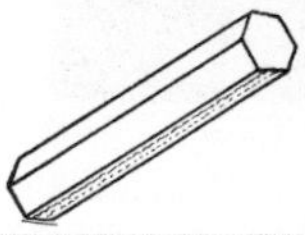	−2.8	−6
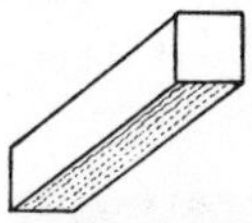	−3.5	−8
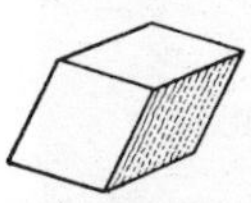	−0.3	−8
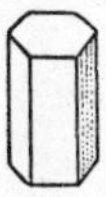	−2.9	−9
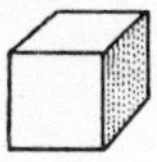	−3.8	−11
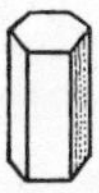	−6.2	−12
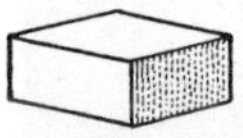	−2.1	−14
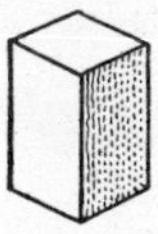	−1.5	−14

occur between uneven multiples of the lattice constants of the crystal and of ice. The best fit for any one dimension of the lattices is shown in the fourth and ninth columns.

ON CADMIUM IODIDE CRYSTAL with spiral growth steps, ice crystals form preferentially at the edges of the steps. Hexagonal sides of ice crystals maintain parallelism.

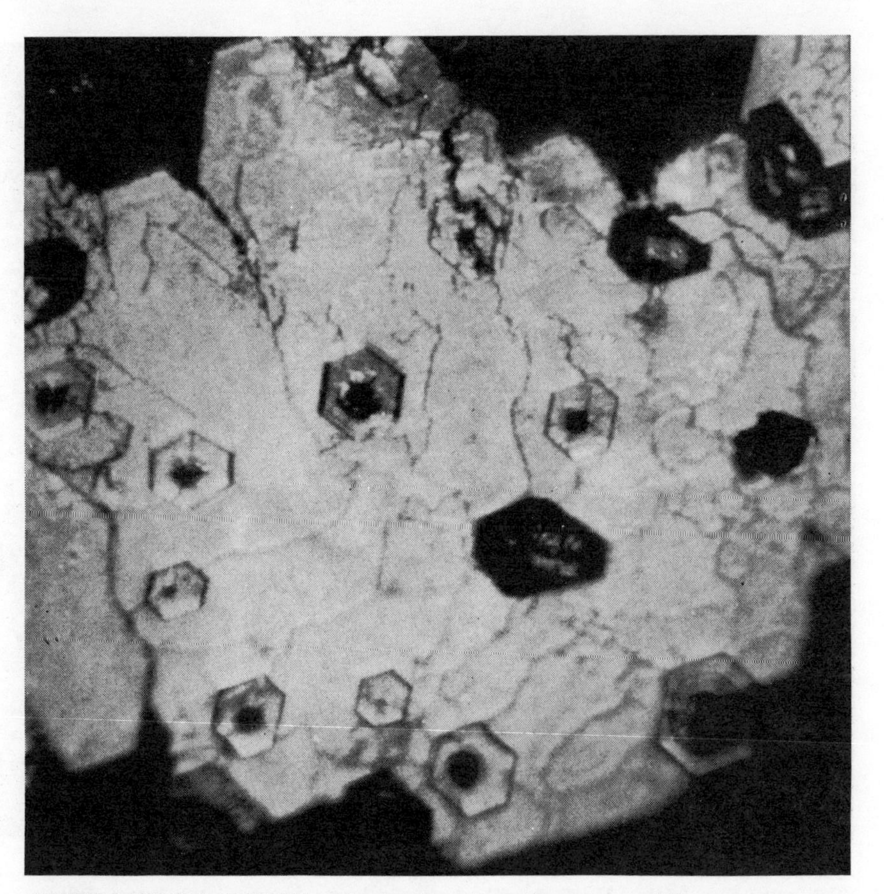

ON LEAD IODIDE CRYSTAL with a variety of growth steps resembling a contour map, ice crystals nucleate at edges of the higher steps. The most effective are about .1 micron high.

TYPICAL HAILSTONES are shown in polarized light. The stone at left, which fell at Boulder, Colo., on June 10, 1969, is 4.8 centimeters in its longest dimension; the stone at right, which fell at Friend, Kan., on June 21, 1969, is slightly smaller. Crystal size indicates temperature of formation; Friend stone, with smaller crystals, was formed in colder environment than the Boulder stone.

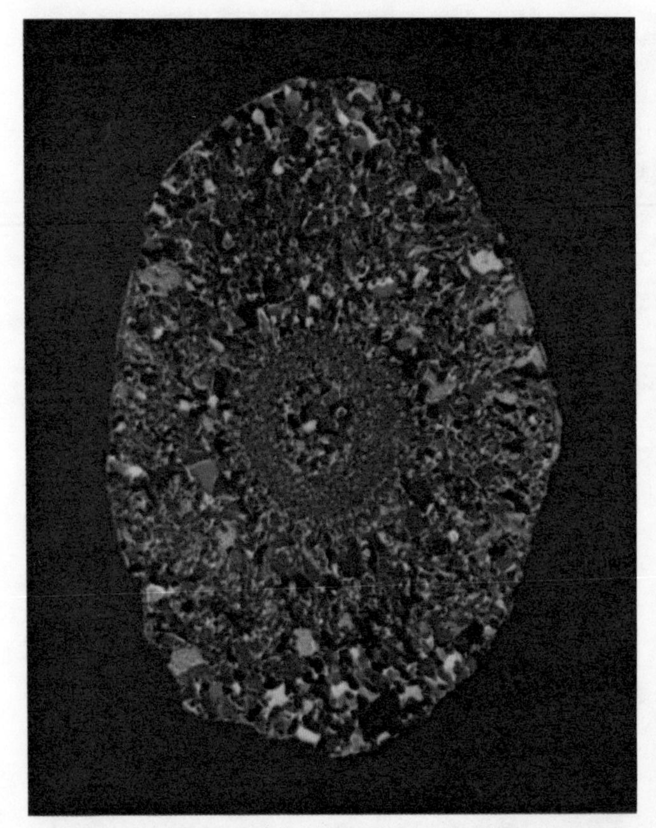

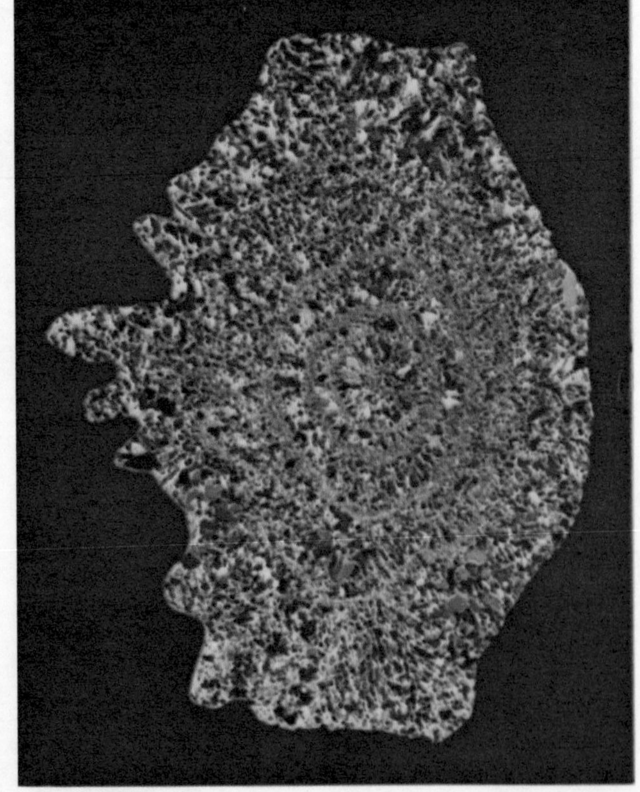

LARGE HAILSTONES fell respectively at Iowa City (*left*) and Coffeyville, Kan. (*right*). The Coffeyville stone, which was almost six inches in diameter and weighed $1\frac{2}{3}$ pounds, is probably the largest authenticated hailstone on record. The Iowa City stone has the flattened spherical shape typical of larger hailstones. Such stones are believed to have tumbled symmetrically as they fell.

Hailstones 5

by Charles and Nancy Knight
April 1971

The internal structure of a hailstone tells the history of its formation. This structure is made visible by cutting a section of the stone and placing it between crossed polarizing filters

A hailstone carries on its surface and in the differences of its internal structure the story of its formation and fall. In the present state of knowledge one cannot reconstruct the entire story from the external and internal evidence, but much progress has been made in that direction. The investigation has its own attractions; it also has economic significance, since hail does a great deal of damage every year. The National Center for Atmospheric Research, where we are studying hailstones, has estimated that in the U.S. the annual damage to crops and property from hail is some $300 million, which in a typical year is higher than the figure for tornado damage. Anything that can be done to mitigate hailstorms would therefore well repay the effort.

Until about 15 years ago no systematic effort had been made to investigate hailstones. One heard reports of memorable hailstorms; they were remembered because of the amount of damage they did, because of the unusual size or shape of the hailstones, or simply because an interested observer happened to experience the storm. There had also been occasional discussions of a particular feature of hailstones, such as the conical shape of many small ones. No one person or group, however, had ever undertaken to find out as much as could be learned about hail.

Between 10 and 15 years ago three small groups—one at the Swiss Federal Institute for Snow and Avalanche Research, another at the Imperial College of Science and Technology in London and the third at the National Physical Research Laboratory of South Africa—did decide to look into the hail problem in detail. These groups did a good share of the pioneering work. Studies are now being conducted in several other countries as well. The results have included a considerable increase in understanding of how hail is formed and a far greater appreciation of the difficulties of the subject.

Several approaches are employed in the study of hailstorms and hail formation. Workers in aircraft map the fields of wind velocity around a storm and within the accessible parts of it. Storms are also studied remotely by radar, which gives three-dimensional displays of the parts of clouds that contain water drops or hail large enough to be detectable by radar, namely particles more than about .2 millimeter in diameter (the lower limit of size for raindrops). Radar can also provide a degree of information about the size of raindrops and hailstones, although these results are often ambiguous.

Another way of studying hailstorms is to examine hailstones. It has long been thought that the intricate layered structure of a hailstone contained a wealth of information about the stone's history. The hope now is that, once the forming of hailstones is properly understood, it will be possible to collect hailstones on the ground at different locations and times, examine them and thereby obtain a clear picture of the storm that produced them.

The environment where hailstones grow is known in a general way from information about convective storms and from observations comparing storms that produce hail with those that do not. Convective storms, of which hailstorms form a small subgroup, occur when the vertical stratification in the temperature of the air is unstable. If the air in the layer at the surface of the earth is much warmer than the air immediately above it, the warmer, less dense air will rise in localized convective currents while the cooler, denser air descends to take its place. The descent of the cooler air and the rise of the warmer air release potential energy. It is this energy that drives the convective currents.

The process is complicated, however, by the vertical pressure gradient in the air. The warm ascending parcel of air rises into an environment where the pressure is lower, so that the parcel expands as it rises. And as it expands it cools. The result is that the convective process tends to be self-limiting.

Expansive cooling leads to the formation of a convective cloud—the familiar cumulus cloud. Warm air can hold more water vapor than cold air, and a current of warm air rising from the surface of the earth can hold a considerable amount of water vapor; a relative humidity of some 60 to 70 percent is not unusual. As the air rises and cools, however, the amount of water it can hold decreases. The relative humidity, which expresses the ratio of the amount of water a body of air holds to the amount it can hold, therefore increases until it reaches 100 percent. Further ascent of the air then causes water vapor to condense into droplets and form a cloud. This process is important to convection, because the condensation produces a large amount of heat (some 550 calories per gram of water condensed). The heat limits further expansive cooling and encourages further convection, enabling the rising currents to attain greater vertical velocities and greater elevations.

Hail grows inside convective storm clouds. The larger hailstones grow only inside the largest and severest storms. The winds in the core of such a storm have high upward velocities; this holds the stones aloft in the hail-growing regions of the storm and allows them to grow larger.

Hail is primarily a phenomenon of temperate climates. It is rare in polar regions because the air is seldom unstable enough to generate strong vertical wind velocities. The surface of the earth is too cold. Hail is also rare in the Tropics because the freezing level in the atmosphere is too high. A warm cloud cannot produce ice.

Clearly a set of conditions necessary for hail to form includes strong vertical wind velocities over a considerable vertical extent, accompanied by temperatures below the freezing point of water. Although these are necessary conditions, they are evidently not sufficient, since not all such clouds produce hail. One objective of investigating hailstones is to learn what set of conditions is sufficient for forming hail.

Hailstones thus grow in the updrafts of convective clouds [*see illustration below*]. While a hailstone is growing in such an updraft, it always falls at its terminal velocity with respect to the air: the velocity at which the wind resistance keeps it from accelerating. Since the density of hailstones does not vary much, terminal velocity is a function of size and shape. If the velocity of the updraft equals the terminal velocity of the hailstone, the stone will remain at a constant altitude until it grows larger (so that its terminal velocity increases) or the updraft velocity changes.

The most important fact about the region of the cloud where hailstones grow is that, although the temperature is below the freezing point, most of the cloud is in the form of liquid water rather than ice. Under certain conditions water can remain for long periods below the freezing temperature without freezing [see "The Undercooling of Liquids," by David Turnbull; SCIENTIFIC AMERICAN, January, 1965]. Water in this condition is called supercooled or undercooled. Since the water in a cloud is typically quite pure, most droplets supercool to −15 degrees Celsius or lower. The freezing of a supercooled drop can be started by certain kinds of dust particle or by collision of the drop with an ice particle. Such a frozen drop of water or alternatively a snow crystal represents the beginning of a hailstone.

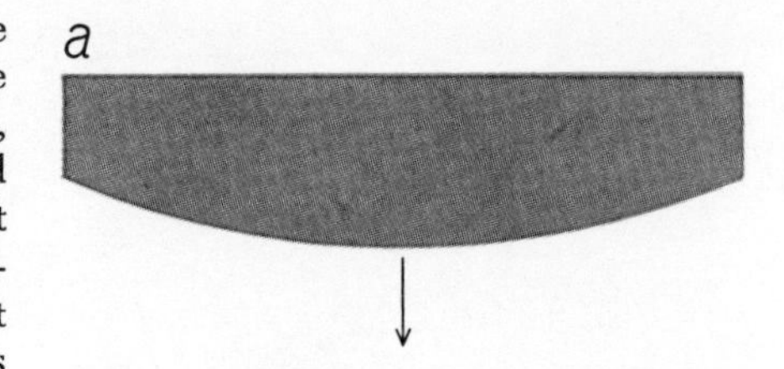

GROWTH OF HAILSTONE results from collisions of a falling stone (*a*) with drop-

Hailstones grow by sweeping up supercooled drops that have not frozen. No doubt they also sweep up occasional ice particles, but this is certainly a minor mechanism of growth. When a supercooled drop strikes an ice surface, two things happen: the drop spreads out on the surface, and it freezes. If it freezes quickly, it will not have time to spread much before it is entirely solid, and it will remain a rounded lump of ice on

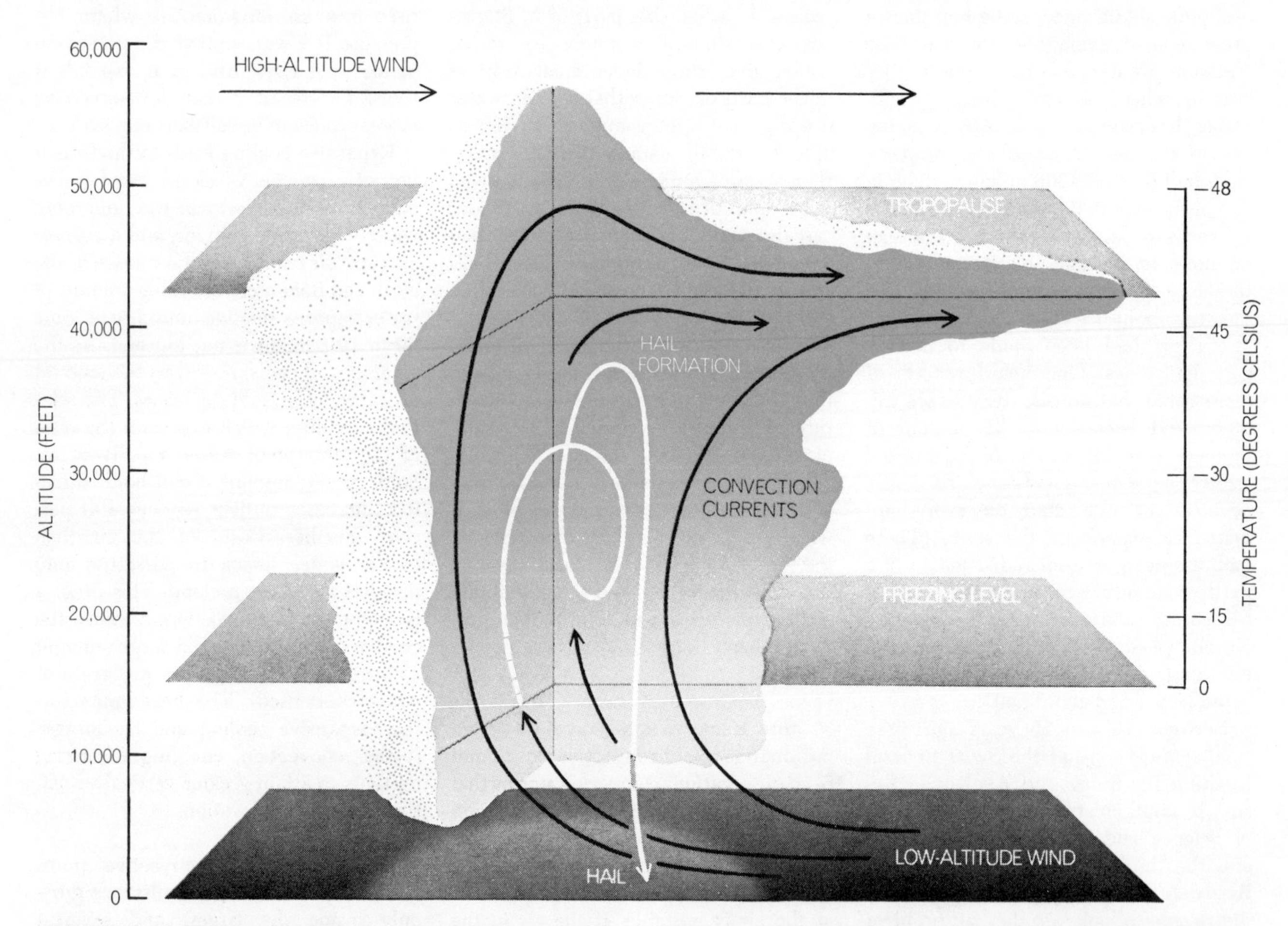

CUMULUS CLOUD provides the environment where hailstones form and grow. The stones usually form in the part of the cloud where the temperature is between −5 and −20 degrees Celsius and supercooled droplets of water are found in the updrafts. Falling stones grow by sweeping up such droplets. If the updrafts are strong, hailstones can grow while they are being carried upward. If such a stone is carried into the anvil of the cloud (*upper right*) and falls out, it can be swept back into the cloud for another circuit.

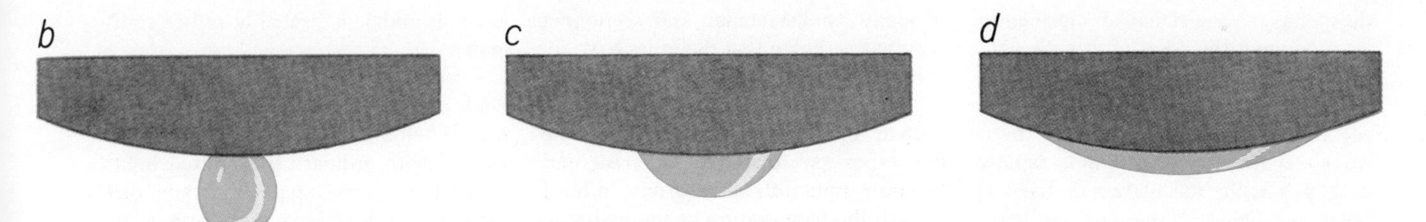

lets of supercooled water. Depending on the conditions, each collision will result in an attached sphere of ice (*b*) or thinner caps (*c, d*). There is more spreading the warmer the drop, the warmer the hailstone, the larger the drop and the harder the impact.

the surface of the stone. If it freezes slowly, it may have time to spread into a thin layer. There are all possible gradations between these extremes.

The freezing of supercooled drops colliding with a hailstone liberates heat. As a supercooled drop freezes its temperature rises to zero degrees C. Only when all the water in the drop is frozen can its temperature again fall below zero.

If the average interval of time between collisions at a given point on a growing hailstone is smaller than the time it takes for one drop to freeze completely, that point will remain at zero degrees C. throughout the growth process. Moreover, the hailstone will include some liquid water from the unfrozen portions of the drops. The amount will depend on the ratio between the collision rate and the rate of heat loss. Hailstones that include a substantial amount of liquid water for this reason are called spongy. They will be observed to splash rather than bounce when they strike the ground. If, on the other hand, the average interval of time between collisions at each point on the hailstone's surface is longer than the freezing time for a drop, the hailstone grows as solid ice at a temperature between zero degrees C. and the temperature of the environment.

Clearly hailstone temperature is one of the fundamentally important factors in any study of hailstone growth. It is also clear that hailstone temperature depends on many interrelated factors. The amount of heat released depends on the growth rate, which in turn depends on the water content of the cloud, the sizes of drops and the falling speed of the hailstone. The rate of transfer of heat to the environment depends on the falling speed of the stone and on the difference in temperature between the stone and the environment. Therefore even if the growth temperature of a hailstone could be determined from the structure of the stone, the meaning in terms of environmental factors would still be far from simple. More direct indicators of environmental conditions would be more useful.

The growth processes of the two forms of atmospheric ice—hailstones and snow crystals—are almost completely different. Snow crystals grow from water vapor by diffusion. Single molecules of water collide with a snow crystal and attach themselves to the crystal lattice. A hailstone collides with water droplets and grows not because of diffusion but because it falls faster than the drops do and so sweeps up the drops from a cylindrical area in its path. Hailstones, far from growing from vapor, are in fact a source of vapor because they are warmer than their environment. Hailstones evaporate as they grow; the evaporation is an important cooling mechanism. Hailstone shapes are determined by complex aerodynamic factors and by heat flow. Snow crystal shapes are determined by diffusion and crystallography.

A close look at hail reveals that the stones differ in appearance. Some are white and others are more nearly clear. By breaking a hailstone in half or melting half of it away, one can observe concentric layers in varying degrees of whiteness or clarity. Close examination shows that these contrasts result from varying amounts of air bubbles in the ice. Ice containing small air bubbles looks white for the same reason that snow looks white: the many air-ice surfaces at different angles form an efficient diffuser and reflector of light.

The layering itself carries a good deal of information about the hailstone. The structures within each layer record changes in the growth environment of the stone. Layering also gives the history of the shape of the stone and information about its falling attitude and whether or not it tumbled or wobbled as it fell.

Thin sections provide the best way to examine a hailstone's layering, air-bubble structure and crystal structure. The method requires only an electric band saw and a cold place to work. The hailstone is first sliced in half with the saw. One sawed surface is smoothed with a file to remove the ridges made by the saw and is pressed onto a warm glass slide. The warmth melts a thin layer of water, which then refreezes, gluing the half-hailstone to the slide. Another saw cut is then made parallel to the first one, leaving a section less than one millimeter thick on the slide. The newly sawed surface is smoothed as before and then polished by vigorous rubbing.

We photograph the sections thus prepared with transmitted light [*see illustration on next page*]. This procedure has the virtue of giving good reproduction of details, although it reverses the normal relation between brightness and air-bubble content. With the source of light behind the thin section, the bubbly layers appear darker than the clear layers.

The simplest story layering can tell is that of a hailstone falling with a constant orientation and consequently growing on only one side. The growth side faces downward, since the hailstone grows by overtaking and sweeping up the water droplets in its path. The conical shape that results is the commonest one among hailstones that measure two centimeters or less in their largest dimension, but it is also found at times in stones as large as five centimeters in diameter.

Hailstones larger than about two centimeters in their largest dimension are usually shaped roughly like flattened spheres. The origin of this shape is the subject of controversy. One hypothesis is that the shape results from aerodynamic molding. The hailstone is thought to grow quite spongily while falling at a constant attitude and to be molded physically by the aerodynamic forces. The airflow around the stone produces pressure gradients such that there is a belt of lowest pressure around the horizontal circumference of the stone and a higher pressure at the top and bottom. The spongy hailstone is thought to become flattened in response to these pressures.

We and several other workers disagree with this view, mainly because the growth of such stones does not appear at all spongy. An alternative mechanism for producing this kind of growth symmetry would be rapid, symmetrical tumbling. We think that is what happens, but the hypothesis has not yet been proved. If hailstones do tumble rapidly as they fall, the calculation of the terminal velocity and the heat-exchange rate—two of the most important factors in hailstone growth—becomes extremely difficult.

Another interesting feature of layering is the evidence it provides that hailstones commonly develop extremely rough shapes, particularly as they grow to the largest sizes. The lumps that form are called lobes. The lobe structure often makes large hailstones look as though they had been formed through the fusion of many smaller stones, but sectioning has always shown that the lobes have resulted from continuous growth.

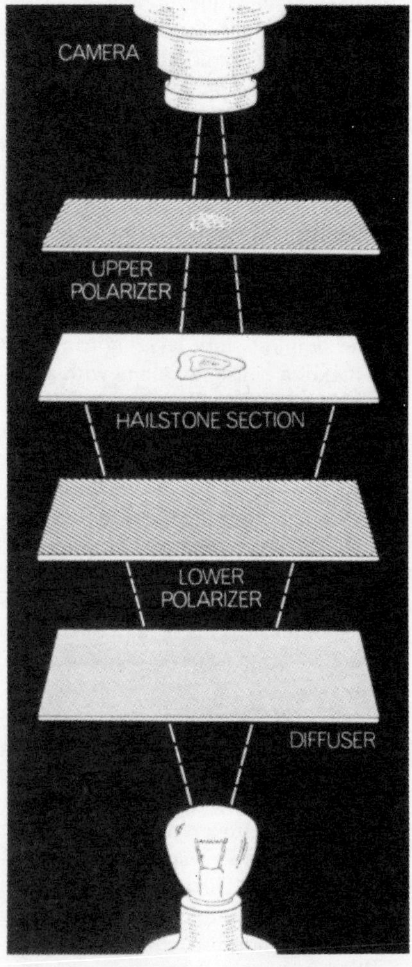

CRYSTAL STRUCTURE of a thin section of a hailstone is viewed or photographed through crossed polarizers, that is, polarizing plates oriented at right angles to each other. When a hailstone section is put between the plates, each crystal in the section rotates the plane of polarization of the light by an amount depending on the orientation of the crystals. Therefore each differently oriented crystal appears to the eye or camera in a different color or shade of gray.

Lobe structures form in hard, cold growth and in spongy growth, but the two types look distinctly different and form for quite different reasons. In hard growth the temperature of the hailstone is below freezing. Each drop that is collected freezes where it hits, and therefore the formation of lobes must result entirely from the collection process.

We believe that a tumbling or oscillating motion of a falling hailstone is necessary for this kind of lobe to develop. A point on the surface of a tumbling hailstone receives droplets from many different directions. If the surface is concave at that point, it is shielded from receiving droplets from certain directions. If the surface is convex, the point can receive droplets from a full hemisphere of directions. Thus any chance bump on the surface of a hailstone will magnify itself by collecting droplets that otherwise would have hit nearby. This process results in the formation of lobes separated by sharp cusps.

The other kind of lobe forms when the surface of the hailstone is at a temperature of zero degrees C. and is covered by a layer of liquid water. This layer is mobile and migrates under the influence of aerodynamic and possibly of centrifugal forces, thereby giving rise to icicles on the surface of the hailstone. These icicle lobes are characterized by the lack of cusps and by bubble and crystal structures indicative of more or less spongy growth.

Hailstone layering can also be used to trace certain events in a stone's history. For example, a discontinuity in layering means that the hailstone has broken in midair. The breaking force may come from the freezing of spongy ice. If a layer grows spongily and includes liquid water, and the hailstone then ascends into colder air or its rate of growth is slowed, the liquid water in it may freeze. Since ice is less dense than water, the freezing creates large internal pressures that commonly cause internal fracturing and sometimes breaking.

Hailstones can also shed entire lobes of the hard-growth type. At times such lobes grow with small necks connecting them to the main body of the stone. We do not know whether the necks are so weak that the wind is able to break off lobes or whether a glancing collision with another hailstone is required. In any case, such a breaking of hard-growth lobes in midair is probably rather common.

Every layer in a hailstone has two kinds of structure: air bubbles and crystals. When a thin section of hailstone is viewed with ordinary transmitted light, all that one sees, apart from any dirt the hailstone may have picked up, is air bubbles. Large numbers of bubbles, too small to be seen individually, give the appearance of continuous shading. Large air bubbles can be seen individually but are often filled with chips of ice from the sectioning process. A few artifactual air bubbles are unavoidably included between the ice and the glass slide, but they are easily identified.

The air bubbles are formed in several ways. The large, radial bubbles typically found in the cusp lines between lobes of the hard-growth type are simply places where no ice grew and air was trapped. The more uniformly dispersed fields of air bubbles can form either in hard, cold growth or in warm, spongy growth but not in conditions between these extremes. Clear, bubble-free ice is produced when the hailstone grows nonspongily at a temperature close to zero degrees C.

In hard, cold growth, particularly if the supercooled cloud droplets are quite small, each drop freezes rapidly on contact with the hailstone and is unlikely to depart much from its original spherical shape. The bulk ice that forms from this kind of growth looks as if it were built up from many little spheres. It includes a large amount of air in the form of dense dispersions of tiny bubbles.

In spongy growth, where liquid water is included within the ice, later freezing (either in the atmosphere or after the hailstone has been collected and stored) produces air bubbles in the same way that freezing ice cubes in a refrigerator does. Air is rather soluble in water and almost insoluble in ice. Water freezing in a confined space must segregate the dissolved air into air bubbles before freezing is completed. At the rather slow freezing rates that are normal these air bubbles can grow considerably larger than the other type. Even if they do not, they always have a characteristic appearance.

Crystal texture in a thin section of a hailstone is examined with two polarizing plates, one between the light source and the section and the other, oriented with its direction of polarization at right angles to that of the first plate, between the section and the eye or the camera. If no hailstone section is in place, no light is transmitted by these crossed

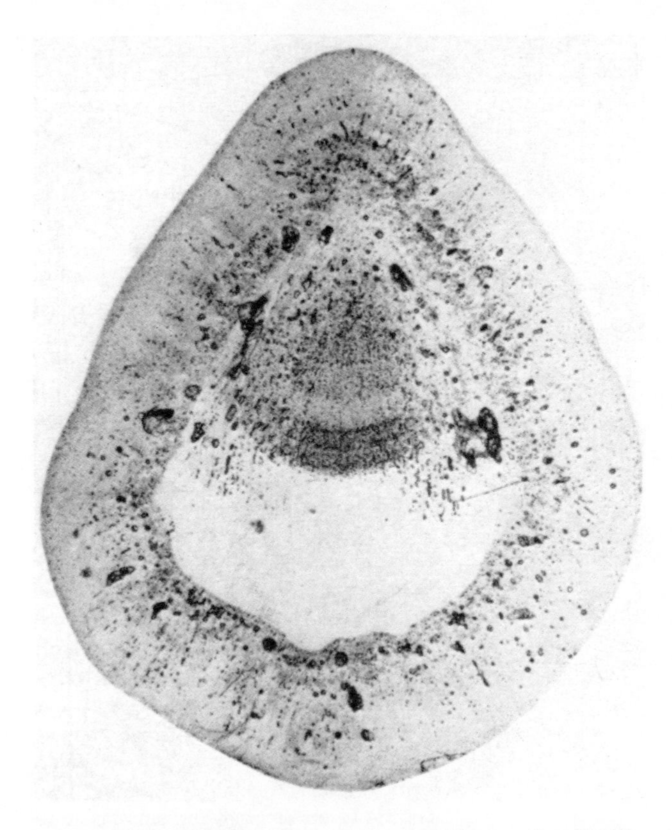

HAILSTONE SHAPE tends to differ according to size. Smaller stones, meaning those less than about two centimeters in their largest dimension, tend to have a conical shape (*left*). The growth side faces downward, since the falling stone grows by sweeping up water droplets. Larger stones tend to be flattened spheres (*right*), probably from tumbling. Stone is shown in largest cross section.

LAYERING OF HAILSTONE is evident in a section of a stone photographed in transmitted light (*left*) and in polarized light (*right*). Darker layers at left are dry-growth areas, where each supercooled droplet of water froze solid as it hit and so maintained something of its spherical shape. Aggregations of such drops include many air bubbles, making the ice appear darker in transmitted light, which reverses the normal appearance. Light layers are spongy-growth areas, in which each water droplet spread over the surface as it hit and froze only partly before the arrival of the next drop. These areas have fewer air bubbles. The close correlation between crystal size and the accumulation of air bubbles is a universal feature of large hailstones that is not understood.

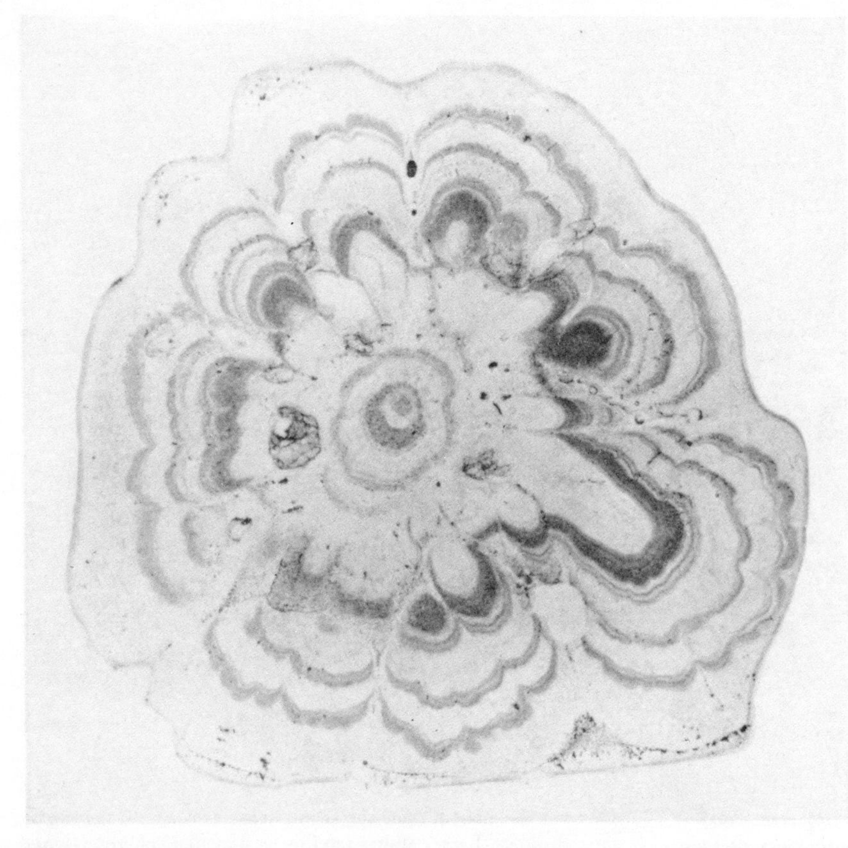

MULTIPLE LOBES appear on a large hailstone. The dry-growth type of lobe formation, represented by the darker areas, followed the development of icicle lobes. Continuity of layering shows that lobes result from growth rather than from the capture of small hailstones.

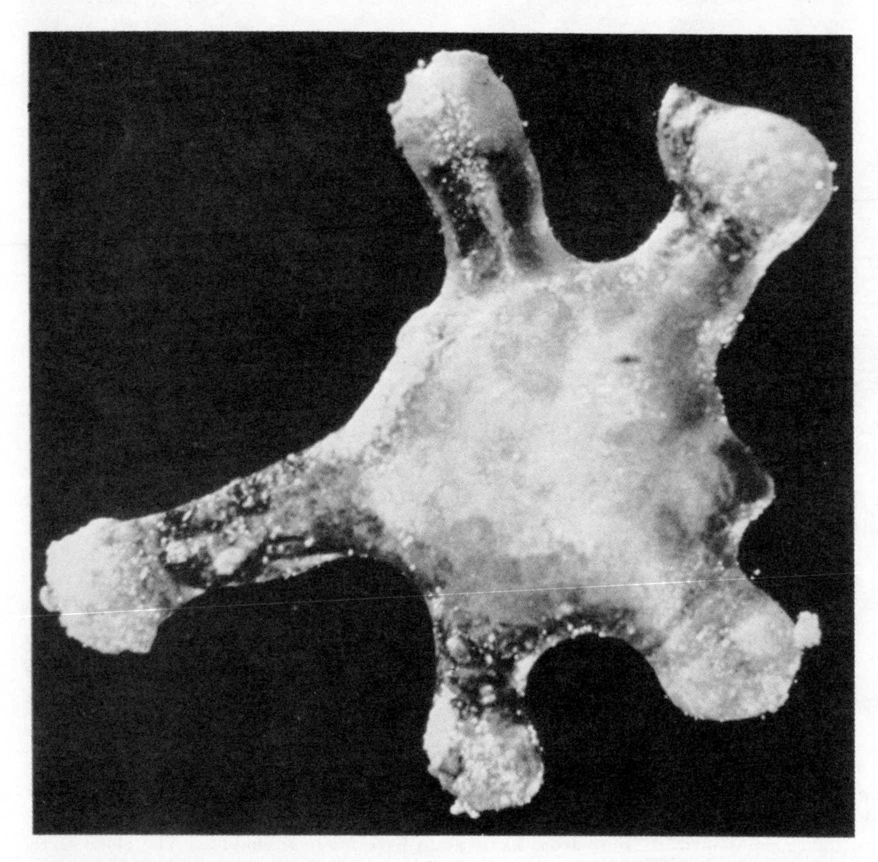

LOBE STRUCTURE is clear in this small hailstone, which is about 2.5 centimeters in its longest dimension. Why the lobes should be in a single plane, as they are here, is not known, but their formation was undoubtedly controlled by the stone's movements as it fell.

polarizers. When a hailstone section is put between the crossed polarizers, each crystal in the section rotates the plane of polarization of the light by an amount depending on the orientation of the crystal. Each differently oriented crystal in the section thus appears in a different shade of gray, and the crystal textures are immediately visible. There are also interference effects that give each crystal a different color, as is evident in the photographs on page 40 of this article. Black-and-white photographs made with crossed polarizers give almost as much information.

Measurement of crystal orientations in hailstones yields a certain amount of information about the growth conditions of a hailstone, but it is difficult and expensive information to get. The procedure has therefore not been widely used. A more useful and much more easily determined factor is crystal size. As one can see from the accompanying photographs made with crossed polarizers, crystal size is highly variable. Concentric layers of large and small crystals are found with (often in correlation with) the layers of air bubbles.

The results of the investigations of the Swiss hail group indicate that crystal size is primarily a function of environmental temperature rather than the hailstone's temperature or rate of growth. The Swiss workers found that this is true except in unusually spongy growth, which is rare in hail and is easily identified. If the environmental temperature is below about −20 to −25 degrees C., the crystals are small. At higher temperatures the crystals are larger. (A crystal is defined as large if it is more than two millimeters in diameter.)

Accepting the Swiss result, one can conclude that wherever a transition appears between large and small crystals in a hailstone, the stone was descending or ascending through the region of the cloud where the temperature was between −20 and −25 degrees C. This is the kind of criterion that is really needed to make effective use of hailstone structure in studying hailstones. Unfortunately two difficulties are met in the routine use of the criterion. First, one must assume that fine-grained layers were not produced by the capture of many small crystals. Most observers believe this assumption is correct, but at present there is no firm proof that it is.

The second difficulty is presented by the almost perfect correlation between small crystal size and bubbly ice when the growth radius of the hailstone is more than about a centimeter. If crys-

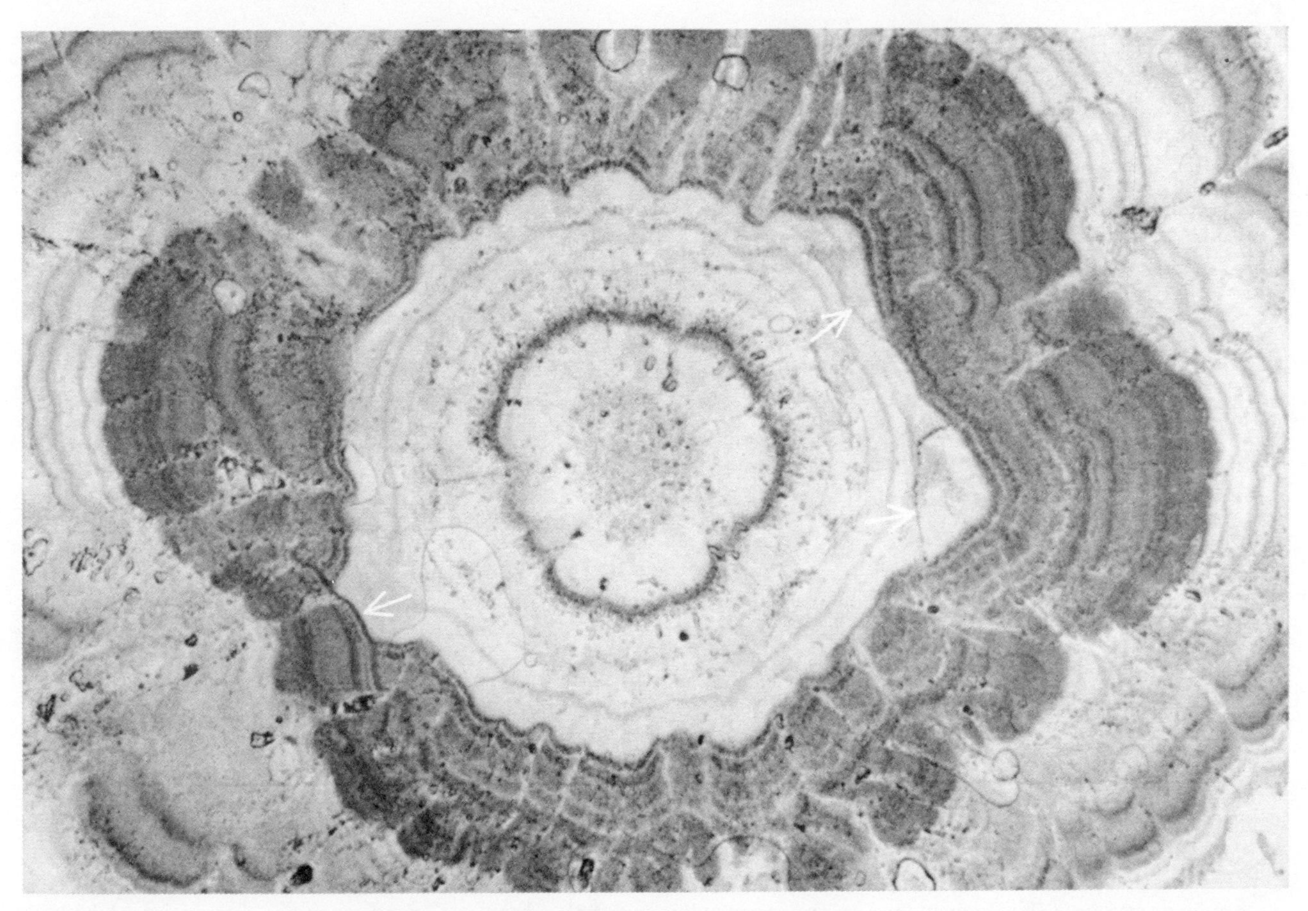

BROKEN LOBE can be traced at the formation marked by arrow at left. Initial spongy growth of the hailstone was followed by cold growth, which continued for about two millimeters before the liquid water in the spongy growth froze and caused the projection to break off. The arrows at right indicate two other places where internal cracking occurred without resulting in any separation.

tal size depends solely on environmental temperature, and if air-bubble content depends on hailstone temperature as well as environmental temperature, there is no evident reason for the close correlation found between crystal size and bubble content. Until the correlation is explained, the criterion of environmental temperature for crystal size cannot be entirely trusted.

Curiously the correlation breaks down for hailstones with a small radius of growth. In such cases one finds large crystals and many air bubbles [*see bottom illustration on page 45*]. The existence of the correlation at large radii of growth and its absence at small radii suggest an influence of the hailstone's speed of fall. No reason for such an influence is known, and the relation of size to growth structure remains a particularly interesting problem.

Once in a while one finds small hailstones that show both a transition in crystal size and an almost uniform bubble structure. Here the criterion of environmental temperature as the determinant of crystal size can surely be trusted, but such hailstones are rare. We have found stones that, according to this criterion, grew in an entirely upward trajectory, continually cooling. Such stones must have emerged from the top of the cloud and fallen to the ground in clear air. We have also found stones that started in a cold environment and warmed up as they grew along a downward trajectory.

Suppressing hailstorms appears to be a possibility because hail grows from supercooled water, which is unstable and can be made to freeze by adding tiny particles of appropriate substances. Silver iodide is one of the best substances for this purpose, being effective at between -5 and -10 degrees C. because its crystal structure is quite similar to the crystal structure of ice. Silver iodide is also easy to disperse in clouds in the form of a fine powder.

In this way it is possible to interfere with the supercooled-water budget of a storm and so affect the production of hail. Whether or not one can interfere in such a way as to produce predictable and beneficial results is a matter of controversy. Probably the only safe thing to say is that real possibilities for suppressing hail exist but that given the present state of knowledge and the rather infrequent occurrence of hail, which makes study difficult, it will take a number of years to ascertain whether or not such suppression is feasible.

It now appears that the more one learns about convective storms in general and hailstorms in particular, the more complicated they seem. This conclusion comes from several lines of evidence, among them the study of the structure of hailstones. If hailstorms conformed to some simple, general plan, one would expect hailstone structures to fall into some general pattern. So far no such pattern has appeared. On the contrary, it seems that hailstones can have almost any conceivable history. Some of them evidently grow during continual ascent, others during continual descent. Others rise and then fall, and still others rise and fall several times.

The late Ukichiro Nakaya of Hokkaido University, who was renowned for his work on the formation of snow crystals, once wrote that they are "hieroglyphics from heaven." The same thing can be said of hailstones, except that the hieroglyphics are much more obscure and have not yet been fully translated.

6 Fog

by Joel N. Myers
December 1968

A kind of grounded cloud, fog can halt sea, air and highway travel. When combined with air pollutants, it can be lethal. Ways are known, however, not only to dissipate fog but also to inhibit its formation.

Fog, once little more than a nuisance except at sea, has become an important hazard to modern man. Its effects on travel are intensified by the speed of the airplane and the automobile. Dense fogs close airports in the U.S. for an average of 115 hours per year, and in 1967 they cost the nation's airlines an estimated $75 million in disrupted schedules as well as inestimable inconvenience to passengers. On present-day turnpikes fog can be disastrous; a single pileup on a fog-shrouded freeway in Los Angeles involved more than 100 vehicles. Above all, fog in combination with air pollution now increasingly afflicts large cities. Its potential was suggested alarmingly by the London smog of December, 1952.

On December 5 a dense fog settled over the city. A strong inversion—warm air lying above cold air—blocked the removal of polluted air by vertical movement, and at the same time the winds were so light that such air was only gradually removed by horizontal movement. Within 24 hours the tons of smoke, dust and chemical fumes given off by the city's furnaces, factories and automobiles turned the fog brown and then black. Two days later the visibility in the city had been reduced to a matter of inches. People fell off wharfs into the Thames and drowned. Others wandered blindly until they died of exposure to the cold. The toxic air afflicted millions of Londoners with smarting eyes, coughing, nausea and diarrhea. It was estimated that the smog killed at least 4,000 persons, mainly from respiratory disorders, and caused permanent injury to tens of thousands.

The principal source of toxicity in most pathological smogs appears to be sulfur dioxide from the smokestacks. The sulfur dioxide is oxidized in the air to sulfur trioxide, which in turn combines with fog droplets to form sulfuric acid. One therefore inhales sulfuric acid with each breath, with resulting acute irritation of the throat and lungs. This seems to have been the chief cause of injury and death in the London smog and in the 1948 smog in Donora, Pa., which killed 20 victims and sickened nearly half of the 14,000 inhabitants.

Strictly speaking, the perennial "smogs" of Los Angeles are misnamed, as they frequently consist of a haze of pollutants that are trapped by an inversion layer of dry air rather than by fog. Polluted air, however, can generate fog and cause it to persist. Once formed, the fog reflects solar radiation back into space, thereby providing an environment favorable to the accumulation of further pollution. In effect fog, itself partly a product of pollution, causes pollution to increase.

Some metropolitan areas in the U.S. suffer from light smogs up to 100 days of the year, and during fall and winter dangerous smog conditions can become frequent. The long-term effects of the urban smogs are believed to include chronic bronchitis, emphysema, asthma and lung cancer. The polluted air corrodes metals, rots wood, causes paint to discolor and flake and may cause extensive damage to vegetation and livestock. In addition to the directly damaging effects, there are indications that the combination of fog and air pollution may be upsetting the delicate balance of man's ecosystem. Smog plays an important role in influencing the earth's gain and loss of radiation, which ultimately determines most of the climatic variables. Smog reduces sunlight, lowers daytime temperature and wind speed, raises humidity and is even suspected of causing a decrease in rainfall. Thus for many reasons—the hazards to travel, the damaging effects of smog, the modification of climate—fog is a growing problem that will demand increasing research attention in the years ahead.

Fog is simply a cloud on the ground, composed, like any cloud, of tiny droplets of water or, in rare cases, of ice crystals, forming an ice fog. Ice fogs usually occur only in extremely cold climates, because the water droplets in a cloud are so tiny they do not solidify until the air temperature is far below freezing, generally 30 degrees below zero Celsius or lower. The droplets of fogs are nearly spherical; they vary in diameter between two and 50 microns and in concentration between 20 and 500 droplets per cubic centimeter of air. The transparency of a fog depends mainly on the concentration of droplets; the more droplets, the denser the fog. A wet sea fog may contain a gram of water per cubic meter; a very light fog may have as little as .02 gram of water per cubic meter.

Since water is 800 times denser than air, investigators were long puzzled as to why fogs did not quickly disappear through fallout of the water particles to the ground. Even allowing for air resistance, a 10-micron water droplet falls in still air with a velocity of .3 centimeter per second, and a 20-micron droplet falls at 1.3 centimeters per second. To explain the persistence of fogs many early investigators concluded that the droplets must be hollow (that is, bubbles). It turns out, however, that the droplets are fully liquid and do fall at the predictable rate, but in fog-creating conditions they either are buoyed up by rising air currents or are continually replaced by new droplets condensing from the water vapor in the air.

The atmosphere always contains

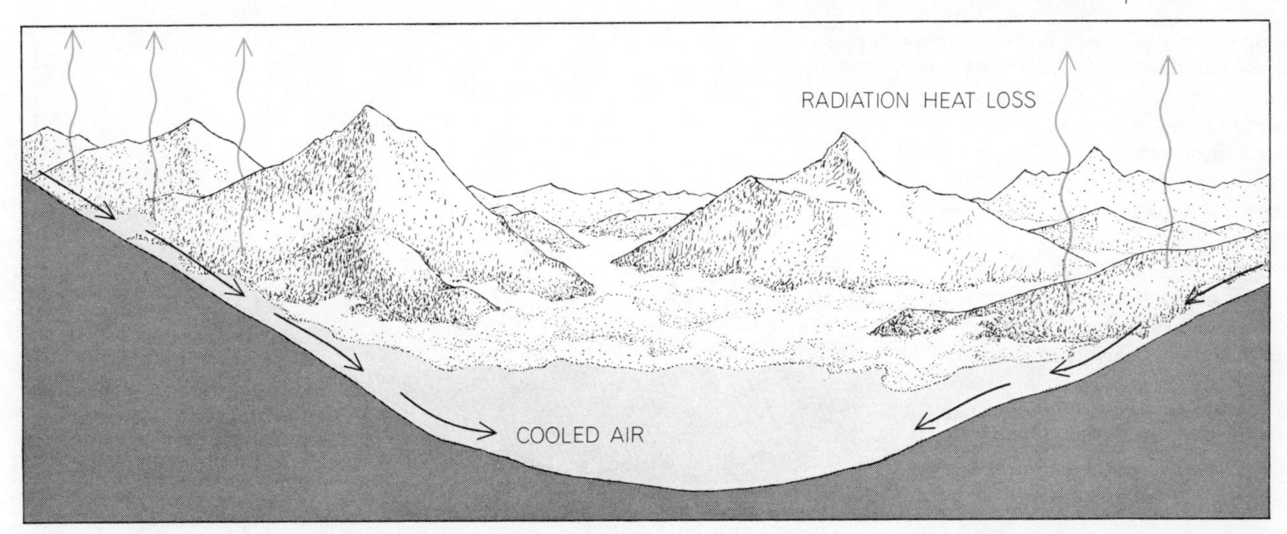

FOG IS FORMED when moist air is cooled; the air then cannot hold as much moisture in the form of water vapor as it can when it is warmer. As the humidity approaches the saturation point tiny water droplets form, obscuring visibility. The diagram indicates how a radiation fog, one of three kinds of "cooling fog," forms. Night cooling has reduced the air's temperature and cause water droplets to appear. The cold, fog-filled air drains downhill and accumulates in low-lying areas (*bottom illustration on page 51*).

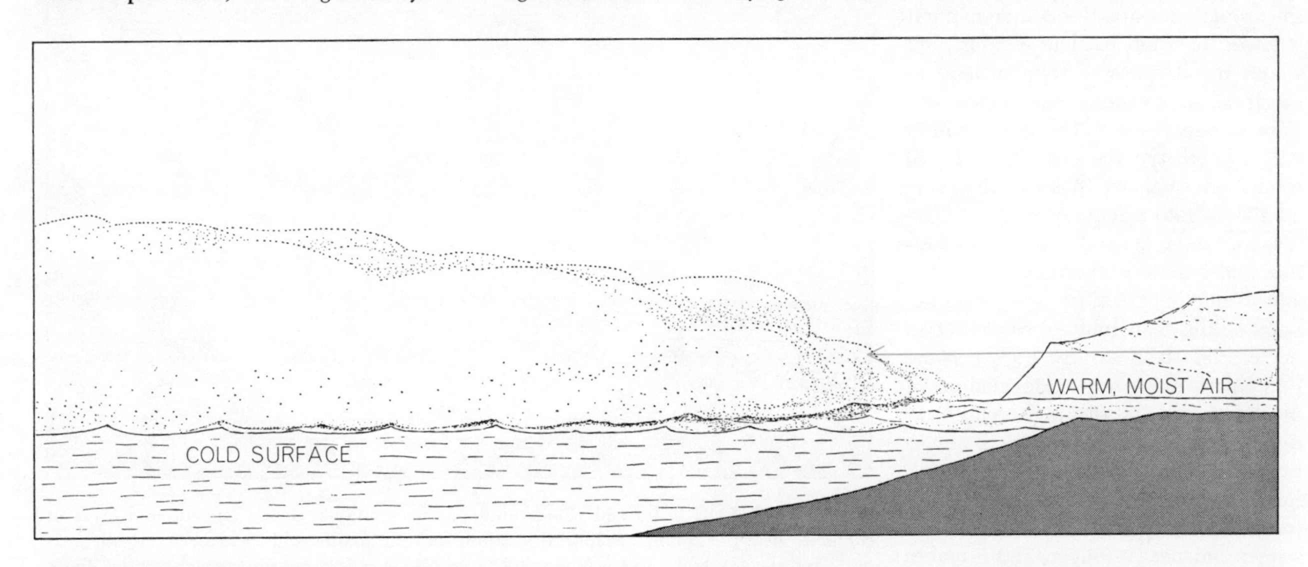

ADVECTION FOG, the second kind of cooling fog, forms when warm, moist air is carried across a cold surface and is cooled to the saturation point. Advection fogs are common at sea where warm air masses come in contact with cold ocean currents (*see top illustration on pages 52–53*) and on land during the cold months when moist tropical air masses are carried across chilled grounds.

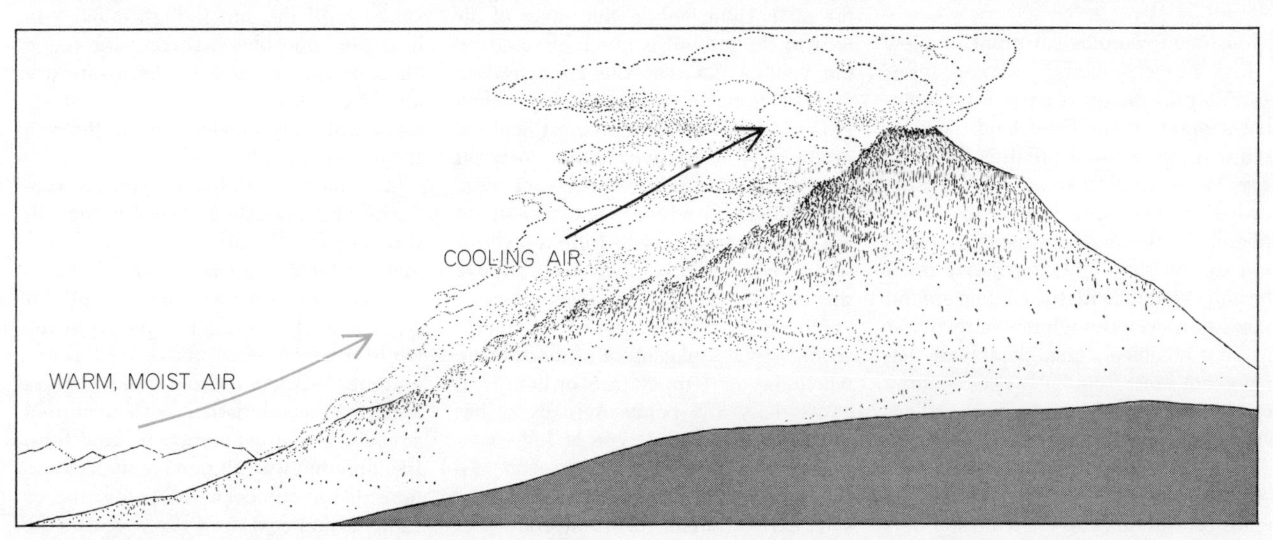

UPSLOPE FOG, the third kind of cooling fog, forms when an air mass is forced upward. In the diagram a topographic barrier forces the air mass to rise. As pressure diminishes the air mass expands, grows cooler and, as in both other cases, soon becomes saturated.

some water vapor, supplied by evaporation from bodies of water, vegetation and other sources. Since the air's capacity for holding water in the form of vapor decreases with falling temperature, even comparatively dry air will reach the saturation point—100 percent relative humidity—when it is cooled sufficiently. At that point the vapor of course begins to condense into liquid water. Fogs often form, however, at humidities well below 100 percent. The droplets condense on tiny particles of dust in the air called condensation nuclei. These are hygroscopic particles, which, because of their affinity for water vapor, initiate condensation at subsaturation humidities—sometimes as low as 65 percent. The nucleus on which the water condenses, which may be a soil particle or a grain of sea salt, a combustion product or cosmic dust, usually dissolves in the droplet. Because the saturation point is lower for solutions than it is for pure water, the droplets of solution tend to condense more water vapor on them and grow in size. A rise in the air's humidity will also enlarge the droplets and will form more of them, thereby thickening a light fog into a dense one.

Given suitable conditions of temperature and humidity, the density of a fog and its microphysical properties will depend on the availability of condensation nuclei and their nature. Fogs become particularly dense near certain industrial plants because of the high concentration of hygroscopic combustion particles in the air. This is not to say that air pollution is generally a primary cause of fog formation, but it does cause fogs to form sooner and persist longer, and it makes them dirtier—hence less transparent—than fogs that develop at higher humidities in relatively clean air.

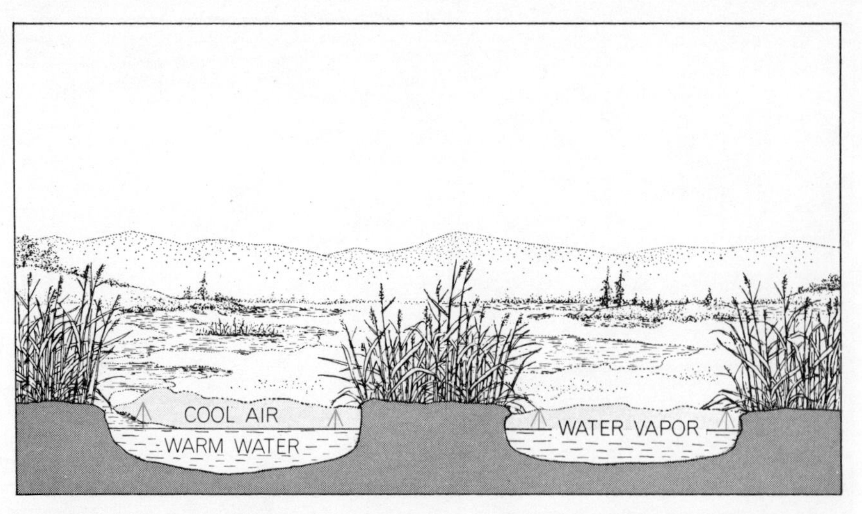

WARM-WATER FOG forms like steam over water that is covered by a much colder air mass. Water vapor from the comparatively warm water rises into the colder air and is rapidly condensed into fog droplets. The fog's intensity depends on the temperature differential.

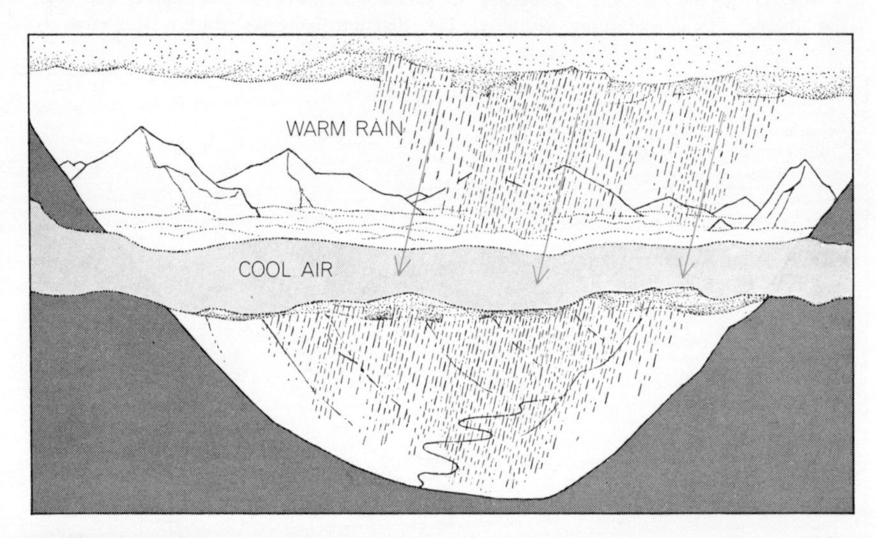

WARM-RAIN FOG is formed when raindrops from higher clouds encounter a layer of cooler air near the ground; evaporating raindrops saturate the cold air layer. As the diagram indicates, the fog looks like a low cloud to an observer looking up from the valley floor.

From a meteorological point of view fogs are classified in several types according to the gross natural processes that generate them. Over land the most common type is the "radiation fog" that arises from nighttime cooling of the earth's surface and the lower atmosphere. As the earth radiates away its heat during the night, fog may form if the air in contact with the cooling ground is moist or, even though it is fairly dry at first, if it is cooled a great deal. Radiation fogs occur most frequently over swampy terrain and in deep, narrow valleys where cold air draining down from the hillsides concentrates in the valley bottom. The likelihood of fog formation depends considerably on the wind speed. A moderate to strong wind, by moving the air about and diluting the cooling effect, tends to prevent the formation of fog. If the air is calm, only a thin layer of air next to the ground is much affected by the cooling, so that the condensation may be restricted to dew or a shallow ground fog. The condition most likely to produce an extensive fog is a slight breeze; by generating turbulence near the ground such a breeze may spread out the cooled surface air to form a layer of fog several hundred feet deep. Above this cold layer the air remains warm.

One might suppose that radiation fogs should reach a maximum around dawn, when the air temperature ordinarily is at its daily low point. Actually it has been found that this type of fog sometimes becomes thickest shortly after sunrise. The reason appears to be that the sun's early rays, not yet strong enough to evaporate the fog droplets, generate turbulence that intensifies and thickens the fog layer. The fog does not begin to dissipate until the sun is high enough to heat the atmosphere, stirring the foggy air so that it mixes with the warm, dry air above. Obviously in pockets that are topographically shielded from the sun the fog will remain longer.

A somewhat different process produces what are called advection fogs. In this case the fog arises from the movement of humid air over a surface that is already cold. Most sea fogs are of this type; indeed, the foggiest places in the world are the areas above cold ocean currents. Advection also commonly plays a part, in combination with nocturnal cooling, in the generation of land fogs. In a different way air moving up a mountainside sometimes produces fog: the air expands and cools as it rises because the atmospheric pressure diminishes with altitude. Advection and upslope fogs can

ICE FOG (*above*) fills the Nenana River valley near Fairbanks, Alaska. Ice fogs consist mainly of ice crystals that form in air cooled to 30 degrees below zero Celsius or lower. They may appear because a temperature drop has produced nearly 100 percent humidity or because the air is saturated by the addition of water vapor.

RADIATION FOG (*below*) spills out of Bald Eagle Valley through a mountain gap near Lock Haven, Pa. Radiation fogs are caused by the night cooling of the earth's surface, which reduces the air temperature and raises the humidity to near-saturation. In Bald Eagle Valley industrial wastes contribute to the fog-forming processes.

provide additions to the water supply. On some coastal mountains in California, for example, the ground receives more water from fog dripping off vegetation than it does from rain. Residents of some arid regions take advantage of this phenomenon by suspending arrays of nylon threads to extract water from drifting fogs.

Fog is also produced by the familiar steaming process we observe above a hot bath or a heated kettle or on a hot roof or parking lot after a summer shower; the vapor rising from the warm water quickly condenses into steam in the cooler air above. In this way the evaporation from a body of water on an unseasonably cold night may generate a shallow fog (up to perhaps 50 feet). Warm rain falling through cool air also can give rise to a steamlike fog; it is a common cause of the fogging in of airfields during rainy weather.

The principal natural agents of fog dispersal are sunshine and brisk wind, but a dense fog has built-in resistance to the sun. Because fog is an excellent reflector of sunlight, only 20 to 40 percent of the impinging solar energy penetrates it to warm the ground and the foggy air, and only part of that heat is available to evaporate the fog droplets. Consequently a dense fog strongly resists dissipation by the sun. Moreover, the resistance is increased when the fog becomes a thick smog.

To what extent has our era of industrialism intensified the fog problem? This is very difficult to determine, because the meteorological records are incomplete and the phenomenon itself is highly variable. The incidence and intensity of fogs vary widely with time and season in any given place and from one place to another, even within a few miles. Furthermore, weather-observation practices have changed considerably over the past century: the observations are much more frequent than they used to be, cover many more locations and are complicated by changes both in the location of observatories and in the observers' nomenclature. The definition of "dense" fog, for example, has been revised repeatedly by the U.S. Weather Bureau. Nevertheless, although reliable comparisons cannot be made in detail, a survey of past records indicates some general trends. It appears that the incidence, duration and probably the density of fogs vary directly with the amount of industrial activity and air pollution in the area involved.

Perhaps the most reliable set of data for a single location is the one for the city of Prague in Czechoslovakia, where consistent observations of fogs have been made for the past century and a half. The records show that in the period since 1881 Prague has had nearly twice as many fogs as in the preceding 80-year period. In general it appears

CLASSIC EXAMPLE of an advection fog is the light, low-lying bank seen obscuring

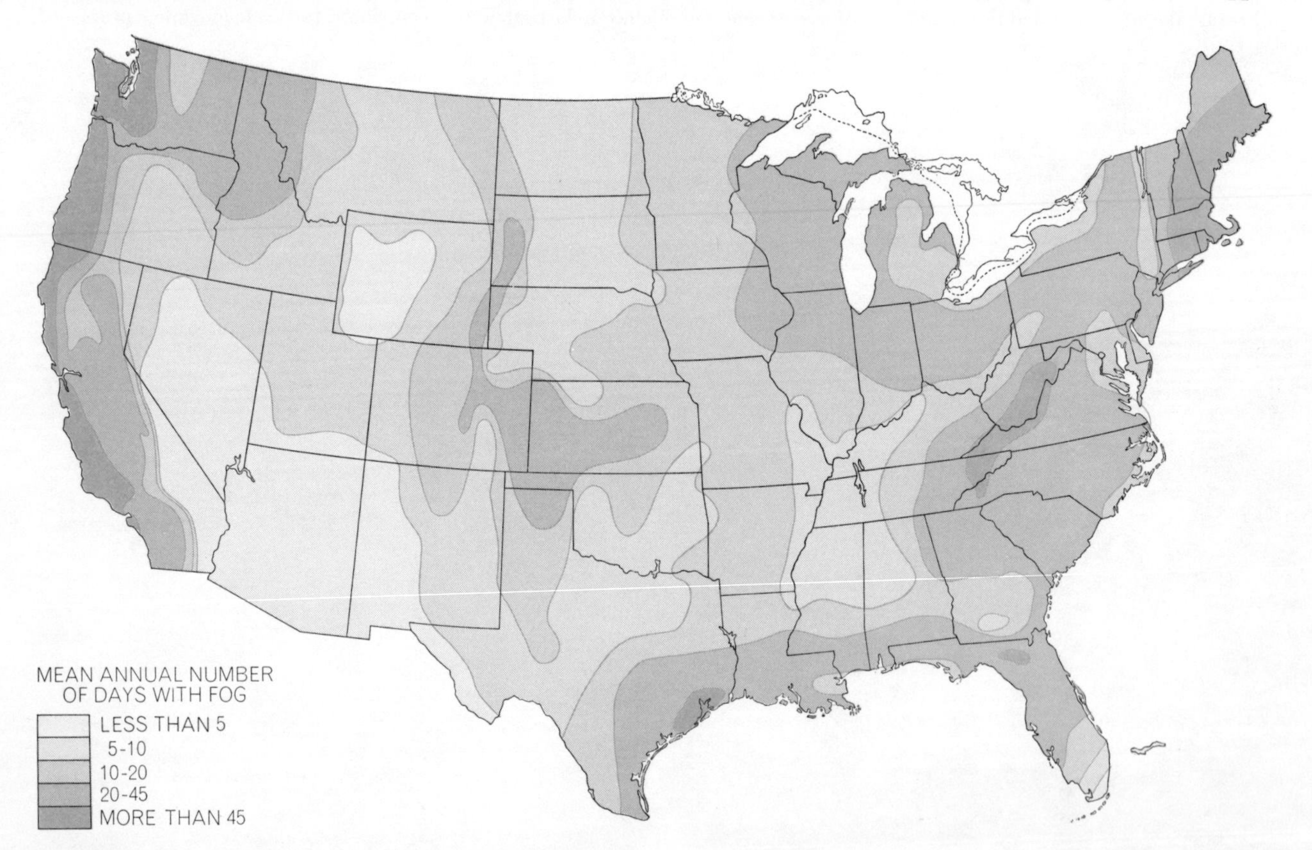

DISTRIBUTION OF FOGS in the continental U.S. is shown by area shading that indicates the days of dense fog per year reported by 251 weather stations from 1900 to 1960. The least foggy parts of the mainland are the desert areas of Arizona, California and Nevada; foggiest are the Pacific and New England coasts. Appalachian hills often have rain fogs; the valleys, radiation fogs.

the Golden Gate Bridge at the entrance to San Francisco Bay in the aerial photograph. When a warm, moist air mass passes over the cold waters of the California Current, it is cooled to the saturation point and some of its water vapor is condensed into fog droplets.

that fog tends to be more frequent in and near cities than in unpolluted rural areas. The available data do not indicate, however, if fogs tend to be denser in cities than they are elsewhere. It is conceivable that the relatively warm microclimate of large cities (due to the solar heating of their streets and buildings and the urban artificial heat) may prevent the strong nocturnal cooling that is usually a prerequisite for dense fogs. If that is so, perhaps the places where dense fogs occur most frequently are highly industrialized small cities and suburban communities that lie downwind of metropolises.

A fog recently studied by Charles L. Hosler, Jr., a meteorologist at Pennsylvania State University, may serve as an illustration of the problem. The fog is seen shortly after sunrise in Bald Eagle Valley in central Pennsylvania on days when little or no fog is observed elsewhere in the state. The valley contains large industrial plants that daily discharge billions of hygroscopic particles and tons of water vapor into the atmosphere. Although fog formation is aided by the overnight drainage of cold air into Bald Eagle Valley, the pollution is a major contributing factor.

What can be done about the growing fog problem? The phenomenon has become a matter of active concern for many interests—government, industry, the airlines, the military—and it is being attacked by investigators in a wide range of disciplines. Clearly one of the prime needs is a vigorous assault on the complex problem of air pollution. Nor is pollution confined exclusively to the atmosphere. A form of pollution that is responsible for some fogs and that has received little attention is the "thermal pollution" of natural water resulting from industrial practices. It gives rise to the formation of fogs by increasing the normal rate of water evaporation.

Enormous quantities of heated water are discharged into our streams and lakes by industrial operations that use water for various cooling purposes. The principal offenders are steel mills, paper factories, sewage-treatment plants, certain chemical and manufacturing plants and particularly electric-power generating stations. Power plants based on nuclear fuel, to which many utilities are now turning, use enormous quantities of water for cooling. The power industry estimates that by 1980 it will be using one-fifth of the total free water runoff in the U.S. for cooling. Thus, ironically, nuclear plants, which are counted on to reduce air pollution by replacing plants that use fossil fuels, may become a major factor in spawning fog and thereby trapping air pollution in the form of smog.

Some industrial plants, notably in the power industry, have taken to getting rid of their heated water by evaporating it in massive cooling towers instead of dumping it into bodies of water. Unfortunately this method is expensive and may still produce fogs. Some towers evaporate hundreds of cubic meters of water per hour, and the tremendous flux of vapor, if trapped under an inversion on a windless night, could generate and maintain a dense fog over a large city. Federal aviation officials believe a cooling tower recently built three miles north of the Morgantown Municipal Airport in West Virginia is responsible for local fogs that have begun to trouble the area. And it appears that fog plumes from other cooling towers may have contributed to automobile accidents on highways near them.

Cooling towers might be turned into assets rather than liabilities in farming areas. The fog they produce could prevent heat from escaping from the soil and thus protect cold-sensitive crops against frost. Some farmers in localities of frequent fogs report that fog blankets extend the frost-free growing season and thereby increase their crop yields. Russian meteorologists have successfully used artificial fogs to protect vineyards from frost. Perhaps power companies should be encouraged to build their new plants in rural regions where farming could benefit from such frost protection.

Among the various possibilities for dealing with the fog problem the approach that has been explored most actively up to now is the idea of dissipating fog by some artificial means. Over the past several decades hundreds of schemes for doing this have been proposed and many have been tried, but so far no universally practicable method has been found. What are the prospects that an effective and not too expensive technique could be developed?

The most direct method would be

simply to blow the fog away, using an artificial wind. This tactic is actually employed around some settlements in the Arctic to disperse ice fogs. In the frigid atmosphere the water vapor emitted from human settlements (indeed, even the exhalations of a reindeer herd) can easily saturate the air and produce a local fog; at a temperature of 30 degrees or more below zero C. the fog consists mainly of ice crystals. Giant fans have been used successfully to free settlements of such fogs. Obviously, however, the method is applicable only on a small scale and in special situations.

Another attack on fog, based on the principle of evaporating the droplets by one means or another, has been applied in several interesting ways. The best-known of these is the FIDO method (Fog Investigation and Dispersal Operation), in which fuel-oil fires have been used to burn off (that is, evaporate) fog on airfields. The British resorted to this technique on military fields in World War II and successfully cleared them for more than 2,000 takeoffs and landings that could not have been undertaken otherwise. The method has important drawbacks, however. It is expensive, requiring hundreds of dollars' worth of fuel to clear a jet runway for 10 to 15 minutes; it creates a fire hazard for planes landing on the field, and it cannot dissipate all dense fogs. Moreover, the smoke and moisture released by the combustion of oil hinder evaporation of the fog. It has been proposed that this problem might be obviated by using electricity, jet-engine exhausts or anthracite coal to provide the heat, but the high cost would still remain a major objection. Drying the air with chemicals rather than heat has been effective in clearing fog in some cases; here again, however, the procedure is too expensive for wide use, and the chemicals employed tend to be corrosive.

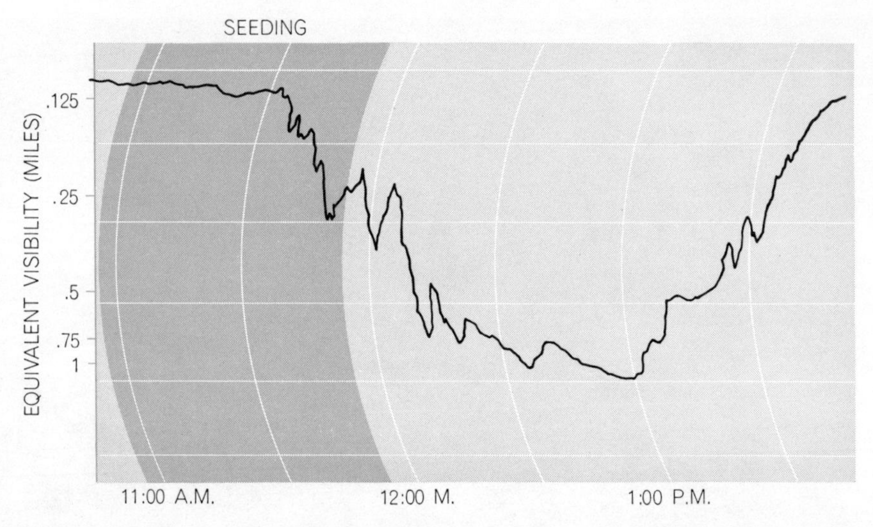

COLD-SEEDING TEST at the airport in Medford, Ore., began shortly before 11:00 A.M.; visibility, as measured by a transmissometer, was less than an eighth of a mile. Thirty minutes after the first seeding run visibility began to improve. By 12:15 P.M. it exceeded the half-mile minimum required at the airport and remained above minimum for over an hour.

Among the various principles of fog dispersal that have been tested, the most promising seems to be the injection of a catalyst or some other agent that will cause the droplets to coalesce and thus grow large enough to fall quickly to the ground. This type of attack has proved its worth in fogs consisting of supercooled droplets. By seeding the fog with particles of a very cold substance such as dry ice or liquid propane, one can cause some of the fog droplets to freeze. Water vapor in the air then condenses onto these ice crystals; the resultant drying of the air turns additional fog droplets back into vapor. The vapor, in turn, accelerates the growth of the ice crystals, which fall to the ground as they enlarge. United Air Lines has been seeding supercooled fogs at fields in the Pacific Northwest and in Alaska for several years and has found the method to be about 80 percent successful in dissipating such fogs. The airline estimates that this investment in fog control has repaid it fivefold by maintaining the regularity of flight schedules. The method has also worked well in other areas where it is applicable. It is effective, however, only for supercooled fogs, which account for only about 5 percent of all the fogs in the U.S. Temperate Zone.

Several ideas for dispersing warm fogs by droplet coalescence are currently being explored. The Air Transport Association is sponsoring a series of tests of a chemical mixture (composition undisclosed) that is said to make droplets combine by an electrical attraction effect. It is reported to have achieved some success in dissipating radiation fogs at Sacramento, Calif., during calm or very light winds. The ability of the method to disperse moving fogs, however, remains to be demonstrated. In any attempt to control fog, the wind condition is crucial. When the air is calm, clearing it of fog presents a comparatively uncomplicated problem because the volume of air that needs to be treated is limited. On the other hand, a breeze can quickly refog a space that has been cleared of fog; when moderate or strong winds are blowing, it is almost impossible to maintain a clearing.

Another scheme that has been proposed for fog dispersal involves dropping carefully controlled doses of salt particles into the fog. The theory is that when these hygroscopic particles deliquesce into solution droplets, they will gain water and grow at the expense of natural fog droplets because the humidity is higher with respect to the solution than it is with respect to natural water. It is hoped that the collection of the water into fewer and larger drops will produce a rapid improvement in visibility and that the fall of the large drops to the ground will maintain the improvement for some time after seeding has ended.

For effective progress toward the economical control or modification of fogs we shall have to learn a great deal more about the basic structure of fog and the chemical, physical and electrical properties of the fog droplets. Intensive studies are going forward on various types of natural fogs and on artificial fogs produced in the laboratory. The results of these investigations will be used to test mathematical models that describe fogs in terms of such quantities as temperature, humidity, wind, condensation nuclei, concentration of droplets and amount of liquid water. It should then be possible to obtain insight into the mechanisms and energy exchanges involved in the formation and maintenance of fogs and allow meteorologists to determine which kind of fog will respond to which dispersal method.

The artificial dissipation of fog will be rather costly in any case. Much might be done to prevent the formation of fogs in the first place. For example, spreading a chemical film over swamps and

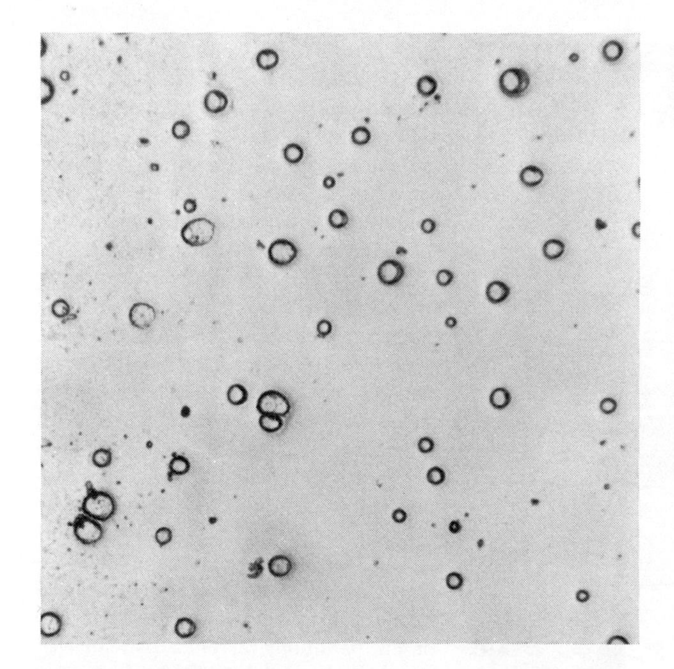

WATER DROPLETS that comprise a fog vary in size from about two microns to as much as 50 microns in diameter; their average size is 20 microns. The droplets in the photomicrograph are from a sample of supercooled fog prepared for a "cold seeding" study.

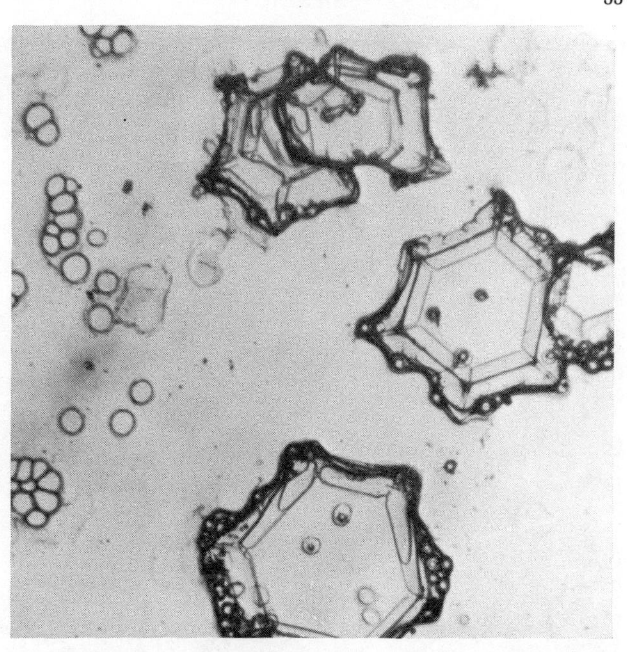

ICE CRYSTALS are formed in a sample of supercooled fog by seeding with propane. The crystals grow as water vapor is deposited on them. Reduced humidity makes the fog droplets evaporate, clearing the air and furnishing more water vapor for crystal growth.

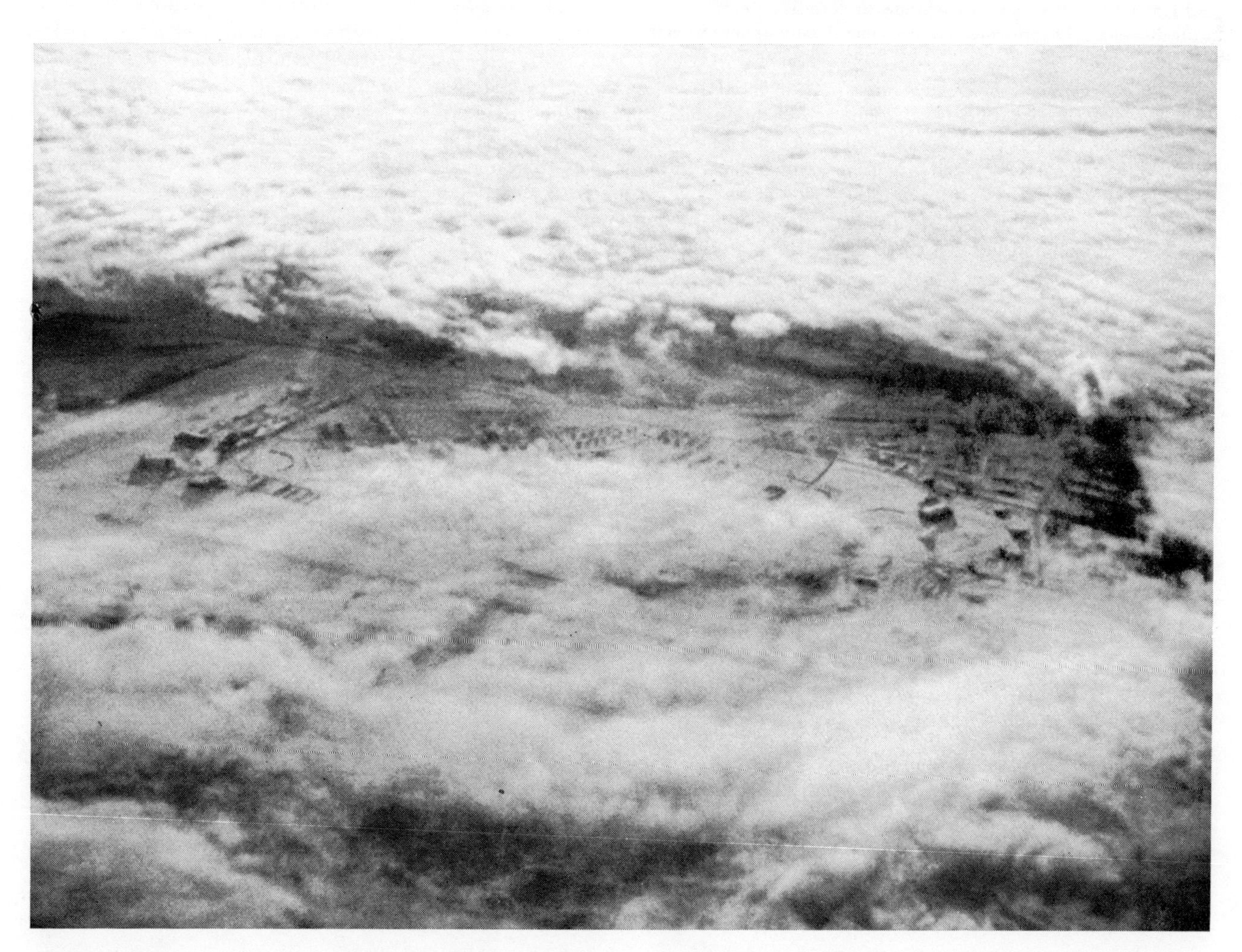

FOG CLEARANCE is achieved by dropping dry ice into a bank of supercooled fog overlying Elmendorf Air Force Base in Anchorage, Alaska. Seeding with dry ice initiated the growth of ice crystals, depriving the air of the liquid water that had made it foggy. Only 5 percent of the fogs in those parts of the U.S. that lie within the Temperate Zone are supercooled and dispersable by cold-seeding.

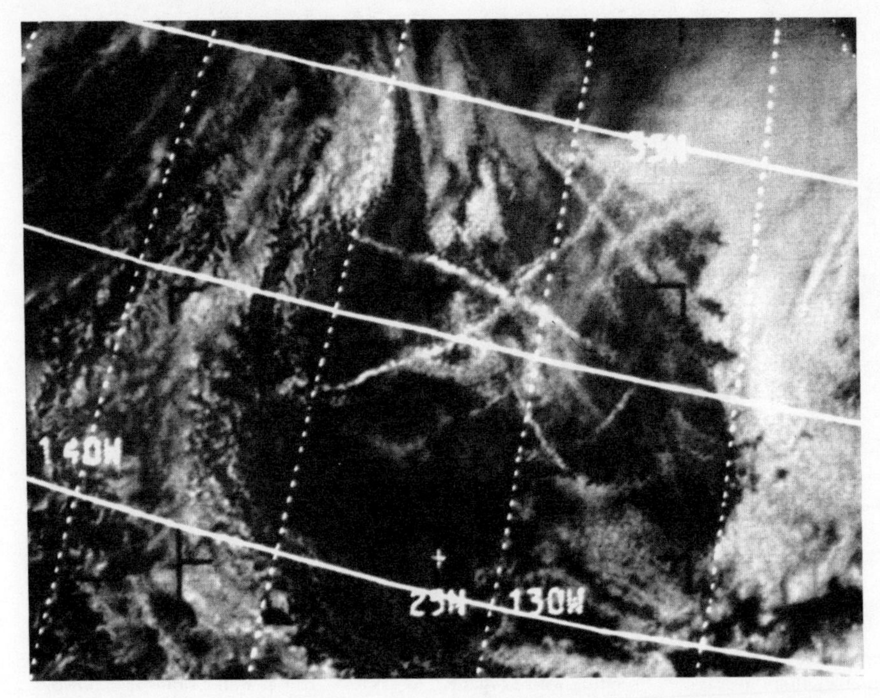

POLLUTION FOGS over the Pacific Ocean off the coast of California are visible in a weather satellite photograph as white trails in the area between 25 and 35 degrees north latitude and 125 and 135 degrees west longitude. Each trail is a narrow fogbank formed in response to the discharge of hygroscopic particles into the atmosphere from the funnels of ships at sea.

lakes in the vicinity of sensitive areas such as airports and highways might considerably reduce evaporation and thus reduce the frequency and intensity of fogs in those areas. There are indications that in places where shallow radiation fogs are common the fogs could be prevented from spreading by planting vegetation thickly around the area of origin. It would be desirable to prohibit pollution-generating factories from operating at times when the air is calm or an inversion is present. And the fog menace could be diminished greatly by giving more careful attention to the selection of proper locations for activities that generate fog and for those that may be troubled by fog.

Factories giving rise to air pollution, for example, should be located at sites where the topography and prevailing winds favor effective dispersal of the pollutants. Plants that must dispose of heated water should not be built near densely populated or well-traveled areas. On the other hand, in the selection of sites for airports, highways, sports stadiums, golf courses and so forth consideration should be given to finding locations where the geography, topography, soil characteristics and other features would tend to minimize the formation and persistence of fogs. For example, ideally an airport should be situated (1) upwind of any nearby source of air pollution, (2) on a plateau that stands high enough to shed cold air into a valley but low enough to avoid hilltop immersion in clouds, (3) away from rivers, lakes or marshy ground and (4) in full, unobstructed exposure to the sun.

By the application of meteorological knowledge to planning and by the development of further methods for fog prevention and dispersal, we may eventually be able to deal with the growing fog and smog problem. The success of these efforts, however, will hinge on the achievement of effective control over the pollution of the air and waters.

II

ATMOSPHERIC PHENOMENA

II ATMOSPHERIC PHENOMENA

INTRODUCTION

Most of the phenomena discussed in this reader are more common than people think. Halos occur 50 to 100 times a year. Rainbows are visible whenever sunlight falls on raindrops, and the once rare glory can be enjoyed on many commercial jet flights. Coronae, the delicate diffraction colors near the sun, can be seen in most low, thin clouds if one blocks out the solar glare. Lightning strikes and thunder rolls an estimated 10 million times a day. Even the distant zodiacal light lingers after dusk and rises again before dawn.

Why don't more people know about these terrific sights? They simply don't know where to look or what to look for. Watch people walking along the street. Where are their eyes? Looking down at their feet, watching the pavement, gazing into shop windows, looking at other people. They are looking everywhere but up! The most brilliant rainbow I ever saw occurred one afternoon in Los Angeles. It was far brighter than I thought a rainbow could be and totally dominated the eastern sky. Yet hundreds of people went about their business, never once glancing up, never once suspecting, never once knowing what they were missing.

As with most things, seeing atmospheric phenomena depends more on an observer's awareness of his or her surroundings than on luck. To demonstrate this, I assigned my "Color and Light" class at UCLA the task of going outside ten times a day on any day they chose, glancing around the sky, and recording what they saw. They were simply amazed. Everyone reported halos, coronas, crepuscular rays, contrast effects, and a number of other phenomena.

One beautiful aspect of physics is its unity. This is no more delightfully illustrated than by the glory. An infrequent sight before the days of airplane travel, the glory can be seen on sunny days if you look at your shadow cast on a fog bank. Appearing as a series of concentric colored rings around the antisolar point (the direction opposite the sun), the glory is usually seen when the sun is low in the sky. Charles T. R. Wilson, who had seen the glory from Ben Nevis in Scotland, set about to study it in the laboratory by recreating the circumstances under which he thought the glory would appear. In 1895 he built a cloud chamber, which could produce a tiny cloud under the controlled lighting conditions in the laboratory. The study was scarcely under way when Wilson discovered that charged particles left their signatures as contrails when they passed through the apparatus. Thus, the cloud chamber found a new mistress in particle physics. Wilson won the Nobel prize in 1927 for his invention, and the delicate glory was swept into a forgotten corner of the physics laboratory.

There it remained until 1965, when H. Moysés Nussenzveig, another particle physicist, developed his theory of the scattering of particles off atomic nuclei, or the complex angular momentum theory. The formulation was intended for nuclear scattering, but it embodied a much more general formalism that included the scattering of light off spherical particles. With only minor modifications, he applied the theory to the glory. The result was our best understanding to date of the glory and its familiar cousin, the rainbow.

For those interested in atmospheric phenomena, there are several popular journals devoted to celestial happenings of all kinds. Magazines such as *Weather* (UK), *The Meteorological Magazine* (UK), *Weatherwise* (USA), *Sky and Telescope* (USA), and *Astronomy* (USA) provide regular reports on atmospheric and extraterrestrial displays as well as regular feature articles on a wide variety of topics. Minnaert's book, mentioned in the preface, is an excellent survey of the field, even though much of it is out of date. A more technical treatment of many of the subjects in this reader can be found in R. A. R. Tricker's *Introduction to Meteorological Optics*. The latest research papers are published in such scientific journals as the *Journal of the Optical Society of America, Applied Optics, Journal of the American Meteorological Society, Journal of the Atmospheric Sciences, Journal of Geophysical Review*, and *Journal of the Royal Meteorological Society*.

Like most scientists, I am at home with journals and scientific equipment. Studying the world around me is what I do best, and it is difficult to resist an appealing research project. Yet there are other ways to experience nature, ways that I rediscover regularly.

At dawn one foggy morning, I was sitting quietly by the ocean. I could see only a little way down the beach, and I felt serene and protected in the arms of the fog. Suddenly the sun broke out, creating a magnificent fogbow, that ghostly white rainbow formed by very small water drops. My first reaction was to dash home for a camera and notebook, but, for reasons I did not understand at the time, I decided to lie back and enjoy the gorgeous bow in the mystical light of the new day. Shutter speeds, f-stops, and troublesome pencils simply would have been out of place. I watched the bow for about 45 minutes, until the fog burned off. Afterwards I expected to feel guilty for letting the opportunity for an easy picture slip by. But I didn't. And now I know why.

Appreciation of our surroundings is as much a part of the human intellect as is curiosity. That day on the beach, sheer appreciation won out over curiosity. I have no way of knowing which state of mind will possess me when I next encounter a fogbow. But I am looking forward to that time, because either way I know I shall benefit from the meeting.

7 The Theory of the Rainbow

by H. Moysés Nussenzveig
April 1977

When sunlight is scattered by raindrops, why is it that colorful arcs appear in certain regions of the sky? Answering this subtle question has required all the resources of mathematical physics

The rainbow is a bridge between the two cultures: poets and scientists alike have long been challenged to describe it. The scientific description is often supposed to be a simple problem in geometrical optics, a problem that was solved long ago and that holds interest today only as a historical exercise. This is not so; a satisfactory quantitative theory of the rainbow has been developed only in the past few years. Moreover, that theory involves much more than geometrical optics; it draws on all we know of the nature of light. Allowance must be made for wavelike properties, such as interference, diffraction and polarization, and for particlelike properties, such as the momentum carried by a beam of light.

Some of the most powerful tools of mathematical physics were devised explicitly to deal with the problem of the rainbow and with closely related problems. Indeed, the rainbow has served as a touchstone for testing theories of optics. With the more successful of those theories it is now possible to describe the rainbow mathematically, that is, to predict the distribution of light in the sky. The same methods can also be applied to related phenomena, such as the bright ring of color called the glory, and even to other kinds of rainbows, such as atomic and nuclear ones.

Scientific insight has not always been welcomed without reservations. Goethe wrote that Newton's analysis of the rainbow's colors would "cripple Nature's heart." A similar sentiment was expressed by Charles Lamb and John Keats; at a dinner party in 1817 they proposed a toast: "Newton's health, and confusion to mathematics." Yet the scientists who have contributed to the theory of the rainbow are by no means insensitive to the rainbow's beauty. In the words of Descartes: "The rainbow is such a remarkable marvel of Nature ...that I could hardly choose a more suitable example for the application of my method."

The single bright arc seen after a rain shower or in the spray of a waterfall is the primary rainbow. Certainly its most conspicuous feature is its splash of colors. These vary a good deal in brightness and distinctness, but they always follow the same sequence: violet is innermost, blending gradually with various shades of blue, green, yellow and orange, with red outermost.

Other features of the rainbow are fainter and indeed are not always present. Higher in the sky than the primary bow is the secondary one, in which the colors appear in reverse order, with red innermost and violet outermost. Careful observation reveals that the region between the two bows is considerably darker than the surrounding sky. Even when the secondary bow is not discernible, the primary bow can be seen to have a "lighted side" and a "dark side." The dark region has been given the name Alexander's dark band, after the Greek philosopher Alexander of Aphrodisias, who first described it in about A.D. 200.

Another feature that is only sometimes seen is a series of faint bands, usually pink and green alternately, on the inner side of the primary bow. (Even more rarely they may appear on the outer side of the secondary bow.) These "supernumerary arcs" are usually seen most clearly near the top of the bow. They are anything but conspicuous, but they have had a major influence on the development of theories of the rainbow.

The first attempt to rationally explain the appearance of the rainbow was probably that of Aristotle. He proposed that the rainbow is actually an unusual kind of reflection of sunlight from clouds. The light is reflected at a fixed angle, giving rise to a circular cone of "rainbow rays." Aristotle thus explained correctly the circular shape of the bow and perceived that it is not a material object with a definite location in the sky but rather a set of directions along which light is strongly scattered into the eyes of the observer.

The angle formed by the rainbow rays and the incident sunlight was first measured in 1266 by Roger Bacon. He measured an angle of about 42 degrees; the secondary bow is about eight degrees higher in the sky. Today these angles are customarily measured from the opposite direction, so that we measure the total change in the direction of the sun's rays. The angle of the primary bow is therefore 180 minus 42, or 138, degrees; this is called the rainbow angle. The angle of the secondary bow is 130 degrees.

After Aristotle's conjecture some 17 centuries passed before further significant progress was made in the theory of the rainbow. In 1304 the German monk Theodoric of Freiberg rejected Aristotle's hypothesis that the rainbow results from collective reflection by the raindrops in a cloud. He suggested instead that each drop is individually capable of producing a rainbow. Moreover, he tested this conjecture in experiments with a magnified raindrop: a spherical flask filled with water. He was able to trace the path followed by the light rays that make up the rainbow.

Theodoric's findings remained largely unknown for three centuries, until they were independently rediscovered by Descartes, who employed the same method. Both Theodoric and Descartes showed that the rainbow is made up of rays that enter a droplet and are reflected once from the inner surface. The secondary bow consists of rays that have undergone two internal reflections. With each reflection some light is lost, which is the main reason the secondary bow is fainter than the primary one. Theodoric and Descartes also noted that along each direction within the angular

DOUBLE RAINBOW was photographed at Johnstone Strait in British Columbia. The bright, inner band is the primary bow; it is separated from the fainter secondary bow by a region, called Alexander's dark band, that is noticeably darker than the surrounding sky. Below the primary bow are a few faint stripes of pink and green; they are supernumerary arcs. The task of theory is to give a quantitative explanation for each of these features.

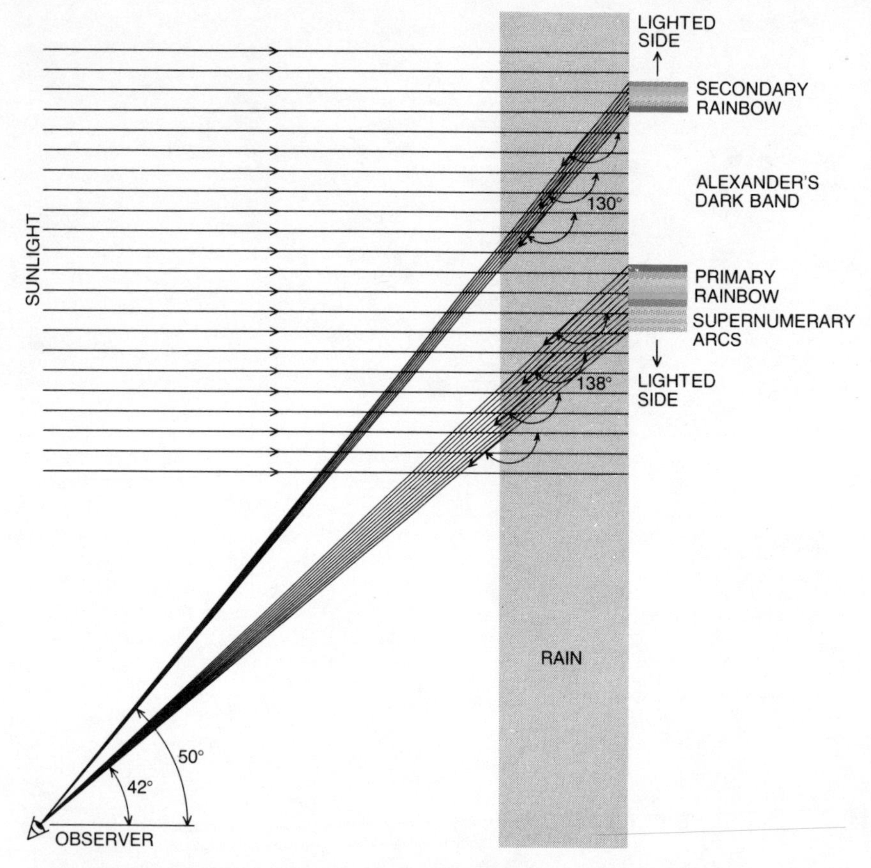

GEOMETRY OF THE RAINBOW is determined by the scattering angle: the total angle through which a ray of sunlight is bent by its passage through a raindrop. Rays are strongly scattered at angles of 138 degrees and 130 degrees, giving rise respectively to the primary and the secondary rainbows. Between those angles very little light is deflected; that is the region of Alexander's dark band. The optimum angles are slightly different for each wavelength of light, with the result that the colors are dispersed; note that the sequence of colors in the secondary bow is the reverse of that in the primary bow. There is no single plane in which the rainbow lies; the rainbow is merely the set of directions along which light is scattered toward the observer.

range corresponding to the rainbow only one color at a time could be seen in the light scattered by the globe. When the eye was moved to a new position so as to explore other scattering angles, the other spectral colors appeared, one by one. Theodoric and Descartes concluded that each of the colors in the rainbow comes to the eye from a different set of water droplets.

As Theodoric and Descartes realized, all the main features of the rainbow can be understood through a consideration of the light passing through a single droplet. The fundamental principles that determine the nature of the bow are those that govern the interaction of light with transparent media, namely reflection and refraction.

The law of reflection is the familiar and intuitively obvious principle that the angle of reflection must equal the angle of incidence. The law of refraction is somewhat more complicated. Whereas the path of a reflected ray is determined entirely by geometry, refraction also involves the properties of light and the properties of the medium.

The speed of light in a vacuum is invariant; indeed, it is one of the fundamental constants of nature. The speed of light in a material medium, on the other hand, is determined by the properties of the medium. The ratio of the speed of light in a vacuum to the speed in a substance is called the refractive index of that substance. For air the index is only slightly greater than 1; for water it is about 1.33.

A ray of light passing from air into water is retarded at the boundary; if it strikes the surface obliquely, the change in speed results in a change in direction. The sines of the angles of incidence and refraction are always in constant ratio to each other, and the ratio is equal to that between the refractive indexes for the two materials. This equality is called Snell's law, after Willebrord Snell, who formulated it in 1621.

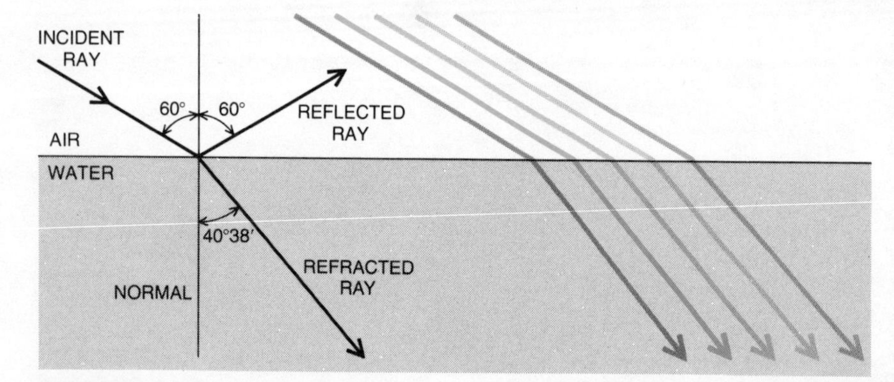

REFLECTION AND REFRACTION of light at boundaries between air and water are the basic events in the creation of a rainbow. In reflection the angle of incidence is equal to the angle of reflection. In refraction the angle of the transmitted ray is determined by the properties of the medium, as characterized by its refractive index. Light entering a medium with a higher index is bent toward the normal. Light of different wavelengths is refracted through slightly different angles; this dependence of the refractive index on color is called dispersion. Theories of the rainbow often deal separately with each monochromatic component of incident light.

A preliminary analysis of the rainbow can be obtained by applying the laws of reflection and refraction to the path of a ray through a droplet. Because the droplet is assumed to be spherical all directions are equivalent and there is only one significant variable: the displacement of the incident ray from an axis passing through the center of the droplet. That displacement is called the impact parameter. It ranges from zero, when the ray coincides with the central axis, to the radius of the droplet, when the ray is tangential.

At the surface of the droplet the incident ray is partially reflected, and this reflected light we shall identify as the scattered rays of Class 1. The remaining light is transmitted into the droplet (with a change in direction caused by refraction) and at the next surface is again partially transmitted (rays of Class 2) and partially reflected. At the next boundary the reflected ray is again split into reflected and transmitted components, and the process continues indefinitely. Thus the droplet gives rise to a series of scattered rays, usually with rapidly decreasing intensity. Rays of Class 1 represent direct reflection by the droplet and those of Class 2 are directly transmitted through it. Rays of Class 3 are those that escape the droplet after one internal reflection, and they make up the primary rainbow. The Class 4 rays, having undergone two internal reflections, give rise to the secondary bow. Rainbows of higher order are formed by rays making more complicated passages, but they are not ordinarily visible.

For each class of scattered rays the scattering angle varies over a wide range of values as a function of the impact parameter. Since in sunlight the droplet is illuminated at all impact parameters simultaneously, light is scattered in virtually all directions. It is not difficult to find light paths through the droplet that contribute to the rainbow, but there are infinitely many other paths that direct the light elsewhere. Why, then, is the scattered intensity enhanced in the vicinity of the rainbow angle? It is a question Theodoric did not consider; an answer was first provided by Descartes.

By applying the laws of reflection and refraction at each point where a ray strikes an air-water boundary, Descartes painstakingly computed the paths of many rays incident at many impact parameters. The rays of Class 3 are of predominating importance. When the impact parameter is zero, these rays are scattered through an angle of 180 degrees, that is, they are backscattered toward the sun, having passed through the center of the droplet and been reflected from the far wall. As the impact parameter increases and the incident rays are displaced from the center of the droplet, the scattering angle decreases. Descartes found, however, that this trend does not continue as the impact parameter is increased to its maximum value, where the incident ray grazes the droplet at a tangent to its surface. Instead the scattering angle passes through a minimum when the impact parameter is about seven-eighths of the radius of the droplet, and thereafter it increases again. The scattering angle at the minimum is 138 degrees.

For rays of Class 4 the scattering angle is zero when the impact parameter is zero; in other words, the central ray is reflected twice, then continues in its original direction. As the impact parameter increases so does the scattering angle, but again the trend is eventually reversed, this time at 130 degrees. The Class 4 rays have a maximum scattering angle of 130 degrees, and as the impact parameter is further increased they bend back toward the forward scattering direction again.

Because a droplet in sunlight is uniformly illuminated the impact parameters of the incident rays are uniformly distributed. The concentration of scattered light is therefore expected to be greatest where the scattering angle varies most slowly with changes in the impact parameter. In other words, the scattered light is brightest where it gathers together the incident rays from the largest range of impact parameters. The regions of minimum variation are those surrounding the maximum and minimum scattering angles, and so the special status of the primary and secondary rainbow angles is explained. Furthermore, since no rays of Class 3 or Class 4 are scattered into the angular region between 130 and 138 degrees, Alexander's dark band is also explained.

Descartes's theory can be seen more clearly by considering an imaginary population of droplets from which light is somehow scattered with uniform intensity in all directions. A sky filled with such droplets would be uniformly bright at all angles. In a sky filled with real water droplets the same total illumination is available, but it is redistributed. Most parts of the sky are dimmer than they would be with uniform scattering, but in the vicinity of the rainbow angle there is a bright arc, tapering off gradually on the lighted side and more sharply on the dark side. The secondary bow is a similar intensity highlight, except that it is narrower and all its features are dimmer. In the Cartesian theory the region between the bows is distinctly darker than the sky elsewhere; if only rays of Class 3 and Class 4 existed, it would be quite black.

The Cartesian rainbow is a remarkably simple phenomenon. Brightness is a function of the rate at which the scattering angle changes. That angle is itself determined by just two factors: the refractive index, which is assumed to be constant, and the impact parameter, which is assumed to be uniformly distributed. One factor that has no influence at all on the rainbow angle is size: the geometry of scattering is the same for small cloud droplets and for the large water-filled globes employed by Theodoric and Descartes.

So far we have ignored one of the most conspicuous features of the rainbow: its colors. They were explained, of course, by Newton, in his prism experiments of 1666. Those experiments demonstrated not only that white light is a mixture of colors but also that the refractive index is different for each color, the effect called dispersion. It follows that each color or wavelength of light must have its own rainbow angle; what we observe in nature is a collection of monochromatic rainbows, each one slightly displaced from the next.

From his measurements of the refractive index Newton calculated that the

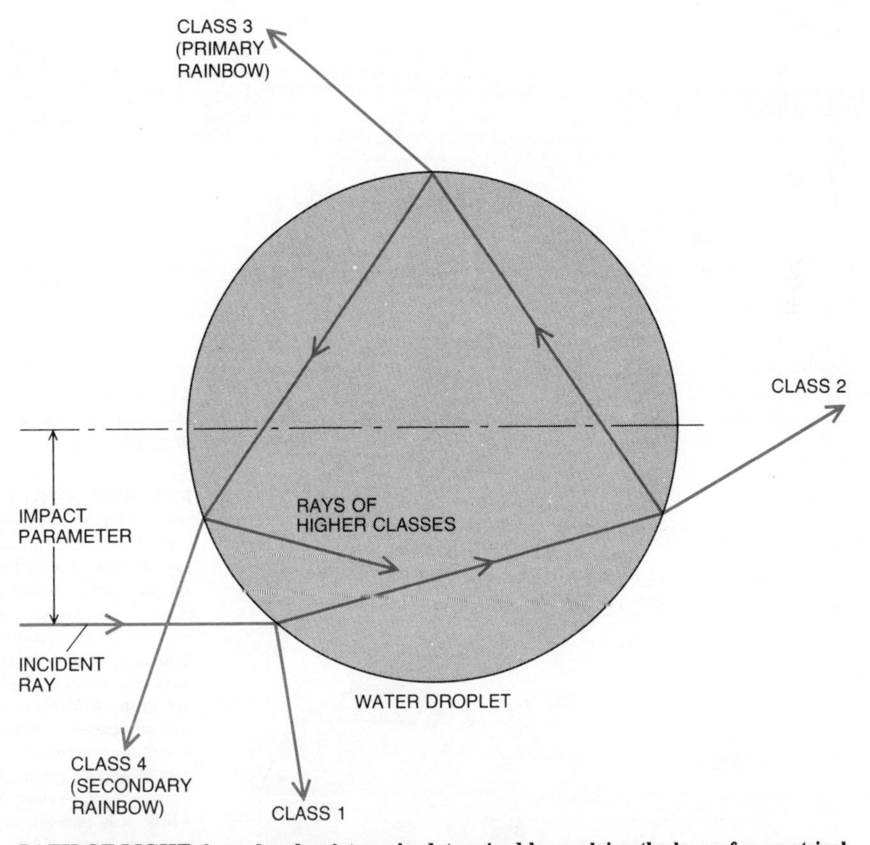

PATH OF LIGHT through a droplet can be determined by applying the laws of geometrical optics. Each time the beam strikes the surface part of the light is reflected and part is refracted. Rays reflected directly from the surface are labeled rays of Class 1; those transmitted directly through the droplet are designated Class 2. The Class 3 rays emerge after one internal reflection; it is these that give rise to the primary rainbow. The secondary bow is made up of Class 4 rays, which have undergone two internal reflections. For rays of each class only one factor determines the value of the scattering angle. That factor is the impact parameter: the displacement of the incident ray from an axis that passes through the center of the droplet.

rainbow angle is 137 degrees 58 minutes for red light and 139 degrees 43 minutes for violet light. The difference between these angles is one degree 45 minutes, which would be the width of the rainbow if the rays of incident sunlight were exactly parallel. Allowing half a degree for the apparent diameter of the sun, Newton obtained a total width of two degrees 15 minutes for the primary bow. His own observations were in good agreement with this result.

Descartes and Newton between them were able to account for all the more conspicuous features of the rainbow. They explained the existence of primary and secondary bows and of the dark band that separates them. They calculated the angular positions of these features and described the dispersion of the scattered light into a spectrum. All of this was accomplished with only geometrical optics. Their theory nevertheless had a major failing: it could not explain the supernumerary arcs. The understanding of these seemingly minor features requires a more sophisticated view of the nature of light.

The supernumerary arcs appear on the inner, or lighted, side of the primary bow. In this angular region two scattered rays of Class 3 emerge in the same direction; they arise from incident rays that have impact parameters on each side of the rainbow value. Thus at any given angle slightly greater than the rainbow angle the scattered light includes rays that have followed two different paths through the droplet. The rays emerge at different positions on the surface of the droplet, but they proceed in the same direction.

In the time of Descartes and Newton these two contributions to the scattered intensity could be handled only by simple addition. As a result the predicted intensity falls off smoothly with deviation from the rainbow angle, with no trace of supernumerary arcs. Actually the intensities of the two rays cannot be added because they are not independent sources of radiation.

The optical effect underlying the supernumerary arcs was discovered in 1803 by Thomas Young, who showed that light is capable of interference, a phenomenon that was already familiar from the study of water waves. In any medium the superposition of waves can lead either to reinforcement (crest on

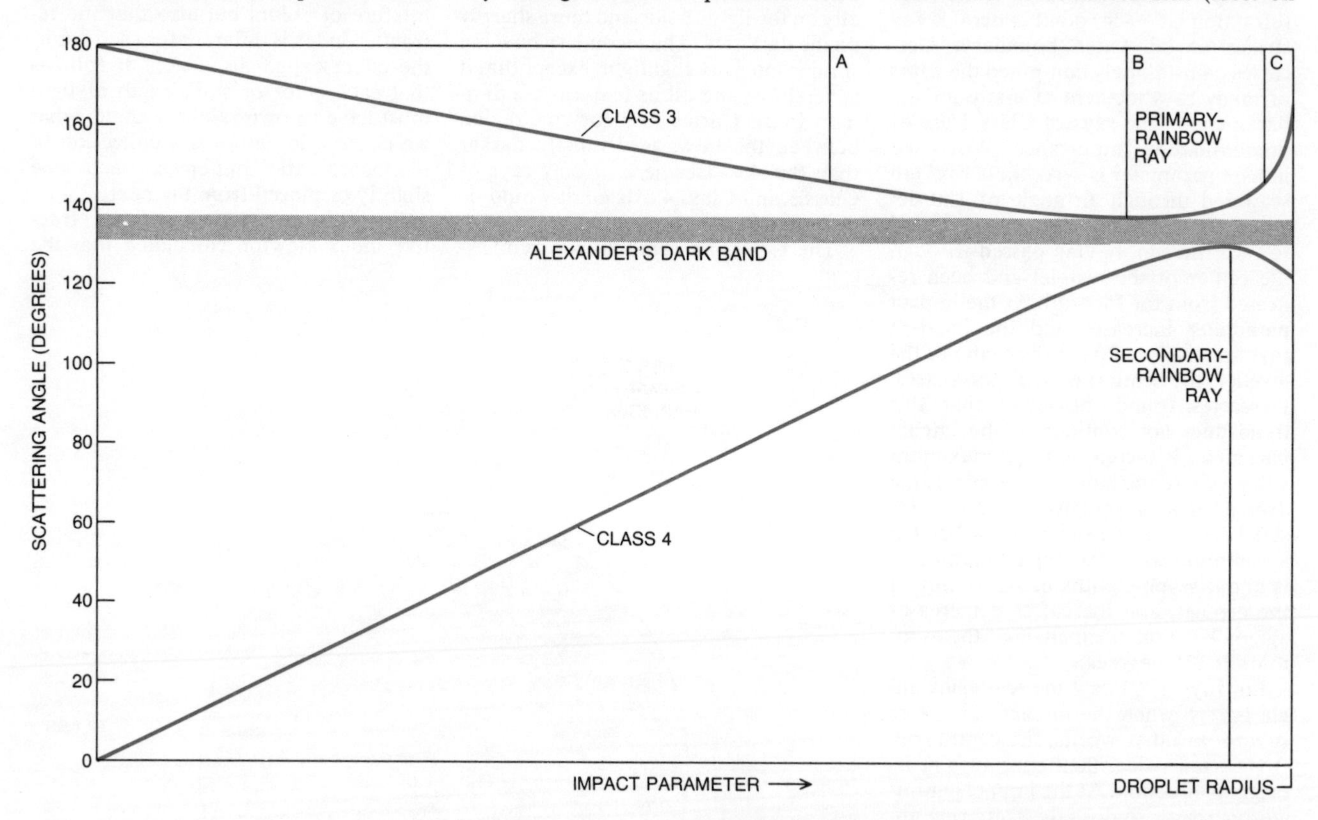

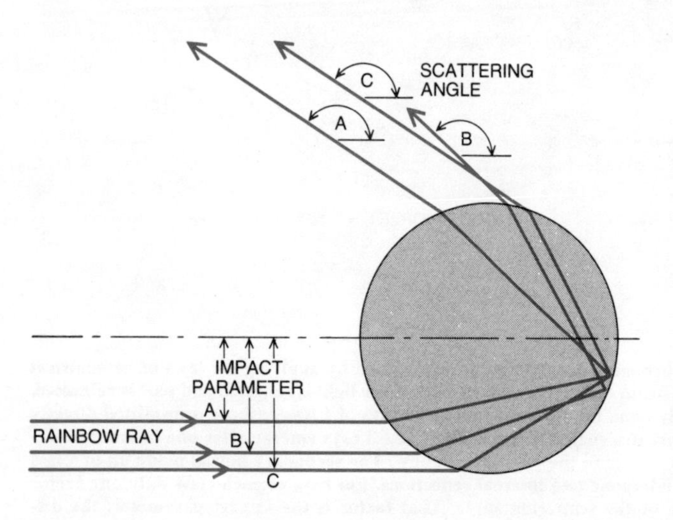

RAINBOW ANGLE can be seen to have a special significance when the scattering angle is considered as a function of the impact parameter. When the impact parameter is zero, the scattering angle for a ray of Class 3 is 180 degrees; the ray passes through the center of the droplet and is reflected by the far surface straight back at the sun. As the impact parameter increases, the scattering angle decreases, but eventually a minimum angle is reached. This ray of minimum deflection is the rainbow ray in the diagram at the left; rays with impact parameters on each side of it are scattered through larger angles. The minimum deflection is about 138 degrees, and the greatest concentration of scattered rays is to be found in the vicinity of this angle. The resulting enhancement in the intensity of the scattered light is perceived as the primary rainbow. The secondary bow is formed in a similar way, except that the scattering angle for the Class 4 rays of which it is composed increases to a maximum instead of decreasing to a minimum. The maximum lies at about 130 degrees. No rays of Class 3 or Class 4 can reach angles between 130 degrees and 138 degrees, explaining the existence of Alexander's dark band. At the left two Class 3 rays, with impact parameters on each side of the rainbow value, emerge at the same scattering angle. It is interference between rays such as these two that gives rise to the supernumerary arcs.

crest) or to cancellation (crest on trough). Young demonstrated the interference of light waves by passing a single beam of monochromatic light through two pinholes and observing the alternating bright and dark "fringes" produced. It was Young himself who pointed out the pertinence of his discovery to the supernumerary arcs of the rainbow. The two rays scattered in the same direction by a raindrop are strictly analogous to the light passing through the two pinholes in Young's experiment. At angles very close to the rainbow angle the two paths through the droplet differ only slightly, and so the two rays interfere constructively. As the angle increases, the two rays follow paths of substantially different length. When the difference equals half of the wavelength, the interference is completely destructive; at still greater angles the beams reinforce again. The result is a periodic variation in the intensity of the scattered light, a series of alternately bright and dark bands.

Because the scattering angles at which the interference happens to be constructive are determined by the difference between two path lengths, those angles are affected by the radius of the droplet. The pattern of the supernumerary arcs (in contrast to the rainbow angle) is therefore dependent on droplet size. In larger drops the difference in path length increases much more quickly with impact parameter than it does in small droplets. Hence the larger the droplets are, the narrower the angular separation between the supernumerary arcs is. The arcs can rarely be distinguished if the droplets are larger than about a millimeter in diameter. The overlapping of colors also tends to wash out the arcs. The size dependence of the supernumeraries explains why they are easier to see near the top of the bow: raindrops tend to grow larger as they fall.

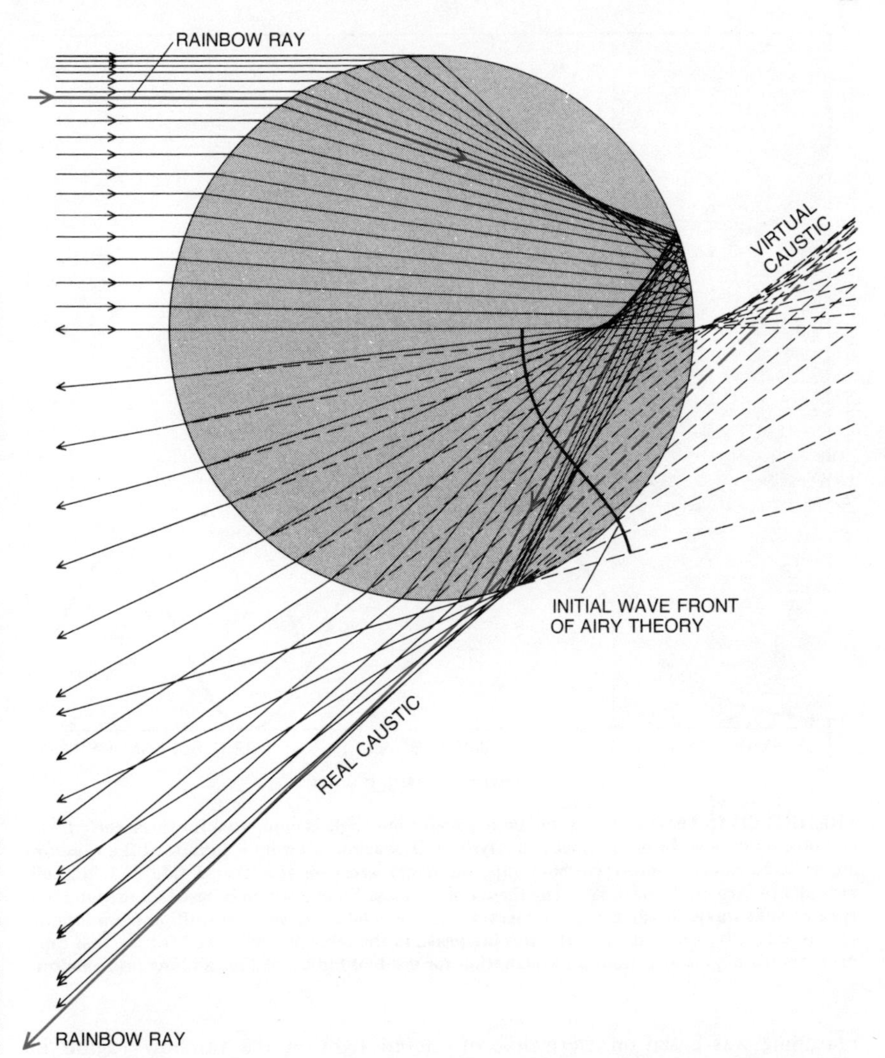

CONFLUENCE OF RAYS scattered by a droplet gives rise to caustics, or "burning curves." A caustic is the envelope of a ray system. Of special interest is the caustic of Class 3 rays, which has two branches, a real branch and a "virtual" one; the latter is formed when the rays are extended backward. When the rainbow ray is produced in both directions, it approaches the branches of this caustic. A theory of the rainbow based on the analysis of such a caustic was devised by George B. Airy. Having chosen an initial wave front—a surface perpendicular at all points to the rays of Class 3—Airy was able to determine the amplitude distribution in subsequent waves. A weakness of the theory is the need to guess the amplitudes of the initial waves.

With Young's interference theory all the major features of the rainbow could be explained, at least in a qualitative and approximate way. What was lacking was a quantitative, mathematical theory capable of predicting the intensity of the scattered light as a function of droplet size and scattering angle.

Young's explanation of the supernumerary arcs was based on a wave theory of light. Paradoxically his predictions for the other side of the rainbow, for the region of Alexander's dark band, were inconsistent with such a theory. The interference theory, like the theories of Descartes and Newton, predicted complete darkness in this region, at least when only rays of Class 3 and Class 4 were considered. Such an abrupt transition, however, is not possible, because the wave theory of light requires that sharp boundaries between light and shadow be softened by diffraction. The most familiar manifestation of diffraction is the apparent bending of light or sound at the edge of an opaque obstacle. In the rainbow there is no real obstacle, but the boundary between the primary bow and the dark band should exhibit diffraction nonetheless. The treatment of diffraction is a subtle and difficult problem in mathematical physics, and the subsequent development of the theory of the rainbow was stimulated mainly by efforts to solve it.

In 1835 Richard Potter of the University of Cambridge pointed out that the crossing of various sets of light rays in a droplet gives rise to caustic curves. A caustic, or "burning curve," represents the envelope of a system of rays and is always associated with an intensity highlight. A familiar caustic is the bright cusp-shaped curve formed in a teacup when sunlight is reflected from its inner walls. Caustics, like the rainbow, generally have a lighted side and a dark side; intensity increases continuously up to the caustic, then drops abruptly.

Potter showed that the Descartes rainbow ray—the Class 3 ray of minimum scattering angle—can be regarded as a caustic. All the other transmitted rays of Class 3, when extended to infinity, approach the Descartes ray from the lighted side; there are no rays of this class on the dark side. Thus finding the intensity of the scattered light in a rainbow is similar to the problem of determining the intensity distribution in the neighborhood of a caustic.

In 1838 an attempt to determine that distribution was made by Potter's Cambridge colleague George B. Airy. His

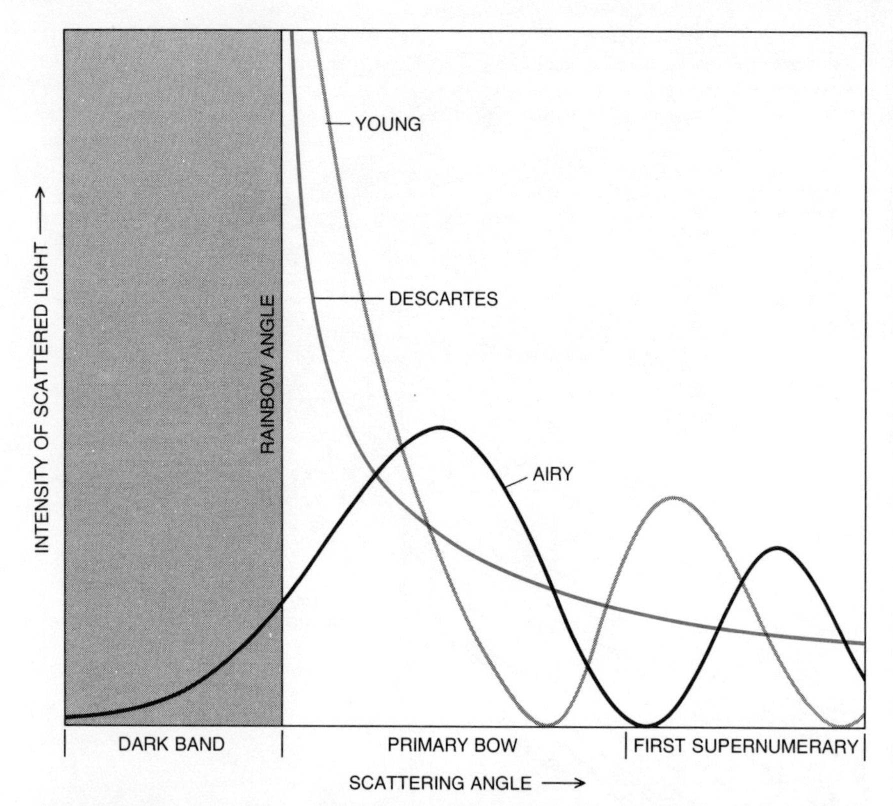

PREDICTED INTENSITY as a function of scattering angle is compared for three early theories of the rainbow. In the geometric analysis of Descartes, intensity is infinite at the rainbow angle; it declines smoothly (without supernumerary arcs) on the lighted side and falls off abruptly to zero on the dark side. The theory of Thomas Young, which is based on the interference of light waves, predicts supernumerary arcs but retains the sharp transition from infinite to zero intensity. Airy's theory relocates the peaks in the intensity curve and for the first time provides (through diffraction) an explanation for gradual fading of the rainbow into shadow.

reasoning was based on a principle of wave propagation formulated in the 17th century by Christiaan Huygens and later elaborated by Augustin Jean Fresnel. This principle regards every point of a wave front as being a source of secondary spherical waves; the secondary waves define a new wave front and hence describe the propagation of the wave. It follows that if one knew the amplitudes of the waves over any one complete wave front, the amplitude distribution at any other point could be reconstructed. The entire rainbow could be described rigorously if we knew the amplitude distribution along a wave front in a single droplet. Unfortunately the amplitude distribution can seldom be determined; all one can usually do is make a reasonable guess for some chosen wave front in the hope that it will lead to a good approximation.

The starting wave front chosen by Airy is a surface inside the droplet, normal to all the rays of Class 3 and with an inflection point (a change in the sense of curvature) where it intersects the Descartes rainbow ray. The wave amplitudes along this wave front were estimated through standard assumptions in the theory of diffraction. Airy was then able to express the intensity of the scattered light in the rainbow region in terms of a new mathematical function, then known as the rainbow integral and today called the Airy function. The mathematical form of the Airy function will not concern us here; we shall concentrate instead on its physical meaning.

The intensity distribution predicted by the Airy function is analogous to the diffraction pattern appearing in the shadow of a straight edge. On the lighted side of the primary bow there are oscillations in intensity that correspond to the supernumerary arcs; the positions and widths of these peaks differ somewhat from those predicted by the Young interference theory. Another significant distinction of the Airy theory is that the maximum intensity of the rainbow falls at an angle somewhat greater than the Descartes minimum scattering angle. The Descartes and Young theories predict an infinite intensity at that angle (because of the caustic). The Airy theory does not reach an infinite intensity at any point, and at the Descartes rainbow ray the intensity predicted is less than half the maximum. Finally, diffraction effects appear on the dark side of the rainbow: instead of vanishing abruptly the intensity tapers away smoothly within Alexander's dark band.

Airy's calculations were for a monochromatic rainbow. In order to apply his method to a rainbow produced in sunlight one must superpose the Airy patterns generated by the various monochromatic components. To proceed further and describe the perceived image of the rainbow requires a theory of color vision.

The purity of the rainbow colors is determined by the extent to which the component monochromatic rainbows overlap; that in turn is determined by the droplet size. Uniformly large drops (with diameters on the order of a few millimeters) generally give bright rainbows with pure colors; with very small droplets (diameters of .01 millimeter or so) the overlap of colors is so great that the resulting light appears to be almost white.

An important property of light that we have so far ignored is its state of polarization. Light is a transverse wave, that is, one in which the oscillations are perpendicular to the direction of propagation. (Sound, on the other hand, is a longitudinal vibration.) The orientation of the transverse oscillation can be resolved into components along two mutually perpendicular axes. Any light ray can be described in terms of these two independent states of linear polarization. Sunlight is an incoherent mixture of the two in equal proportions; it is often said to be randomly polarized or simply unpolarized. Reflection can alter its state of polarization, and in that fact lies the importance of polarization to the analysis of the rainbow.

Let us consider the reflection of a light ray traveling inside a water droplet when it reaches the boundary of the droplet. The plane of reflection, the plane that contains both the incident and the reflected rays, provides a convenient geometric reference. The polarization states of the incident light can be defined as being parallel to that plane and perpendicular to it. For both polarizations the reflectivity of the surface is slight at angles of incidence near the perpendicular, and it rises very steeply near a critical angle whose value is determined by the index of refraction. Beyond that critical angle the ray is totally reflected, regardless of polarization. At intermediate angles, however, reflectivity depends on polarization. As the angle of incidence becomes shallower a steadily larger portion of the perpendicularly polarized component is reflected. For the parallel component, on the other hand, reflectivity falls before it begins to increase. At one angle in particular, reflectivity for the parallel-polarized wave vanishes entirely; that wave is totally transmitted. Hence for sunlight incident at that angle the internally reflected ray is completely polarized perpendicular to the plane of reflection. The angle is called Brewster's angle, after David Brewster, who discussed its significance in 1815.

Light from the rainbow is almost completely polarized, as can be seen by looking at a rainbow through Polaroid sunglasses and rotating the lenses around the line of sight. The strong polarization results from a remarkable coincidence: the internal angle of incidence for the rainbow ray is very close to Brewster's angle. Most of the parallel component escapes in the transmitted rays of Class 2, leaving a preponderance of perpendicular rays in the rainbow.

With the understanding that both matter and radiation can behave as waves, the theory of the rainbow has been enlarged in scope. It must now encompass new, invisible rainbows produced in atomic and nuclear scattering.

An analogy between geometrical optics and classical particle mechanics had already been perceived in 1831 by William Rowan Hamilton, the Irish mathematician. The analogues of rays in geometrical optics are particle trajectories, and the bending of a light ray on entering a medium with a different refractive index corresponds to the deflection of a moving particle under the action of a force. Particle-scattering analogues exist for many effects in optics, including the rainbow.

Consider a collision between two atoms in a gas. As the atoms approach from a large initial separation, they are at first subject to a steadily increasing attraction. At closer range, however, the electron shells of the atoms begin to interpenetrate and the attractive force diminishes. At very close range it becomes an increasingly strong repulsion.

As in the optical experiment, the atomic scattering can be analyzed by tracing the paths of the atoms as a function of the impact parameter. Because the forces vary gradually and continuously, the atoms follow curved trajectories instead of changing direction suddenly, as at the boundary between media of differing refractive index. Even though some of the trajectories are rather complicated, each impact parameter corresponds to a single deflection angle; moreover, there is one trajectory that represents a local maximum angular deflection. That trajectory turns out to be the one that makes the most effective use of the attractive interaction between atoms. A strong concentration of scattered particles is expected near this angle; it is the rainbow angle for the interacting atoms.

A wave-mechanical treatment of the atomic and nuclear rainbows was formulated in 1959 by Kenneth W. Ford of Brandeis University and John A. Wheeler of Princeton University. Interference between trajectories emerging in the same direction gives rise to supernumerary peaks in intensity. A particle-scattering analogue of Airy's theory has also been derived.

An atomic rainbow was first observed in 1964, by E. Hundhausen and H. Pauly of the University of Bonn, in the scattering of sodium atoms by mercury atoms. The main rainbow peak and two supernumeraries were detected; in more recent experiments oscillations on an even finer scale have been observed. The rainbows measured in these experiments carry information about the interatomic forces. Just as the optical rain-

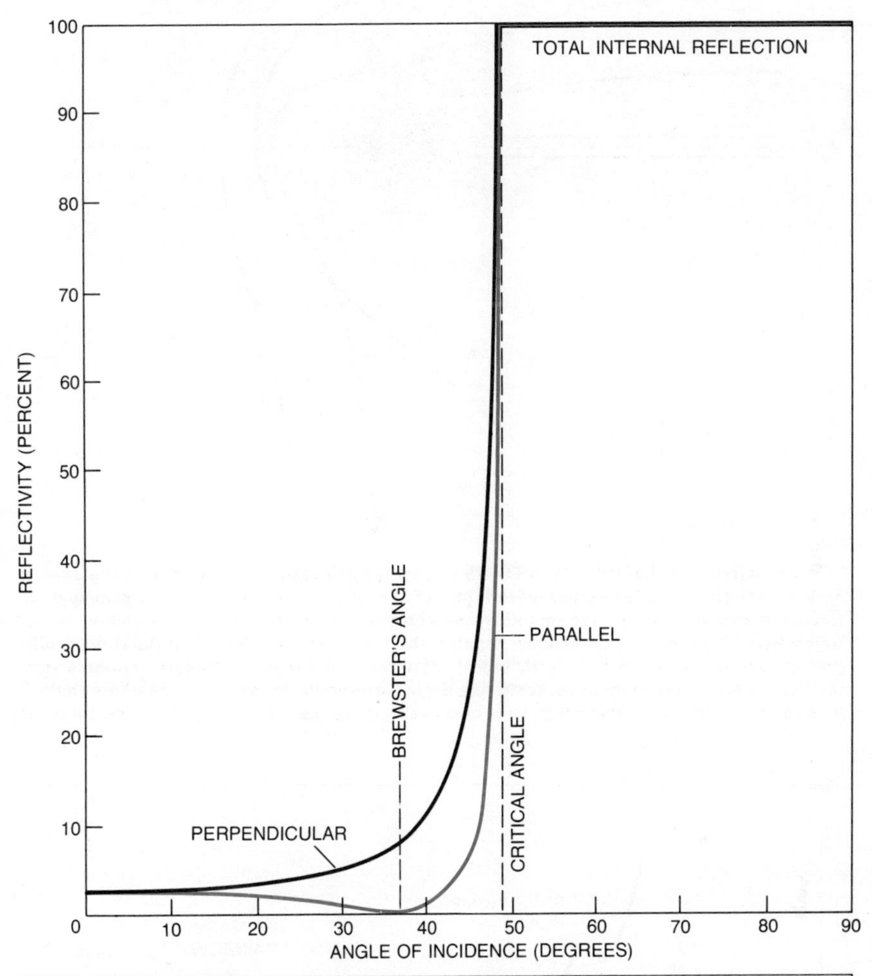

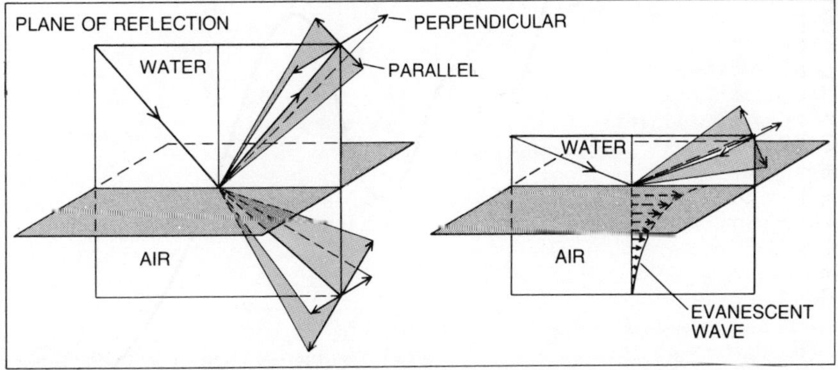

POLARIZATION OF THE RAINBOW results from differential reflection. An incident ray can be resolved into two components polarized parallel to and perpendicular to the plane of reflection. For a ray approaching an air-water boundary from inside a droplet the reflectivity of the surface depends on the angle of incidence. Beyond a critical angle both parallel and perpendicular components are totally reflected, although some light travels parallel to the surface as an "evanescent wave." At lesser angles the perpendicular component is reflected more efficiently than the parallel one, and at one angle in particular, Brewster's angle, parallel-polarized light is completely transmitted. The angle of internal reflection for the rainbow ray falls near Brewster's angle. As a result light from the rainbow has a strong perpendicular polarization.

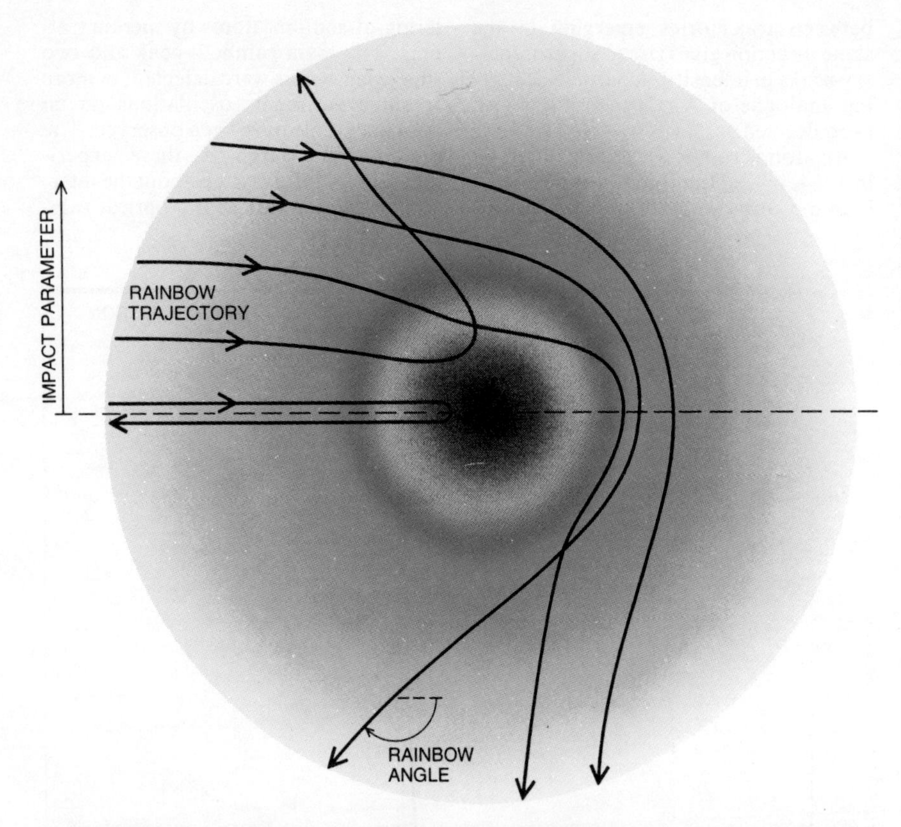

SCATTERING OF ATOMS BY ATOMS creates a particulate rainbow. The role played in optical scattering by the refractive index is played here by interatomic forces. The principal difference is that the forces vary smoothly and continuously, so that the atoms follow curved trajectories. As one atom approaches another the force between them is initially a steadily growing attraction (*colored shading*), but at close range it becomes strongly repulsive (*gray shading*). A local maximum in the scattering angle corresponds to the optical rainbow angle. It is the angle made by the trajectory most effective in using the attractive part of the potential.

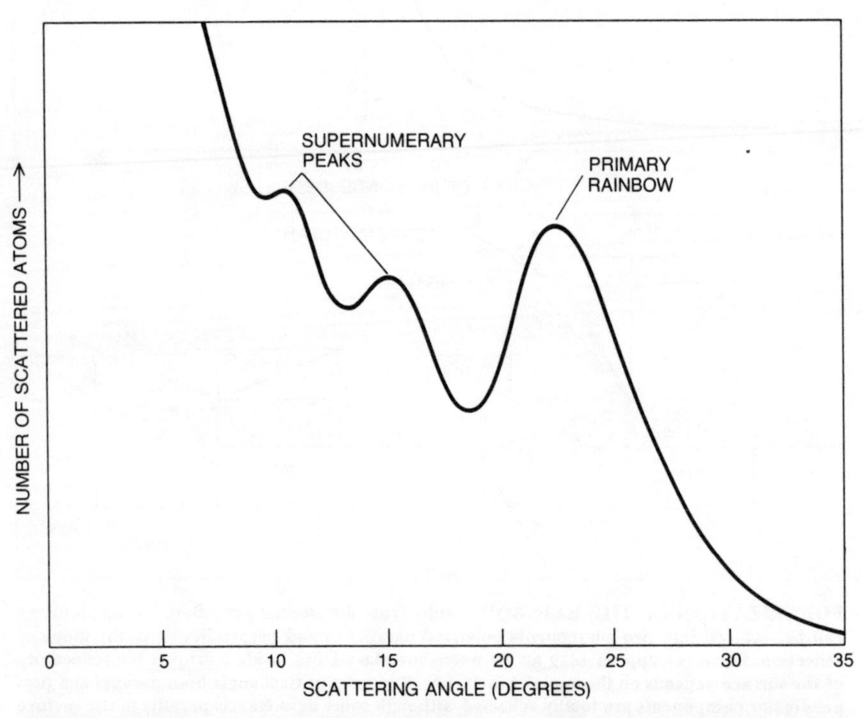

ATOMIC RAINBOW was detected by E. Hundhausen and H. Pauly of the University of Bonn in the scattering of sodium atoms by mercury atoms. The oscillations in the number of scattered atoms detected correspond to a primary rainbow and to two supernumerary peaks. A rainbow of this kind embodies information about the strength and range of the interatomic forces.

bow angle depends solely on the refractive index, so the atomic rainbow angle is determined by the strength of the attractive part of the interaction. Similarly, the positions of the supernumerary peaks are size-dependent, and they provide information about the range of the interaction. Observations of the same kind have now been made in the scattering of atomic nuclei.

The Airy theory of the rainbow has had many satisfying successes, but it contains one disturbing uncertainty: the need to guess the amplitude distribution along the chosen initial wave front. The assumptions employed in making that guess are plausible only for rather large raindrops. In this context size is best expressed in terms of a "size parameter," defined as the ratio of a droplet's circumference to the wavelength of the light. The size parameter varies from about 100 in fog or mist to several thousand for large raindrops. Airy's approximation is plausible only for drops with a size parameter greater than about 5,000.

It is ironic that a problem as intractable as the rainbow actually has an exact solution, and one that has been known for many years. As soon as the electromagnetic theory of light was proposed by James Clerk Maxwell about a century ago, it became possible to give a precise mathematical formulation of the optical rainbow problem. What is needed is a computation of the scattering of an electromagnetic plane wave by a homogeneous sphere. The solution to a similar but slightly easier problem, the scattering of sound waves by a sphere, was discussed by several investigators, notably Lord Rayleigh, in the 19th century. The solution they obtained consisted of an infinite series of terms, called partial waves. A solution of the same form was found for the electromagnetic problem in 1908 by Gustav Mie and Peter J. W. Debye.

Given the existence of an exact solution to the scattering problem, it might seem an easy matter to determine all its features, including the precise character of the rainbow. The problem, of course, is the need to sum the series of partial waves, each term of which is a rather complicated function. The series can be truncated to give an approximate solution, but this procedure is practical only in some cases. The number of terms that must be retained is of the same order of magnitude as the size parameter. The partial-wave series is therefore eminently suited to the treatment of Rayleigh scattering, which is responsible for the blue of the sky; in that case the scattering particles are molecules and are much smaller than the wavelength, so that one term of the series is enough. For the rainbow problem size parameters up to several thousand must be considered.

A good approximation to the solution by the partial-wave method would require evaluating the sum of several thousand complicated terms. Computers have been applied to the task, but the results are rapidly varying functions of the size parameter and the scattering angle, so that the labor and cost quickly become prohibitive. Besides, a computer can only calculate numerical solutions; it offers no insight into the physics of the rainbow. We are thus in the tantalizing situation of knowing a form of the exact solution and yet being unable to extract from it an understanding of the phenomena it describes.

The first steps toward the resolution of this paradox were taken in the early years of the 20th century by the mathematicians Henri Poincaré and G. N. Watson. They found a method for transforming the partial-wave series, which converges only very slowly onto a stable value, into a rapidly convergent expression. The technique has come to be known as the Watson transformation or as the complex-angular-momentum method.

It is not particularly hard to see why angular momentum is involved in the rainbow problem, although it is less obvious why "complex" values of the angular momentum need to be considered. The explanation is simplest in a corpuscular theory of light, in which a beam of light is regarded as a stream of the particles called photons. Even though the photon has no mass, it does transport energy and momentum in inverse proportion to the wavelength of the corresponding light wave. When a photon strikes a water droplet with some impact parameter greater than zero, the photon carries an angular momentum equal to the product of its linear momentum and the impact parameter. As the photon undergoes a series of internal reflections, it is effectively orbiting the center of the droplet. Actually quantum mechanics places additional constraints on this process. On the one hand it requires that the angular momentum assume only certain discrete values; on the other it denies that the impact parameter can be precisely determined. Each discrete value of angular momentum corresponds to one term in the partial-wave series.

In order to perform the Watson transformation, values of the angular momentum that are conventionally regarded as being "unphysical" must be introduced. For one thing the angular momentum must be allowed to vary continuously, instead of in quantized units; more important, it must be allowed to range over the complex numbers: those that include both a real component and an imaginary one, containing some multiple of the square root of -1. The plane defined by these two components is referred to as the complex-angular-momentum plane.

Much is gained in return for the mathematical abstractions of the complex-angular-momentum method. In particular, after going over to the complex-angular-momentum plane through the Watson transformation, the contributions to the partial-wave series can be redistributed. Instead of a great many terms, one can work with just a few points called poles and saddle points in the complex-angular-momentum plane. In recent years the poles have attracted great theoretical interest in the physics of elementary particles. In that context they are usually called Regge poles, after the Italian physicist Tullio Regge.

Both poles and saddle points have physical interpretations in the rainbow problem. Contributions from real saddle points are associated with the ordinary, real light rays we have been considering throughout this article. What about complex saddle points? Imaginary or complex numbers are ordinarily regarded as being unphysical solutions to an equation, but they are not meaningless solutions. In descriptions of wave propagation imaginary components are usually associated with the damping of the wave amplitude. For example, in the total internal reflection of a light ray at a water-air boundary a light wave does go "through the looking glass." Its amplitude is rapidly damped, however, so that the intensity becomes negligible within a depth on the order of a single wavelength. Such a wave does not propagate into the air; instead it becomes attached to the interface between the water and the air, traveling along the surface; it is called an evanescent wave. The mathematical description of the evanescent wave involves the imaginary components of a solution. The effect called quantum-mechanical tunneling, in which a particle passes through a potential barrier without climbing over it, has a similar mathematical basis. "Complex rays" also appear on the shadow side of a caustic, where they describe the damped amplitude of the diffracted light waves.

Regge-pole contributions to the transformed partial-wave series are associated with surface waves of another kind. These waves are excited by incident rays that strike the sphere tangentially. Once such a wave is launched, it travels around the sphere, but it is continually damped because it sheds radiation tangentially, like a garden sprinkler. At each point along the wave's circumferential path it also penetrates the sphere at the critical angle for total internal reflection, reemerging as a surface wave after taking one or more such shortcuts. It is interesting to note that Johannes Kepler conjectured in 1584 that "pin-

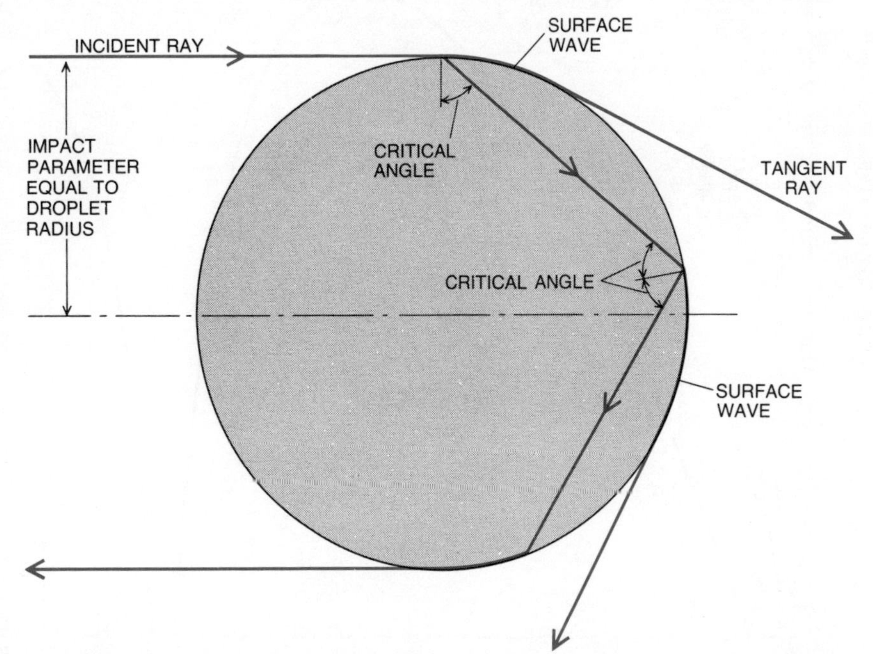

COMPLEX-ANGULAR-MOMENTUM theory of the rainbow begins with the observation that a photon, or quantum of light, incident on a droplet at some impact parameter (which cannot be exactly defined) carries angular momentum. In the theory, components of that angular momentum are extended to complex values, that is, values containing the square root of -1. The consequences of this procedure can be illustrated by the example of a ray striking a droplet tangentially. The ray stimulates surface waves, which travel around the droplet and continuously shed radiation. The ray can also penetrate the droplet at the critical angle for total internal reflection, emerging either to form another surface wave or to repeat the shortcut.

wheel" rays of this kind might be responsible for the rainbow, but he abandoned the idea because it did not lead to the correct rainbow angle.

In 1937 the Dutch physicists Balthus Van der Pol and H. Bremmer applied Watson's transformation to the rainbow problem, but they were able to show only that Airy's approximation could be obtained as a limiting case. In 1965 I developed an improved version of Watson's method, and I applied it to the rainbow problem in 1969 with somewhat greater success.

In the simple Cartesian analysis we saw that on the lighted side of the rainbow there are two rays emerging in the same direction; at the rainbow angle these coalesce into the single Descartes ray of minimum deflection and on the shadow side they vanish. In the complex-angular-momentum plane, as I have mentioned, each geometric ray corresponds to a real saddle point. Hence in mathematical terms a rainbow is merely the collision of two saddle points in the complex-angular-momentum plane. In the shadow region beyond the rainbow angle the saddle points do not simply disappear; they become complex, that is, they develop imaginary parts. The diffracted light in Alexander's dark band arises from a complex saddle point. It is an example of a "complex ray" on the shadow side of a caustic curve.

It should be noted that the adoption of the complex-angular-momentum method does not imply that earlier solutions to the rainbow problem were wrong. Descartes's explanation of the primary bow as the ray of minimum deflection is by no means invalid, and the supernumerary arcs can still be regarded as a product of interference, as Young proposed. The complex-angular-momentum method simply gives a more comprehensive accounting of the paths available to a photon in the rainbow region of the sky, and it thereby achieves more accurate results.

In 1975 Vijay Khare of the University of Rochester made a detailed comparison of three theories of the rainbow: the Airy approximation, the "exact" solution, obtained by a computer summation of the partial-wave series, and the rainbow terms in the complex-angular-momentum method, associated with the collision of two saddle points. For the dominant, perpendicular polarization the Airy theory requires only small corrections within the primary bow, and its errors become appreciable only in the region of the supernumerary arcs. For the scattered rays polarized parallel to the scattering plane, however, Airy's approximation fails badly. For the supernumerary arcs the exact solution shows minima where the Airy theory has maximum intensity, and vice versa. This serious failure is an indirect result of the near coincidence between the angle of internal reflection for the rainbow rays and Brewster's angle. At Brewster's angle the amplitude of the reflected ray changes sign, a change the Airy theory does not take into account. As a result of the change in sign the interference along directions corresponding to the peaks in

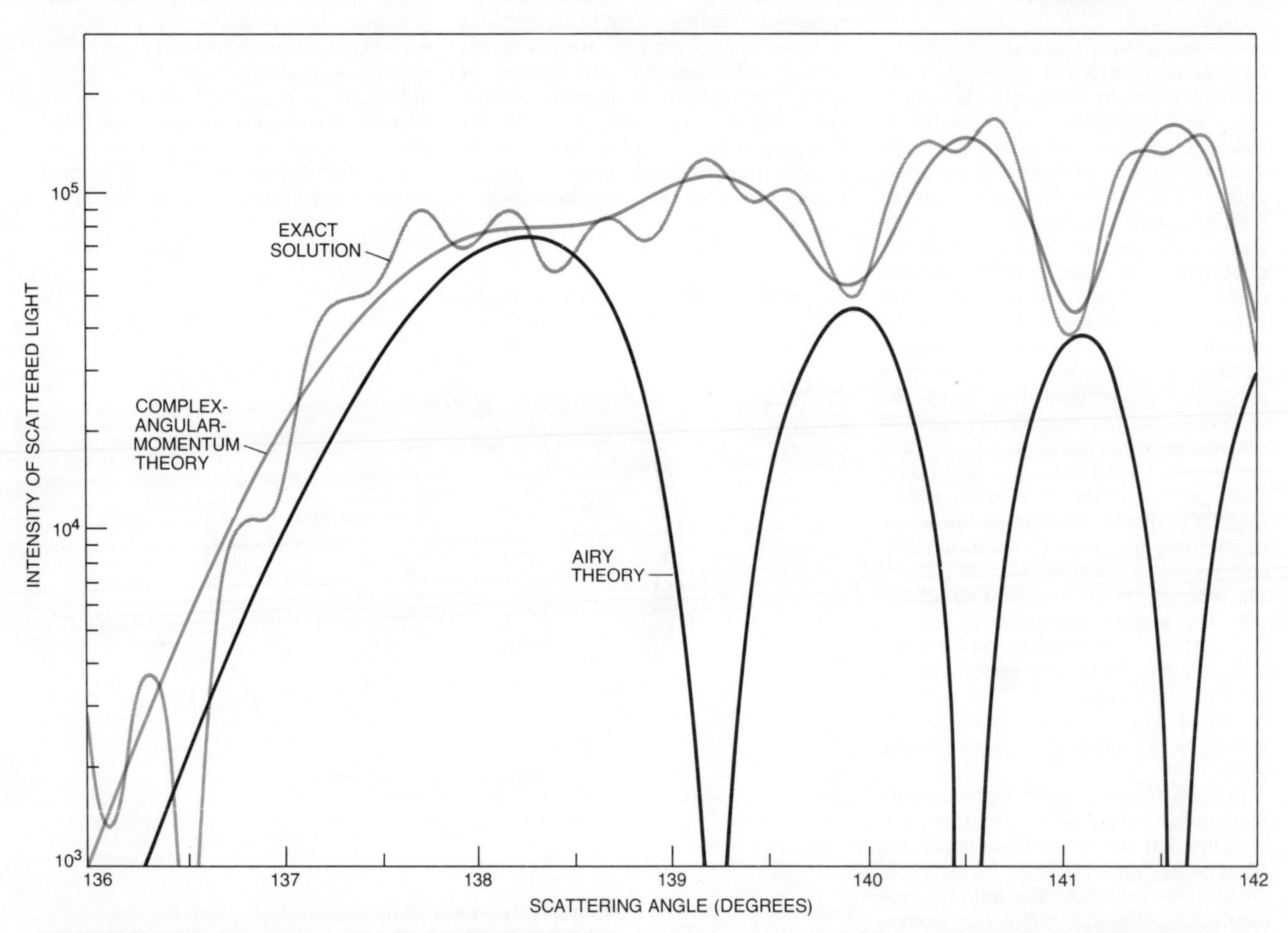

QUANTITATIVE THEORIES of the rainbow predict the intensity of the scattered light as a function of the scattering angle and also with respect to droplet size and polarization. Here the predictions of three theories are presented for parallel-polarized light scattered by droplets with a circumference equal to 1,500 wavelengths of the light. One curve represents the "exact" solution to the rainbow problem, derived from James Clerk Maxwell's equations describing electromagnetic radiation. The exact solution is the sum of an infinite series of terms, approximated here by adding up more than 1,500 complicated terms for each point employed in plotting the curve. The Airy theory is clearly in disagreement with the exact solution, particularly in the angular region of the supernumerary arcs. There the exact solution shows troughs at the positions of Airy's peaks. The results obtained by the complex-angular-momentum method, on the other hand, correspond closely to the exact solution, failing only to reproduce small, high-frequency oscillations. These fluctuations are associated with another optical phenomenon in the atmosphere, the glory, which is also explained by the complex-angular-momentum theory.

the Airy solutions is destructive instead of constructive.

In terms of large-scale features, such as the primary bow, the supernumerary arcs and the dark-side diffraction pattern, the complex-angular-momentum result agrees quite closely with the exact solution. Smaller-scale fluctuations in the exact intensity curve are not reproduced as well by the rainbow terms in the complex-angular-momentum method. On the other hand, the exact solution, for a typical size parameter of 1,500, requires the summation of more than 1,500 complicated terms; the complex-angular-momentum curve is obtained from only a few much simpler terms.

The small residual fluctuations in the exact intensity curve arise from higher-order internal reflections: rays belonging to classes higher than Class 3 or Class 4. They are of little importance for the primary bow, but at larger scattering angles their contribution increases and near the backward direction it becomes dominant. There these rays are responsible for another fascinating meteorological display: the glory [see the article "The Glory," by Howard C. Bryant and Nelson Jarmie, beginning on the following page].

The glory appears as a halo of spectral colors surrounding the shadow an observer casts on clouds or fog; it is most commonly seen from an airplane flying above clouds. It can also be explained through the complex-angular-momentum theory, but the explanation is more complicated than that for the rainbow. One set of contributions to the glory comes from the surface waves described by Regge poles that are associated with the tangential rays of Kepler's pinwheel type. Multiple internal reflections that happen to produce closed, star-shaped polygons play an important role, leading to resonances, or enhancements in intensity. Such geometric coincidences are very much in the spirit of Kepler's theories.

A second important set of contributions, demonstrated by Khare, is from the shadow side of higher-order rainbows that appear near the backward direction. These contributions represent the effect of complex rays. The 10th-order rainbow, formed only a few degrees away from the backward direction, is particularly effective.

For the higher-order rainbows Airy's theory would give incorrect results for both polarizations, and so the complex-angular-momentum theory must be employed. One might thus say the glory is formed in part from the shadow of a rainbow. It is gratifying to discover in the elegant but seemingly abstract theory of complex angular momentum an explanation for these two natural phenomena, and to find there an unexpected link between them.

8 The Glory

by Howard C. Bryant and Nelson Jarmie

July 1974

This halo of prismatic colors is most often seen around the shadow of an airplane on a cloud. Its cause is not the same as that of the common rainbow, and involves phenomena at the frontier of physics

"If it be shortly after sunup of a morning when the fog has obliterated the highway below, I am then rewarded with a spectacle rare to witness. Looking up the coast toward Nepenthe... the sun rising behind me throws an enlarged shadow of me into the iridescent fog below. I lift my arms as in prayer, achieving a wingspan no god ever possessed, and there in the drifting fog a nimbus floats about my head, a radiant nimbus such as the Buddha himself might proudly wear. In the Himalayas, where the same phenomenon occurs, it is said that a devout follower of the Buddha will throw himself from a peak—'into the arms of Buddha.'"

So does Henry Miller, in *Big Sur and the Oranges of Hieronymus Bosch*, describe his observation of the meteorological phenomenon known as the glory. He conveys his feeling of apotheosis on viewing his shadow on a fog bank, "glorified" with colored rings similar to those of the rainbow. The spectacle is indeed a rare one for ground-based observers such as the solitary hiker in Miller's narrative, because it requires an unusual configuration of the sun, the observer and a cloud composed of droplets of uniform size. It is seen regularly, however, by air travelers, particularly those who know where to look. In fact, it is sometimes called the pilot's bow. Other names for the glory are the anticorona and the brocken bow.

Like the common rainbow, the glory is caused by the scattering of sunlight by droplets of water. Like the primary rainbow, the brightest in a series of rainbows, it consists of concentric rings of color, with red the outermost and violet the innermost, encircling a bright central region in the direction opposite to that of the sun. Unlike the primary rainbow, whose red ring is invariably at an angle of 42 degrees from the direction of the shadow cast by the observer, the glory has rings whose angular diameter varies inversely with the diameter of the droplets that give rise to them. The primary rings are often accompanied by as many as four similar sets of rings of larger angular diameter. Typically the innermost red ring has a diameter of two or three degrees.

To see a glory close up you must view the cloud of uniform water droplets in such a way that your shadow is projected on the cloud. You will be rewarded by a vision of the shadow of your head surrounded by a series of colored haloes. Moreover, a feeling of uniqueness may be oddly enhanced: if someone else is with you, his shadow will not appear to be so endowed. One may even speculate that the artistic practice of rendering the heads of holy or powerful personages with luminous and sometimes colored haloes or nimbuses could have arisen from the observation of such haloes on fog banks by solitary mystics on well-illuminated heights. The use of the halo is not restricted to Christian iconography: glorylike structures can be seen surrounding the heads of Roman emperors and Greek gods, and of icons from China, Burma and India, suggesting that such representations may have a universal natural origin.

The first scientific record of the glory was a drawing made by Antonio de Ulloa during a French expedition to Peru in 1735. Both he and Pierre Bouguer also wrote descriptions, translations of which can be found in R. A. R. Tricker's *Introduction to Meteorological Optics*. Balloonists in the 19th century were able to see the glory encircling the shadow of the basket of the balloon, as Gaston Tissandier relates in his *Observations météorologiques en ballon*. These sightings and others are described in detail in the classic work *Meteorologische Optik*, by Josef M. Pernter and Felix M. Exner, published in Vienna in 1910.

C. T. R. Wilson built the first cloud chamber in 1895 for the purpose of recreating the glory in the laboratory. He put aside his original objective when he found that energetic charged particles leave visible tracks of water droplets in the moist air. In his Nobel prize lecture of 1927 he reported: "In September, 1894, I spent a few weeks in the observatory which then existed on the summit of Ben Nevis, the highest of the Scottish hills. The wonderful optical phenomena shown when the sun shone on the clouds surrounding the hilltop, and particularly the colored rings surrounding the sun (coronas) or surrounding the shadow cast by the hilltop or observer on mist or cloud (glories), greatly excited my interest and made me wish to imitate them in the laboratory."

What is the explanation of this lovely apparition? We have indicated that the glory is caused by the scattering of light from water droplets back toward the source of the light. In the process the scattered light is enhanced. To be sure, light is enhanced when it is scattered backward from many things other than fog or clouds, including plowed fields, foliage, the eyes of many animals and dewy grass. Let us first discuss some of these nonglory effects to illustrate the variety of the mechanisms at work.

The "cornfield effect," or backscattering from a plowed field or foliage, can actually be observed on any rough surface that casts small shadows. When we look at the surface from the direction of the illumination, we do not see the shadows and the surface looks unusually bright. The reason is that the brightness is contrary to our expectation: from other directions parts of the surface are darkened by the shadows, and the eye

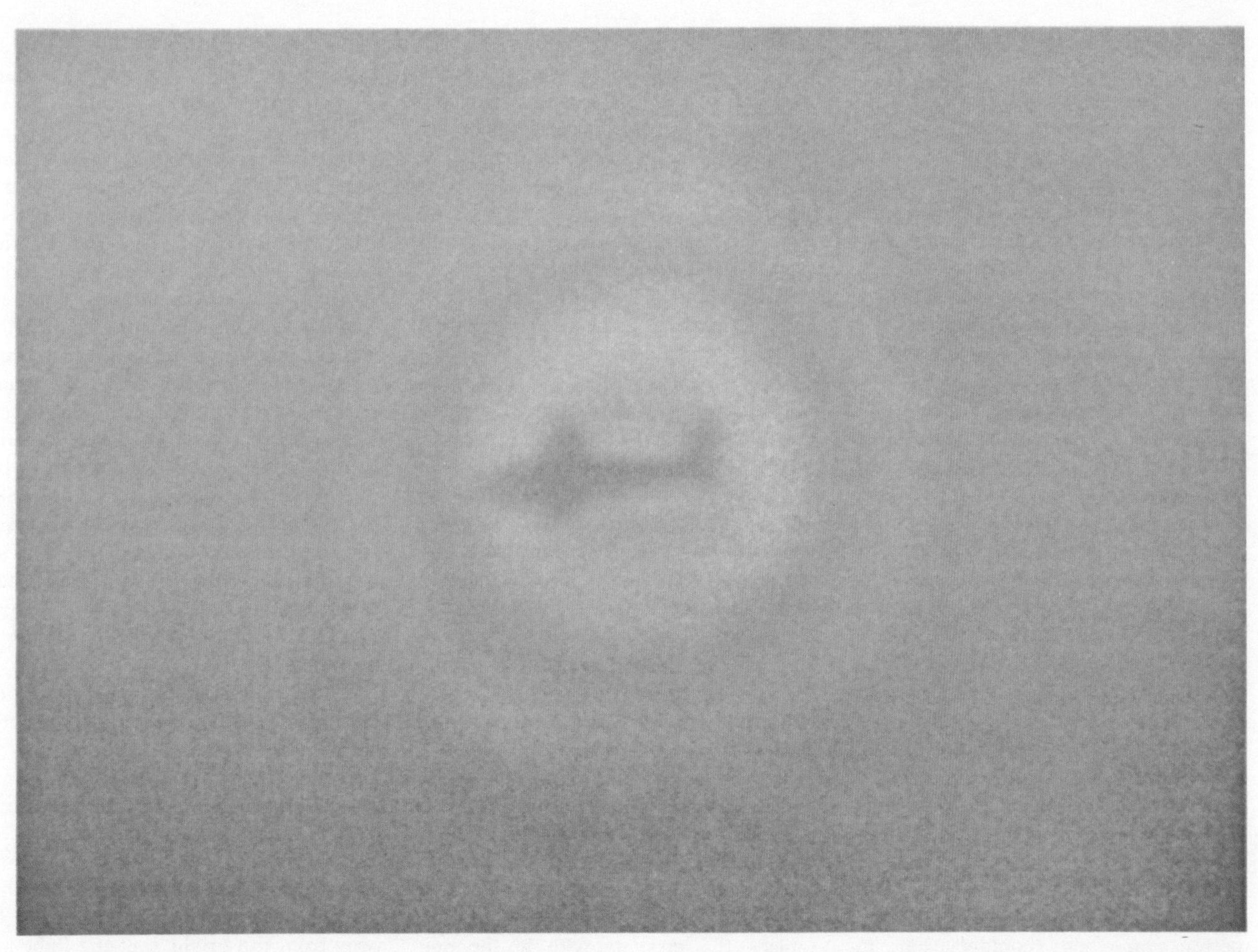

GLORY FROM THE AIR surrounding the shadow of an airplane projected on a cloud below is seen in this photograph made by Fritz Goro. Since the glory is often visible under these circumstances, it is sometimes called the pilot's bow. Around the shadow there is a glow covering a circular area half-obscured by the shadow. Surrounding the central area are two sets of concentric colored rings with blue on the inside and red on the outside. To unaided eye the colors are sometimes much brighter than they appear here.

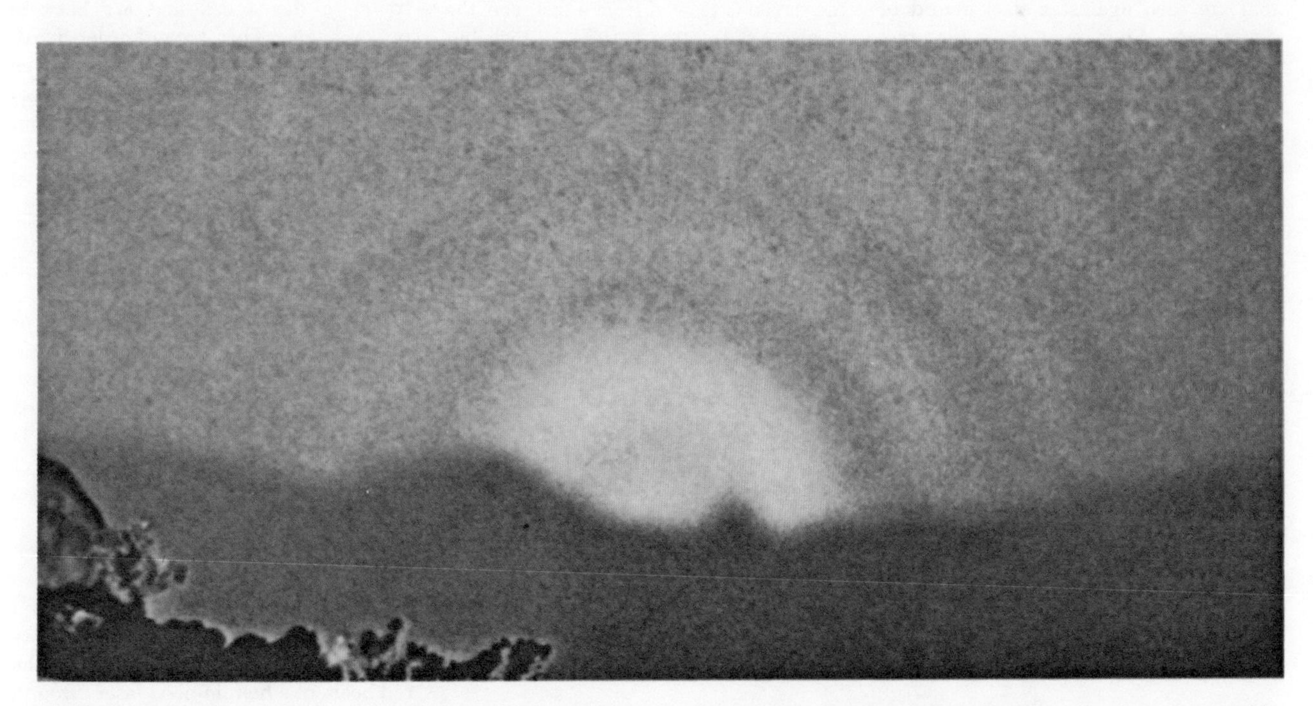

GLORY FROM THE GROUND was photographed by John C. Brandt of the National Aeronautics and Space Administration at the Zodiacal Light Observatory on Haleakala Crater on the Hawaiian island of Maui. The picture was taken before sunset as the crater was filled with cloud or fog. Five glory rings are visible; their angular radii, measured from the red rings, are 1.2 degrees, 3.0 degrees, 4.9 degrees, 6.7 degrees and 8.3 degrees. Normally the shadow of the person viewing the glory is visible, but here it is not.

integrates the image so that the surface appears uniformly darker.

Backscattering from animal eyes arises from the fact that the illuminating light ray is reversed with high precision. Light entering the animal's eye from the direction of the observer is brought to a focus on the animal's retina; then some portion of it is scattered back toward the lens of the animal's eye and refracted back in the direction from which it came. The result is that the animal's eye appears to be illuminated from within.

The effect is enhanced in cats, dogs, rabbits and other animals because they have a reflecting layer behind the retina. Man lacks this layer, but on occasion his eyes backscatter very well. David L. MacAdam, editor of *The Journal of the Optical Society of America,* comments: "Any photographer who has taken many close-up color pictures with a flash lamp built into his camera, close to the lens, will recall his dismay when some of his pictures were ruined by brilliant red spots coincident with the pupils of the eyes of some of his subjects. A considerable portion of fair-haired, light-eyed persons have such strong reflection from the fundus of the eye as to produce this Heiligenschein [the German for halo]. It can also be seen clearly by another person over whose shoulder a bare incandescent tungsten bulb is shining directly into the eyes of the subject, in an otherwise poorly lighted room."

It is effects of the Heiligenschein type that are seen against a background of dewy grass. When one looks at the grass from the direction of illumination, the shadow of one's head appears to be surrounded by a bright area. As with animal eyes, the water droplets, which are more or less spherical, serve as miniature converging lenses that collect the light and focus it on the blades of grass on which they rest. Much of the focused light is scattered in all directions by the leaf, but some of it reenters the droplet and is refracted backward in the direction from which it came.

A completely different mechanism is needed to account for the glory. In 1947 the Dutch astronomer H. C. van de Hulst put forward the explanation that the light of the glory is sent back from the edges of the spherical water droplets in the cloud. For the moment let us set aside the question of how the light is returned from the droplet's edge and concentrate on understanding how the glory would be produced in that way by a random distribution of uniform droplets.

Since each droplet is returning the light from its edge, each is effectively a ring-shaped light source sending light back toward the sun. One way to simulate examining the optics of a field of backward-scattering water droplets is to replace the droplets with an opaque screen that has ring-shaped apertures in it and illuminate the screen from behind with a beam whose rays are parallel [*see bottom illustration on page 76*]. When the screen is viewed from a distance, the resulting diffraction pattern, or distribution of the intensity of the scattered light, will be very much like the diffraction pattern of the glory.

Why should this be so? The diffraction pattern from a single ring can be understood by regarding each point on the ring as a separate source of coherent light waves. This means that all the wavelets coming from each point on the ring will be in phase with one another. The light arriving at any particular point on another screen at some distance from the aperture will consist of contributions from all the points around the ring. Only on the part of the screen that is directly in front of the ring will the wavelets be exactly in phase, since it is only there that the distance each wavelet has traveled from the ring's edge is the same as the distance every other wavelet has traveled. At that point on the screen there will be a circular spot of maximum brightness.

In a region at a small angle away from the spot of maximum brightness the path each wavelet takes is either longer or shorter than the path of its neighbor, and the wavelets begin to interfere destructively with one another. The intensity of the light falls to a minimum, and there is a ring of minimum brightness around the central bright spot. At a greater angle away from the bright spot the light intensity begins to increase again to a second-order maximum: wavelets from opposite sides of the ring have a difference of one wavelength in the distance they travel to the screen and are back in phase again. This second-order maximum (and successive maximums at even greater angles, for which the differences in path length differ by two, three, four or any other whole number of wavelengths) is not as intense as the principal maximum in the center because it is only at that one central spot that all the wavelets are in phase.

So far we have described what happens with only one ring. To illustrate what happens in a glory we prepared a random array of many ring sources. First we drew 241 circles two millimeters in diameter with black ink on a piece of white paper about 12 centimeters on a side. Then we photographed the piece of paper to get a 35-millimeter negative; the negative showed transparent rings on a black background reduced in size by a factor of 16. We set up a parallel-ray source of coherent light by passing the red beam of a helium-neon laser through a shutter into a microscope lens with a pinhole at its focal point. The beam was then directed through a converging lens whose focal point was also at the pin-

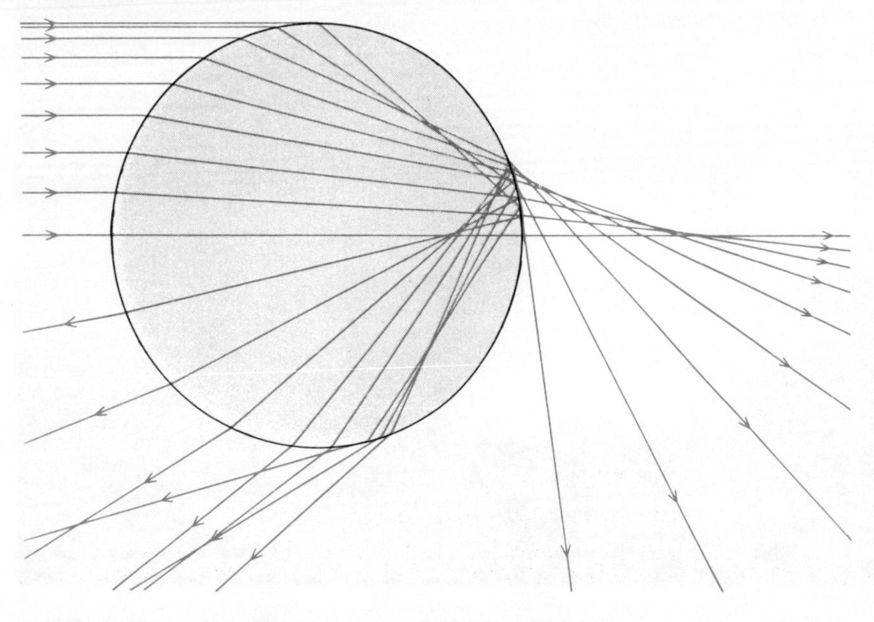

LIGHT THAT ACCOUNTS FOR THE COMMON RAINBOW is not sent straight back toward the observer. The primary rainbow is caused by parallel light rays that are incident from the left being refracted off to lower left by a spherical water droplet. The rays to the right of the droplet are those that go straight through without contributing to the rainbow.

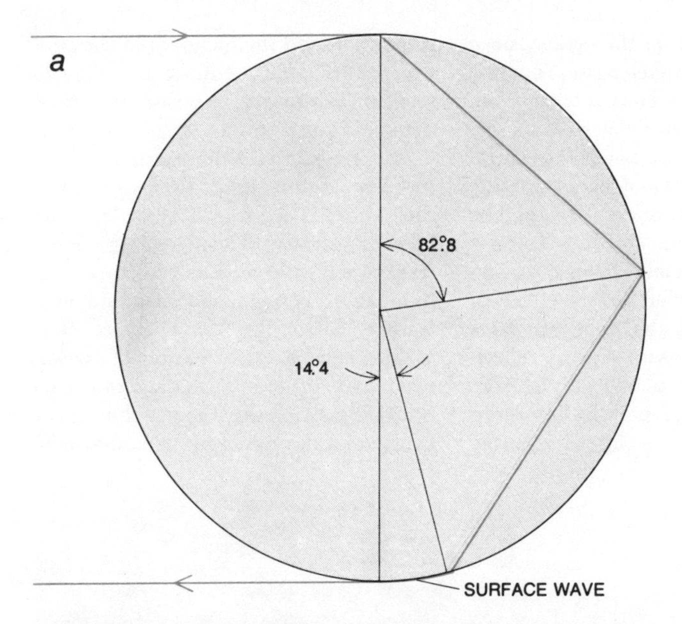

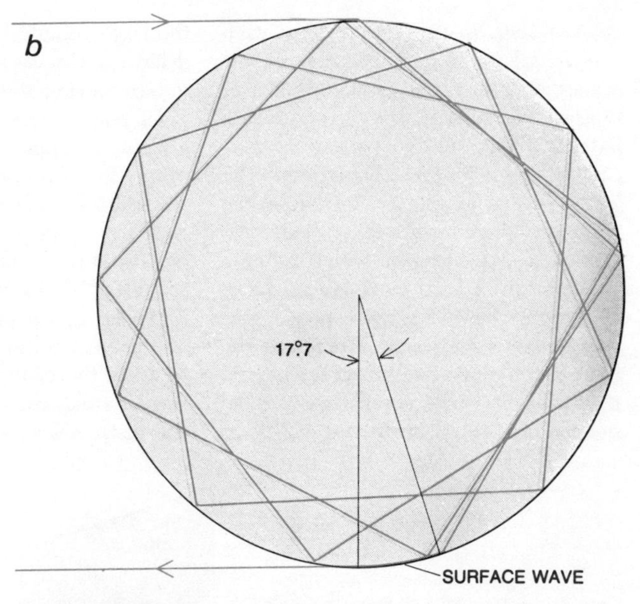

LIGHT THAT ACCOUNTS FOR THE GLORY follows paths that are different from those that are responsible for the common rainbow. The paths consist of light that is reflected repeatedly at an angle of 82.8 degrees within the droplet, together with small segments of surface waves in which the light clings to the surface of the droplet and is conveyed the rest of the way around the droplet to the backward direction. When a number of different paths give rise to light waves that are in phase, there is a resonance, or enhancement, in the backscattered light. The glory is believed to be due principally to the paths in which the light travels halfway around the droplet (*a*) and those in which it travels three and a half times around the droplet (*b*) before being sent straight back.

hole. This train of optical devices gave us a parallel-ray beam of coherent light two centimeters in diameter, into which we inserted the photographic negative of the array of rings.

We now had a field of coherent ring sources, each about 125 micrometers across, which together simulated the effect of a uniform field of spherical droplets that are backscattering light. In order to record the diffraction pattern as it would appear at a distance from the array we put a lens with a focal length of one meter in front of the array and placed a sheet of photographic film in its focal plane. The resulting photograph of concentric bright and dark rings approximately represents the appearance of a glory seen through a red filter [*see illustration on page 77*].

We can now explain the presence of

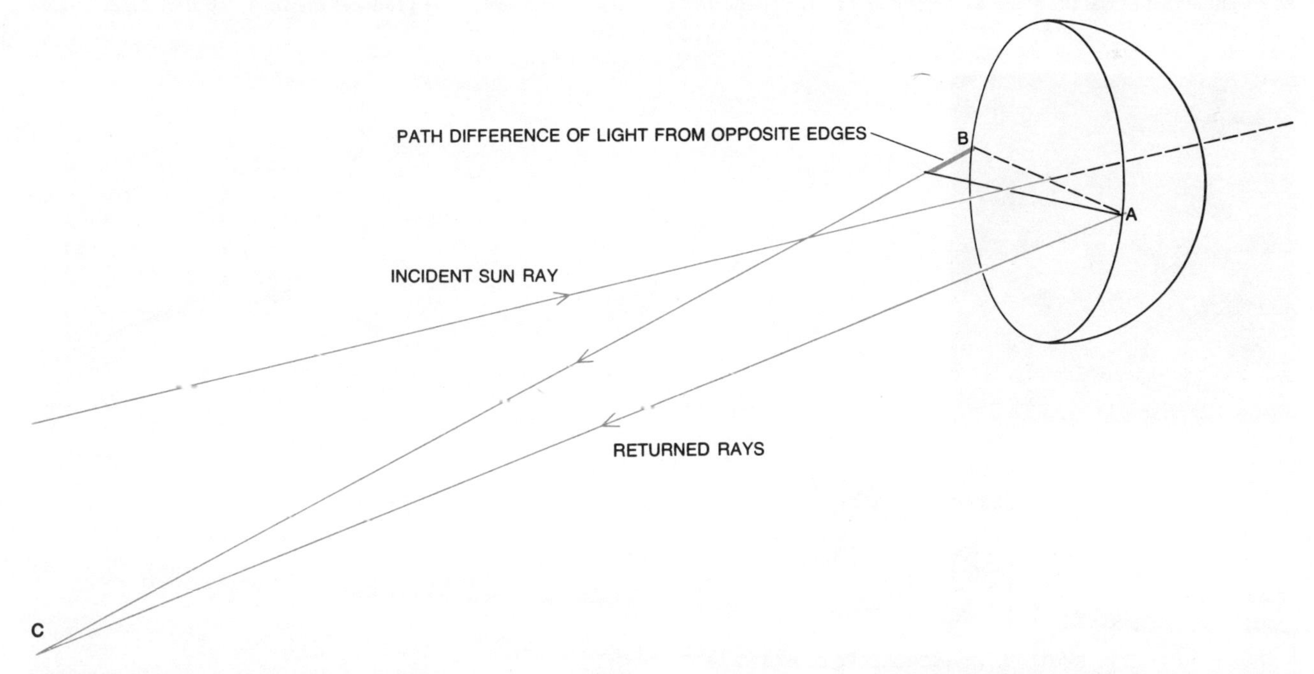

EDGES OF WATER DROPLETS of uniform size scatter sunlight back toward the observer in the formation of a glory. Here the path of rays returning from a single droplet is traced; the droplet is cut in half to show the geometry of the rays. If the observer is straight back from the droplet, the waves on every ray from the edge of the droplet will be in phase and will reinforce one another. As the angle of the observer to the path of the light falling on a droplet increases, the waves from opposite edges begin to go out of phase and interfere. As the angle increases further, they come back into phase and reinforce again. A secondary maximum in brightness is reached near the point where the difference between paths *AC* and *BC* is one wavelength. The rings of the glory are colored because the many different wavelengths in sunlight go through their cycles of maximum and minimum brightness at different angles.

colored rings in the glory. When each wavelength in the solar spectrum is backscattered from an array of droplets of uniform diameter, it will give rise to a pattern similar to the pattern of concentric rings in the photograph. The longer the wavelength, the larger the diameter of the diffraction rings. Although all wavelengths will contribute to the bright central maximum, at other angles only light of certain wavelengths will be at a maximum. Therefore the light in the glory away from the central maximum will be colored. In every sequence of spectral colors red will be at the largest angle from the center, since it has the longest wavelength. Thus the outermost ring of the glory is always red.

Of course, our artificial ring sources are only an approximation of the optical properties of an array of backscattering water droplets. One difference is that the laser light giving rise to the diffraction pattern in the artificial array is not backscattered. We have left out the effects of the polarization of the light; we have also neglected the beam directly reflected from the center of the droplet. The correct treatment of polarization alters the distribution of the light's intensity somewhat, and the inclusion of the central point source produces an enhancement or diminution of every other ring, depending on its phase.

We now return to the question of how droplets scatter light backward from their edges. The fact that they do so can be demonstrated by constructing a large transparent sphere and directing sunlight at it. Permanently mounted in a lecture hall at the University of New Mexico is a heliostat, an optical mechanism made up of a series of mirrors that tracks the sun across the sky during the course of a day and sends a sunbeam 35

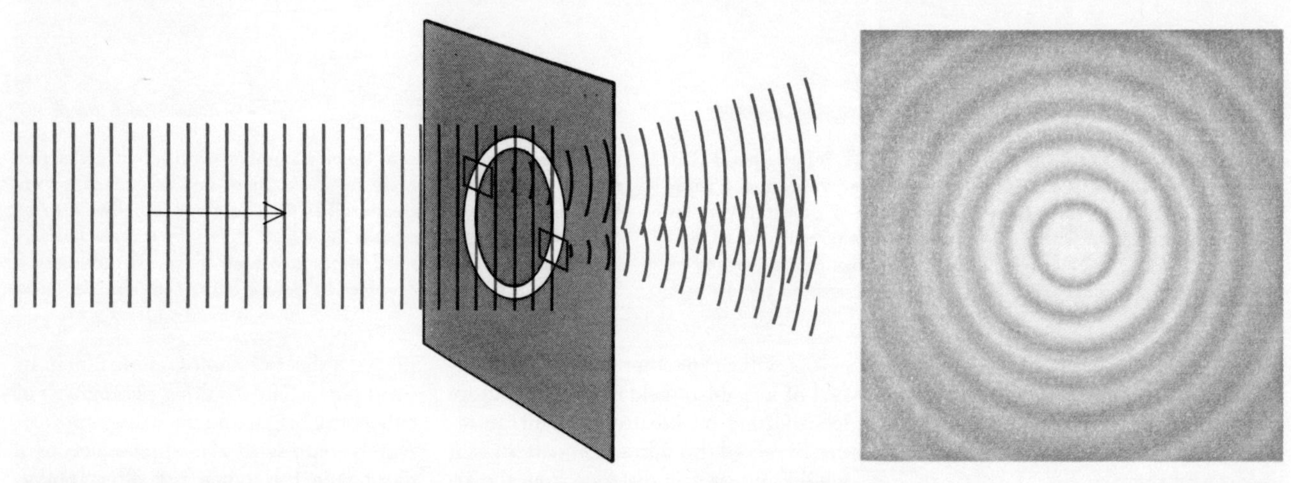

SINGLE RING APERTURE (*left*) illuminated by plane-parallel light (*color*) gives rise to a diffraction pattern that is a series of concentric circles. The ring aperture can be regarded as one droplet in the cloud that causes the glory. Each point on the ring (*indicated by the two rectangles*) acts as a separate source of wavelets of coherent light. Wavelets interfere either constructively or destructively at different angles away from direction of light coming straight back. This interference produces the light and dark rings (*right*).

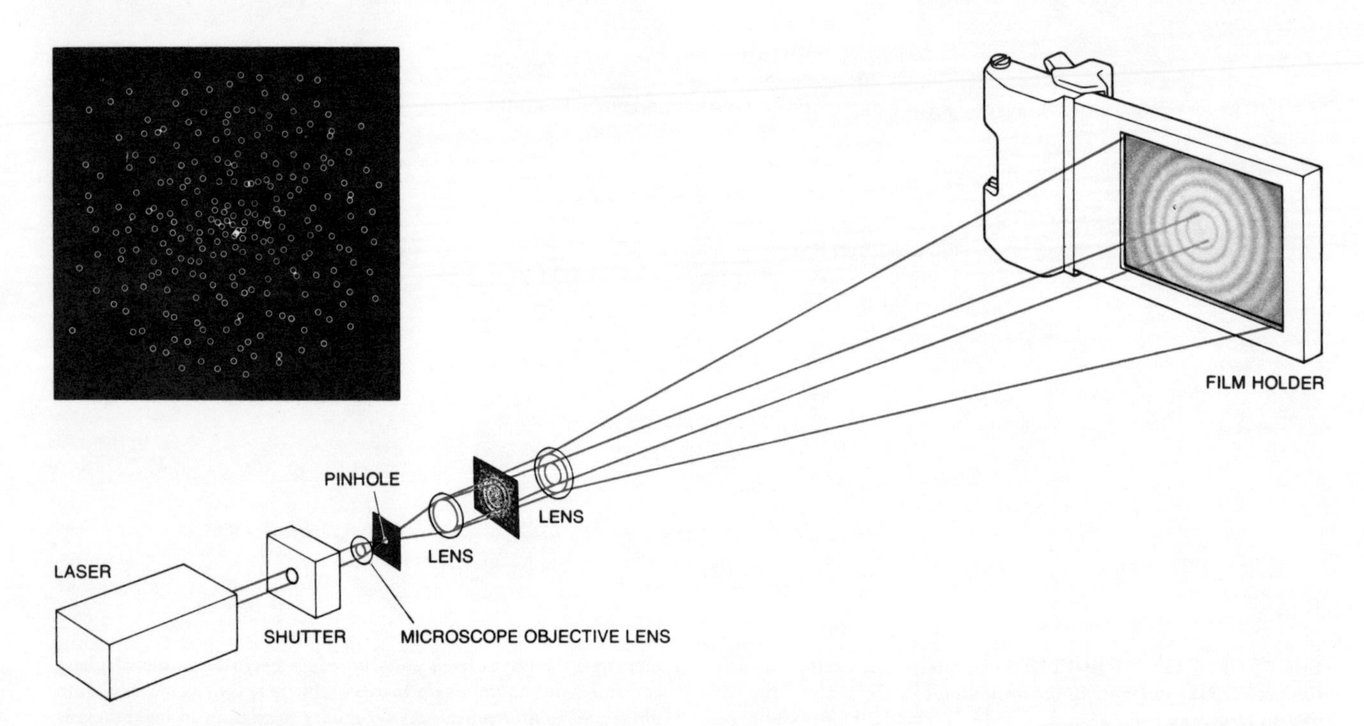

MANY RING APERTURES together (*insert at top left*) act as a cloud of uniform water droplets and generate a diffraction pattern that resembles the glory. First a series of 241 randomly placed circles were drawn on white paper. Then a much reduced photographic negative was made from the drawing. A parallel-ray beam of red light from a helium-neon laser was directed through the array of rings; a lens formed an image of the diffraction pattern that was recorded on Polaroid film (*see illustration on opposite page*).

centimeters in diameter across the front of the hall. Into this beam we can insert a Lucite sphere with a diameter of 30.48 centimeters (12 inches).

In addition to producing a marvelous rainbow that covers one of the white walls of the lecture hall, the system demonstrates the "backward" optics of transparent spheres. We can view the sphere from the direction of the sunbeam by inserting a small mirror into the beam. It can be seen that the backward-directed light indeed comes principally from the edge of the sphere, with a small additional contribution from specular reflection at the center of the sphere's face.

When the sphere is viewed from a point that is even at a slight angle to directly backward, its appearance changes considerably. Its edge is still well illuminated, and the equatorial region of the edge is particularly bright. The specular reflection from the face is off center.

If a plant leaf or a piece of paper is placed five centimeters behind the sphere, one can see the Heiligenschein. The sphere becomes very bright. The presence of the leaf greatly enhances the amount of light sent backward in the direction from which it came, to the extent that the glow reflected back to the mirror of the heliostat can be seen throughout the lecture hall.

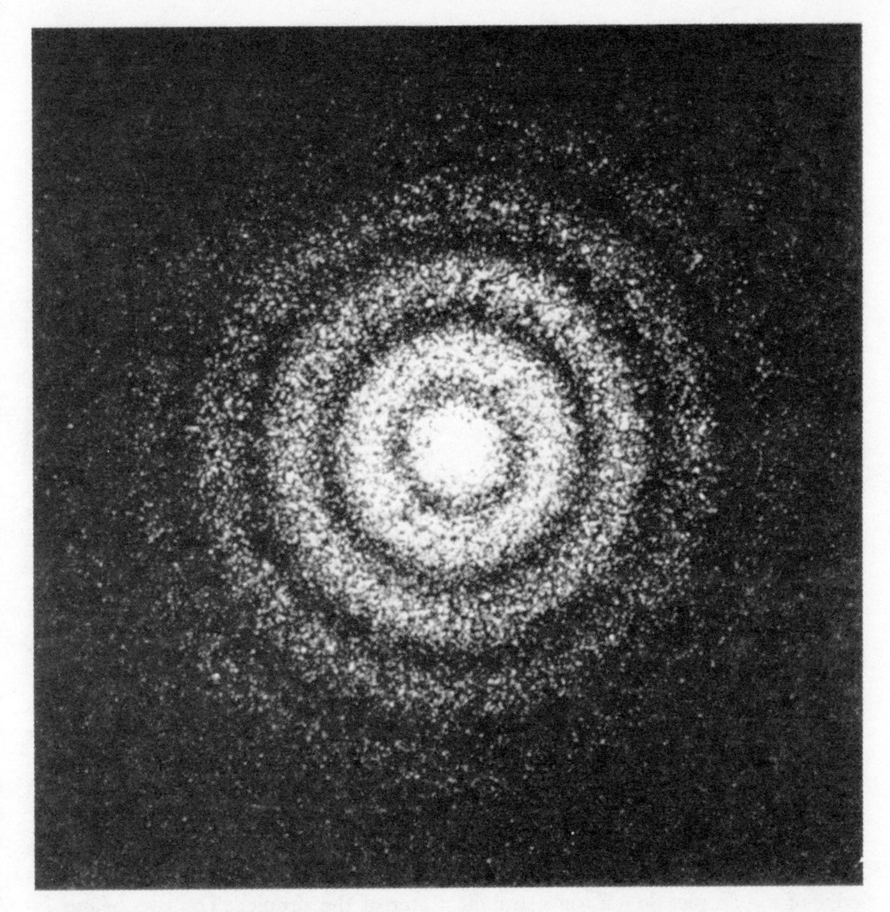

DIFFRACTION PATTERN that approximately represents the appearance of a glory seen through a red filter was photographed with the apparatus that is shown in the illustration at the bottom of the opposite page. Graininess in the picture is caused by interference among individual apertures; if the number of apertures were increased, graininess would decrease.

This procedure with sunlight and the Lucite sphere is instructive, but more relevant studies have been made in the laboratory with laser light and real droplets of water suspended in midair. Theodore S. Fahlen, who was then working at the university, found that droplets could be suspended in two different ways. In the first method a resonator made from a cylindrical piezoelectric crystal and a circular watch glass produced an acoustical standing wave. Droplets as large as a millimeter in diameter could be suspended in midair at the region of maximum acoustical pressure for up to several minutes without appreciable vibration.

The second method simply entailed suspending the droplet by its own surface tension from a glass fiber whose end had been enlarged to form a tiny bead; the fiber could hold droplets two or three millimeters in diameter. This method proved to be more useful than the other one. If the optical effects are to be observed, the surface of the droplet must be exceedingly quiet and smooth, much quieter and smoother than the droplets that were levitated in the acoustical resonator. M. J. Saunders of Bell Laboratories has used the fiber method to make and study water droplets as small as nine microns in diameter. (His fibers were bits of spider web.) Saunders has verified that even with such small droplets the backscattered light comes from the edge of the sphere.

Why should the light be scattered mostly from the edge? In trying to answer that question we approach some current frontiers of optics and particle physics, even though the precise solution of how an electromagnetic wave (in this case light) is scattered from a transparent sphere has been known since the beginning of the century.

In 1908 the German physicist Gustav Mie showed that the intensity of an electromagnetic wave scattered from a sphere can be calculated as precisely as one wants for any angle, including angles in the backward direction. He showed that the intensity of the light scattered at any angle can be represented as a sum of a series of algebraic terms, each composed of involved mathematical expressions. These terms are not of the type that can be easily computed on the back of an envelope. In addition the number of terms that must be evaluated in order to arrive at the intensity at a given angle for a specific wavelength is somewhat larger than the circumference of the sphere in nanometers divided by the wavelength in nanometers. For a droplet one millimeter in diameter, for example, and green light of a wavelength of 500 nanometers, some 6,300 terms have to be evaluated and added together. And to get the entire intensity pattern for just one wavelength of light requires repeating the process for a number of angles.

Thus in addition to a desire to learn the details of Mie's predictions for a given case one must have access to the services of a high-speed computer. It is only within the past decade that Mie calculations for large spheres have been conducted to any extent. To make matters worse, the intensity of the backscattered light is extremely variable. As A. J. Cox has demonstrated at the University of New Mexico, a very small change in the wavelength of the light can result in a change of intensity as large as a factor of 100.

There are less exact but perhaps more instructive ways to treat the scattering

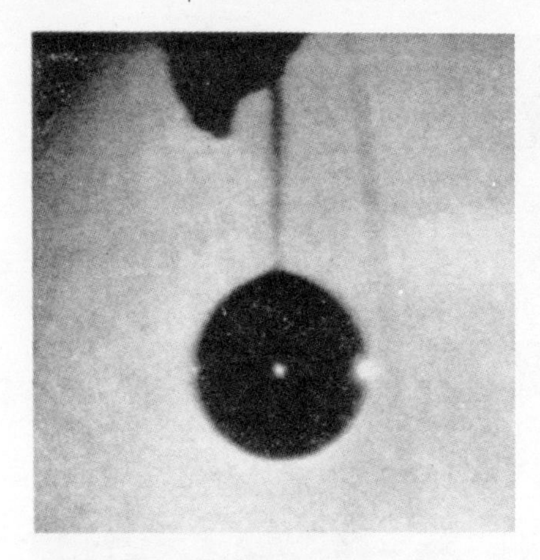

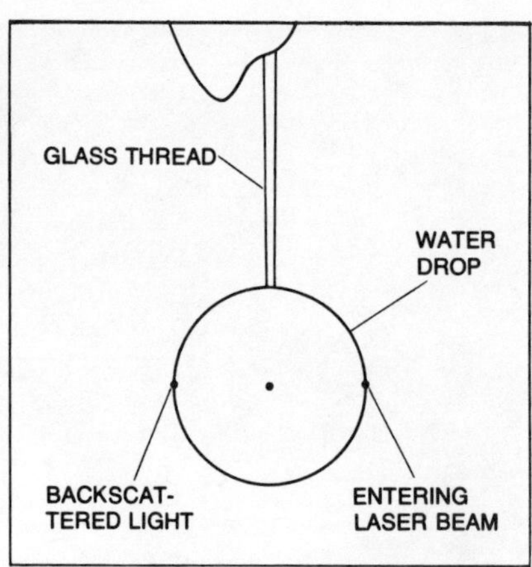

WATER DROPLET SUSPENDED from a glass thread returns a beam of laser light straight back toward the observer. The beam entered at the right edge of the droplet and was sent back at the left edge. Experiment demonstrates that in giving rise to the glory, light waves must be conveyed around the surface of the droplets. Droplet is two millimeters in diameter; thread is 40 micrometers thick. Central bright spot on droplet is due to backlighting.

of light. For instance, primary and secondary rainbows can be explained in terms of geometrical optics: the primary bow is produced by light that is reflected once inside a droplet of water and the secondary bow is produced by light that is reflected twice. The glory cannot be explained on this basis, because in an ordinary rainbow the rays striking the edge of the droplet do not come straight back. It is nonetheless possible to understand in terms of ray optics the small contribution made to the glory along the axis of the incident light beam.

Rays are reflected not only from the outer surface of the drop but also from the inner surface. Since only some 2 percent of the light falling perpendicularly on an interface between air and water is reflected, the amount of light traversing paths that involve more than one such reflection will be negligible compared with the two direct reflections. The relative phase of the two rays comprising the axial contribution varies with the diameter of the droplet. The two beams interfere with each other to give rise to an intensity that varies on a sine curve as the diameter of the droplet varies. This property has been used, in fact, to make precise determinations of the rate at which droplets of various sizes evaporate.

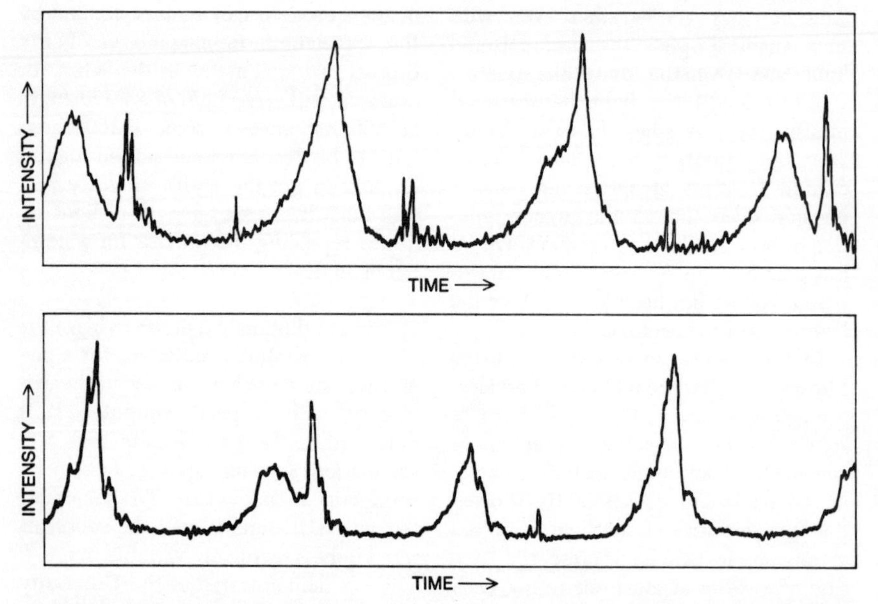

OSCILLOGRAPH TRACINGS of the intensity of a laser beam emerging from the margin of an evaporating droplet over a period of 20 minutes show a repetitive pattern of three humps on which are imposed spikes. The tracing at the top was made when the droplet had a diameter of 1,153 micrometers; the tracing at the bottom was made when it had shrunk to 741 micrometers. Humps are due principally to paths taken by light traveling halfway around droplet and 3½ times around it. Spikes are resonances from hundreds of paths.

The explanation of why the light is returned from the edge of the droplet is surprising even to many people familiar with physical optics. There must be a process that takes light impinging on one edge of the droplet and sends it back in the opposite direction from the opposite edge. The mechanism that produces the rainbow, in which rays are refracted into the sphere, are reflected internally once and are refracted back out again at a preferred angle of 42 degrees from the backward direction, at first appears to be inadequate to explain the glory. Van de Hulst proposed that an internally reflected ray is important but that it is combined with another mechanism we have not yet considered: the surface wave.

When a light beam strikes the air-water interface from the water side, there is a critical angle at which the beam refracted into the air is parallel to the interface. At that angle the beam has a long "tail" on the "downstream" side. This tail is the surface wave, or more properly the lateral wave, and it carries some of the light farther around the circumference of the droplet than one would expect from geometrical optics alone. Thus the surface wave provides the necessary ingredient that makes van de Hulst's hypothesis viable.

As the diameter of the droplets that give rise to the glory is increased, the intensity of the glory shows periodic resonances, or fluctuations. These resonances indicate that the effect must be due to two or more beams whose difference in phase depends on the size of the droplet. When the two or more beams interfere constructively, the intensity of the glory will be high; when the beams interfere destructively, the intensity will be low.

The existence of such multiple-beam resonances has been confirmed experimentally. Fahlen suspended a very quiet evaporating droplet and focused light from a helium-neon laser on it through a small hole in a mirror set at an angle of 45 degrees. The returning light was reflected by the same mirror through a system of variable-aperture lenses onto the photocathode of a photomultiplier, which converts the light into an electric current. When the laser beam was directed along a tangent to one side of the droplet, the returning beam was received coming back along a tangent to the opposite side of the droplet. The amplitude of the current produced by the photomultiplier was recorded with a light-beam oscillograph, an instrument that writes on a rapidly moving strip of photosensitive paper with a narrow ultraviolet beam. The deflection of the beam along the paper was made proportional to the amount of laser light received by the photomultiplier [*see bottom illustration on opposite page*].

The oscillograph tracings clearly showed the highly structured periodic fluctuations in the intensity of the returning ray as the droplet evaporated. The signal was composed of a periodic set of humps with a series of sharp spikes on top of them; the time that elapses between repetitions of a certain spike structure is about three times longer than the time between adjacent humps.

Although it is not apparent from short sections of the traces, the oscillograph record covering the 20-minute lifetime of the droplet shows striking long-term changes in the basic three-hump structure. The spikes slowly change their magnitude and their position on the humps. A new spike rises to prominence at about every 72nd hump. If we number the humps in cycles of three, we find that the most prominent spike of all occurs on the same hump in the cycle every 73rd cycle. From this Fahlen and one of us (Bryant) concluded that the spike period is about 73/74 of the three-hump cycle.

What is the significance of these humps and spikes? The elapsed time between humps can be directly related to the change in the diameter of the droplet by monitoring this intensity of the axial ray. When the intensity goes from a maximum to a minimum and back to a maximum again, wave optics tells us that the diameter of the water droplet has changed by about three-eighths of the wavelength of the light. This fact enables us to determine that the hump period and the spike period respectively correspond to changes in the droplet diameter of .09 and .26 wavelength. The fluctuations in intensity therefore reflect very small changes indeed. For example, for the red laser beam, which has a wavelength of 630 nanometers, the hump period corresponds to a change in the droplet diameter of only .056 micrometer. For a droplet one millimeter in diameter that is a change of only 56 parts per million.

Needless to say, great pains are required to obtain a droplet quiet enough to yield these results, which indicates how complicated a complete explanation of the glory must be. Although the droplets studied in the laboratory have diameters that are 50 times greater than those of the droplets in the clouds where the glory occurs in nature, Mie calculations undertaken by Cox and Fahlen show

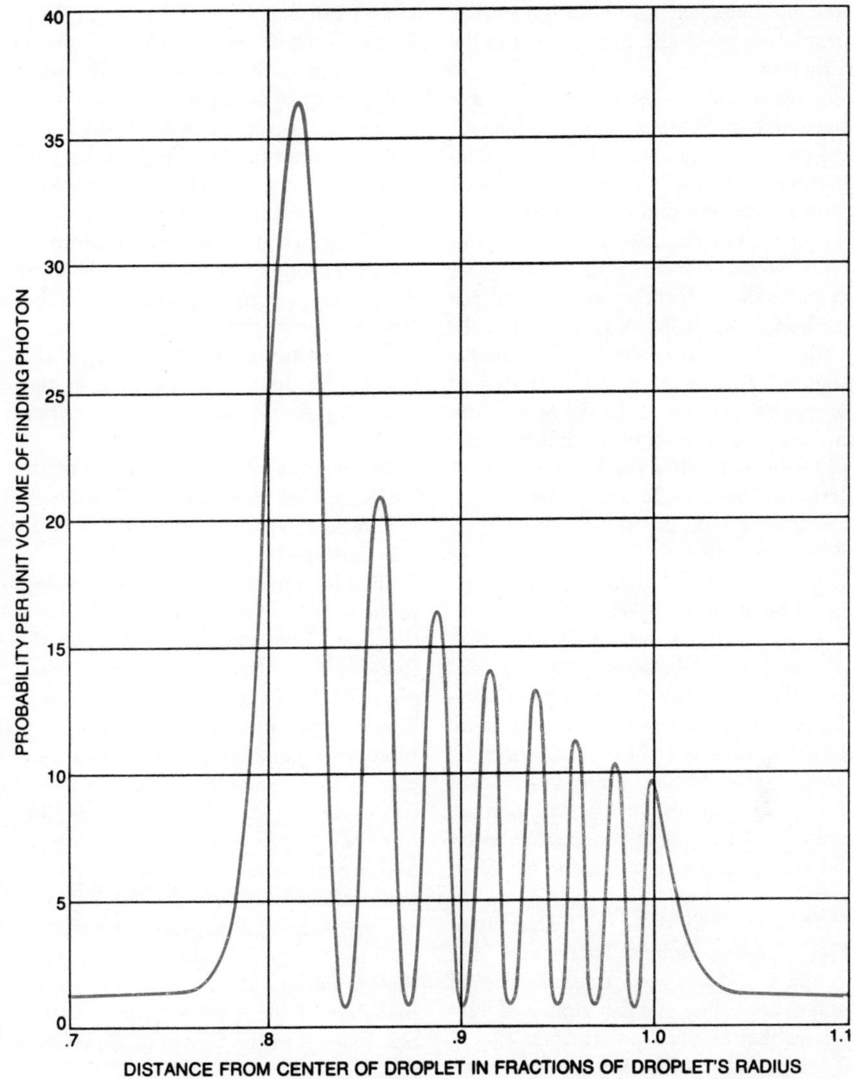

PROBABILITY OF FINDING A PHOTON at a given distance from the center of a droplet shows a set of maximums and minimums that one would expect from rays traveling around the droplet at a resonance similar to that producing one of the narrower spikes seen in the oscillograph tracings in the bottom illustration on page 72. Example corresponds to a droplet with a diameter of 40.44 micrometers illuminated by light from a helium-neon laser.

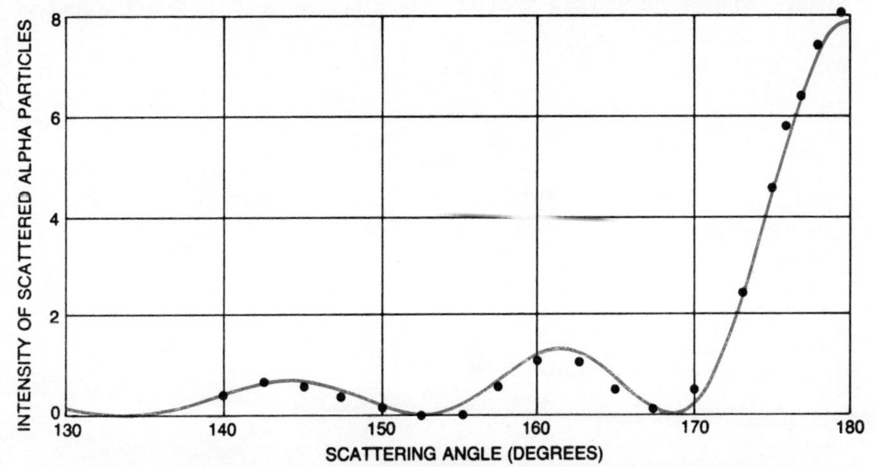

SUBATOMIC PARTICLES PRODUCE A GLORY, as can be seen in this distribution of alpha particles (helium nuclei) at an energy of 29 million electron volts backscattered by nuclei of calcium 40. The solid line is the prediction based on a theory of the glory developed by the authors; the points are actual measurements made by A. Budzanowski and colleagues at the Institute of Nuclear Physics at Cracow in Poland. The maximums at the angles of 180 degrees, 162 degrees and 144 degrees are analogous respectively to the central bright region and the side rings of the optical glory. Intensity is given in arbitrary units.

that the behavior of light in the smaller droplets is much the same as it is in the larger ones.

Fahlen and one of us (Bryant) have been able to devise a simple mathematical model that can predict intensity fluctuations that are in qualitative agreement with experiment and with the exact predictions from the Mie theory. In principle the model includes an infinite series of paths through and around the surface of the sphere. H. M. Nussenzveig of the University of Rochester has developed a rigorous treatment of the scattering of waves by a sphere. Like the Mie theory his analysis consists of an infinite series of terms, but unlike the Mie theory each term can be directly interpreted as representing a light ray being internally reflected one, two, three or more times as it passes around and through the droplet. The humps observed in the oscillograph traces are principally the result of interference between a strong rainbowlike ray (a ray that is internally reflected only once) and a ray that is internally reflected 14 times, traveling around the droplet three and a half times. The spikes, being very narrow, must be the result of constructive interference among a large number of rays of similar intensities.

In master's theses at the University of New Mexico, Robert Thede and later Jaime Wong analyzed a certain class of spikes according to a description of light waves that neglects polarization effects. Their results can be interpreted to obtain the probability of finding a particle of light (a photon) at a given point in space. Thede and Wong found that there are several different types of spike that can be classified according to the probability of finding a photon at a given distance from the center of the water droplet. One such probability plot is shown in the top illustration at the left. We call it a Type 8 spike because it has eight maximums. Spikes of Type 6, Type 7 and Type 9 were found to be prominent for droplets with a diameter of 30 microns. The peaks and troughs in the probability curve can be interpreted in terms of waves reflecting around the inside of the droplet and interfering with one another constructively (the peaks) and destructively (the troughs). These waves correspond to light rays bouncing around inside the droplet at close to the critical angle.

In 1971 Lawrence Sromovsky of the University of Wisconsin, following on the work of Nussenzveig, showed in detail that the spikes are efficiently described by a mathematical concept close to the hearts of some elementary-particle theorists. The concept is that of the Regge pole, introduced by the Italian physicist Tullio Regge. The Regge pole is an abstract mathematical way of representing an entire class of physical situations. For example, all our Type 8 spikes can be described as originating from one Regge pole, so that instead of needing an infinite number of mathematical terms to describe the behavior of Type 8 resonances, one can make do with the Regge pole alone. Thus we find that an important component of the explanation of the glory is quite similar to one concept in the theory of elementary particles.

Since the wave-particle duality exhibited by light is shared by all elementary particles, beams of particles scattered in accelerator experiments may also show optical effects such as the rainbow and the glory. In fact, the concept of a surface wave in a water droplet can be applied to the behavior of high-energy particles bombarding an atomic nucleus. There is no direct correspondence, however, between the scattering of light by a transparent sphere and the scattering of particles by a nucleus. In the nuclear optical model the nucleus is rather like a muddy droplet with a poorly defined surface, so that surface waves would be strongly absorbed and not as clearly defined as they are in a droplet of clear water. Nevertheless, a simple extension of the glory mechanism serves very well in describing certain instances of the backscattering of high-energy particles by nuclei.

Like elementary particles and nuclei, entire atoms and molecules exhibit wave behavior when they are accelerated and scattered. In such experiments nothing comparable to the backward-scattered glory is observed, but wave effects resembling the rainbow have been. Moreover, there is a distinct interference pattern in the forward direction; it is often called the forward glory. Changes in such patterns produced by molecular beams have been utilized by chemists to investigate the energetic threshold of chemical reactions.

We hope that, having read this article, the reader who has never seen a glory will now be on the lookout for one. The easiest way to see a glory in this age of air travel is to watch for the small shadow of one's airplane on a layer of cloud below. If the conditions are right, the shadow will be surrounded by the bull's-eye pattern of the glory—a striking demonstration of the wave nature of light and a colorful reminder of the underlying unity of the physical world.

Miragesㅤ9

by Alistair B. Fraser and William H. Mach
January 1976

There are several types of apparition to be seen across the sea or the land. Each has an explanation in terms of the optical properties of a fluid medium: the atmosphere

In 1906 Robert E. Peary, striking for the North Pole, stood on the summit of Cape Thomas Hubbard at the north end of Axel Heiberg Land. To the northwest, at a distance he believed to be about 120 miles, he saw "snow-clad summits above the ice horizon." Later he saw them again, this time from Cape Columbia on Ellesmere Island, and wrote: "My heart leaped the intervening miles of ice as I looked longingly at this land and, in fancy, I trod its shores and climbed its summits, even though I knew that that pleasure could be only for another in another season." That other man turned out to be Donald B. MacMillan, leader of the expedition to "Crocker Land" in 1913. As the group approached the supposed location of Crocker Land (83 degrees north, 103 degrees west) it obligingly appeared before them. MacMillan wrote: "There could be no doubt about it. Great heavens, what a land! Hills, valleys, snow-capped peaks extending through at least 120 degrees of the horizon." They then tramped 30 miles "inland" over the Arctic ice without seeing a thing. Crocker Land was a mirage!

Both Peary and MacMillan had undoubtedly witnessed one of the most spectacular types of mirage, the Fata Morgana. It has been named after the Fairy Morgan (Fata Morgana in Italian), who appears in some of the Arthurian legends as King Arthur's sister. She was credited with the magical power of creating castles in the air. In retrospect that is an apt description of what happens, because the images seen in the Fata Morgana bear no resemblance to the object from which they were formed. Fantastic sights can appear in spite of the fact that the only object in the distance is a barren surface of snow or water.

One of the best early descriptions of the Fata Morgana was written by an Italian priest, Father Angelucci, who related his experience in a letter to a colleague. On the morning of August 14, 1643, he was looking out over the Strait of Messina from the city of Reggio on the southern tip of Italy. As he watched, "the ocean which washes the coast of Sicily rose up and looked like a dark mountain range." In front of the mountain "there quickly appeared a series of more than 10,000 pilasters which were a whitish-gray color," but then "the pilasters shrank to half their height and built arches like those of Roman aqueducts." Before it all vanished castles appeared above the aqueduct, each with towers and windows.

To properly understand how the atmosphere can give rise to such strange apparitions it is first necessary to examine the much simpler types of mirage. A particularly striking example of a common type of mirage was seen and photographed in the spring of 1972 by one of us (Fraser), who watched with fascination as two boys strolled off a beach at Seattle and casually walked out onto the waters of Puget Sound among the sailboats. The scene was so compellingly real that it was easy to believe the records that describe observations of other men apparently walking on water, examples of which occur not only in Christian writings but also in the literature of Buddhism and of the Greeks of the Hellenistic period.

FERRYBOAT "ILLAHEE" is operated on Puget Sound by the Washington State Ferry System. Appearance of Puget Sound ferries is distorted in mirages shown on the following page.

This simpler type of mirage differed from the Fata Morgana in that whereas the mountains of Crocker Land and the castles and aqueducts seen by Angelucci were nonexistent, the boys who were perceived walking on the water actually existed. They had, however, been seen by observation through a portion of the atmosphere that, acting as a giant lens, was bending the light rays that passed through it. The image of people walking on water was certainly real, but it was an image, not an object. One could not assume that because the image showed people walking on water the

INFERIOR MIRAGE WITH TOWERING results when temperature and temperature gradient are greatest at the surface and decrease with height. In an inferior mirage the image is displaced downward from the object. Towering means image is magnified.

TWO-IMAGE INFERIOR MIRAGE of a ferryboat appears as a series of towers. Everything below the ship's bridge has vanished. Requirements for this mirage are the same as for an inferior mirage with towering except that surface-temperature gradient is larger.

SUPERIOR MIRAGE results when the temperature increases with height so that the image is displaced up from the object. Magnified portholes and the squashed passenger deck and bridge result when boat is at *A* in the image space of diagram at the bottom of page 84.

MISSHAPEN FERRYBOAT appears to be squashed and to be on the top of a high wall. This is a superior mirage that results because the ferryboat was in approximately the position that is marked *C* in the image space of the diagram at the bottom of page 84.

people were actually doing anything of the sort. Usually in viewing one takes for granted this distinction between an image and an object. One knows that the somewhat distorted-looking person who appears on the television screen is just an image and that the object, a perfectly normal individual, is a considerable distance away in front of a television camera. Images seen through the atmospheric lens are no more illusory than the images seen through a telescope or even through a pair of eyeglasses. With the atmospheric lens or any other lens, however, the image may not look the same as the object would if it were seen without the lens.

The word mirage comes from the French verb *se mirer*, to be reflected, and although many of the images seen in mirages resemble those seen in irregular mirrors, the concept of reflection plays no part in mirages. Acting as a lens rather than as a mirror, the atmosphere produces mirages by refraction. The lens is obviously not the kind of lens that would be found in a camera or a pair of eyeglasses. Such lenses, made of glass, have a uniform index of refraction, and the light is made to bend and generate images by their curvature. In the atmosphere the lens has no shape, since both the observer and the object are inside it. The atmosphere causes light to bend as a result of gradual variations of the index of refraction within it. As part of our meteorological research at Pennsylvania State University we have examined this fascinating question of the image-forming ability of the atmospheric lens in an attempt to learn more about the structure and behavior of the atmosphere and to determine how the information carried by light is altered as it passes through the air.

The index of refraction of air depends on the density of the air and the amount of moisture in it. The contribution of moisture to phenomena involving visible light is so small, however, that it can be ignored. The density of the air depends on its temperature and pressure. Since most of the mirages we shall discuss are caused by shallow layers of air over which the change in pressure is slight, we can pretend that the index of refraction depends only on the temperature.

A high temperature corresponds to a low density and a low index of refraction. The stronger the temperature gradient (the greater the change of temperature with distance), the stronger the gradient of the index of refraction (the greater the change of the index of refraction with distance) and the greater the amount of refractive bending of the

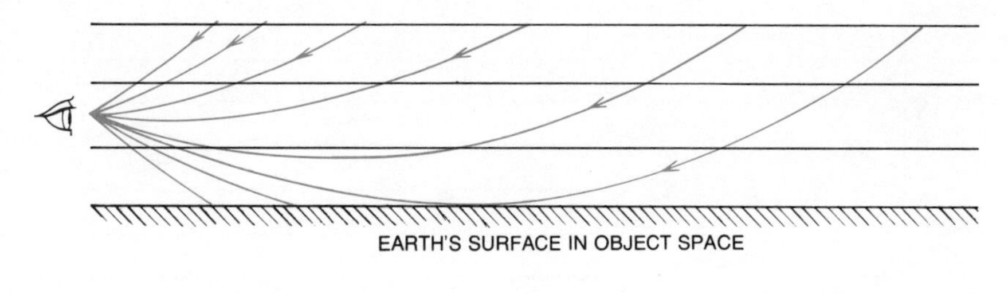

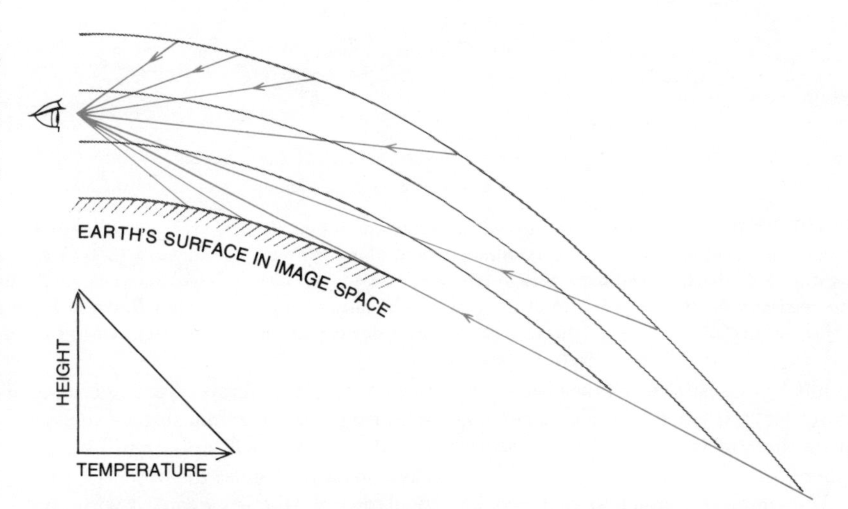

OBJECT SPACE AND IMAGE SPACE are contrasted. If light rays (*color*) pass through a region of the atmosphere where the change of temperature with height is constant, as indicated in the temperature profile at bottom left, the rays take a parabolic track. This is what happens in real space, that is, the space occupied by the object one is seeing (*top*). One's interpretation of what one is seeing, however, almost always rests on the assumption that light is traveling in a straight line. Therefore it is best to visualize a mirage by deforming the real space as if the rays were taking straight paths (*bottom*). Then surfaces that are actually flat are represented as being bent (*gray*). That is how space appears to the eye in a mirage; it is the "image space" in terms of which mirages are discussed in this article.

light. If the temperature in the air is the same everywhere, the light will travel in a straight line. The nature of the atmospheric lens and therefore of the mirage images it produces thus depends on the way the temperature varies in the atmosphere, primarily with height.

Light will take a parabolic path when it passes through a region of the atmosphere where the change of temperature with height is constant [*see illustration above*]. The curvature of the ray is proportional to the temperature gradient measured perpendicularly to the ray, so that the ray is curved most strongly when it is traveling parallel to the lines of constant temperature. The curvature of the light ray will cause an image, such as a view of a distant boat, to be displaced from the position of the object (the boat itself). Since the ray always bends so that the cold (denser) air is on the inside of the curve, the image is displaced in the direction of the warm (less dense) air.

Most mirages involve viewing over distances of anywhere from about half a kilometer to about five kilometers, so that a mirage is only weakly influenced by the curvature of the earth. For the purposes of this discussion it is therefore convenient to assume that the earth is flat. Over a flat earth any other surfaces drawn at uniform heights would also be flat. If there is a temperature gradient in the air, the light rays that pass through the atmosphere will be not straight but curved.

The space through which the rays pass is the "real" or "object" space, but it is not the space that is perceived by the eye. One's interpretation of what one sees in one's surroundings is almost always predicated on the implicit assumption that light travels in a straight line. To comprehend a mirage it is better to imagine that the "real" space has been deformed so that the rays passing through it are traveling in a straight line and the previously flat surfaces, such as the surfaces of the earth, have become bent. We now have a drawing of space as it appears to the eye; the drawing can be described as a representation of the "apparent" or "image" space. The mirages described here will be discussed mainly in terms of the image space that

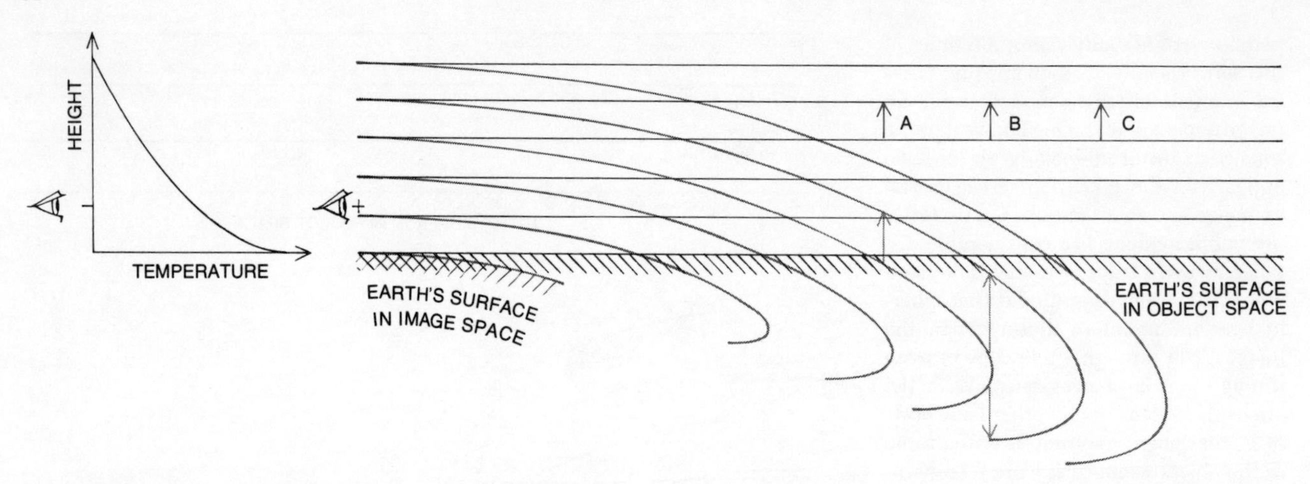

IMAGE SPACE for a two-image mirage is portrayed. When the temperature profile (*left*) has its maximum gradient where the temperature is highest, a two-image mirage can be produced. Because the maximum temperature is also at the ground, the mirage is inferior, so that all image-space surfaces (*gray*) bend down with increasing distance from the observer. Corresponding object-space surfaces (*black*) are horizontal. An object *A* would be seen in image space as a single image, displaced downward and magnified. An object *B* would be seen twice in image space, once erect and once inverted. Object *C* could not be seen unless it was a bit higher.

results from a particular type of temperature (or refraction) profile occurring in the bottom few meters of the atmosphere.

The simplest distinction that can be made is between a superior (literally upper) mirage and an inferior (lower) one. When the temperature increases with height, a horizontal surface such as a body of water will appear to be concave upward. It gives the observer, particularly one who is viewing it with binoculars, the impression of being inside a large, shallow bowl. It is a superior mirage because the image is displaced upward from the position of the object. The phenomenon was known to English sailors as looming.

An inferior mirage occurs when the temperature decreases with height. Then a horizontal surface will appear to be

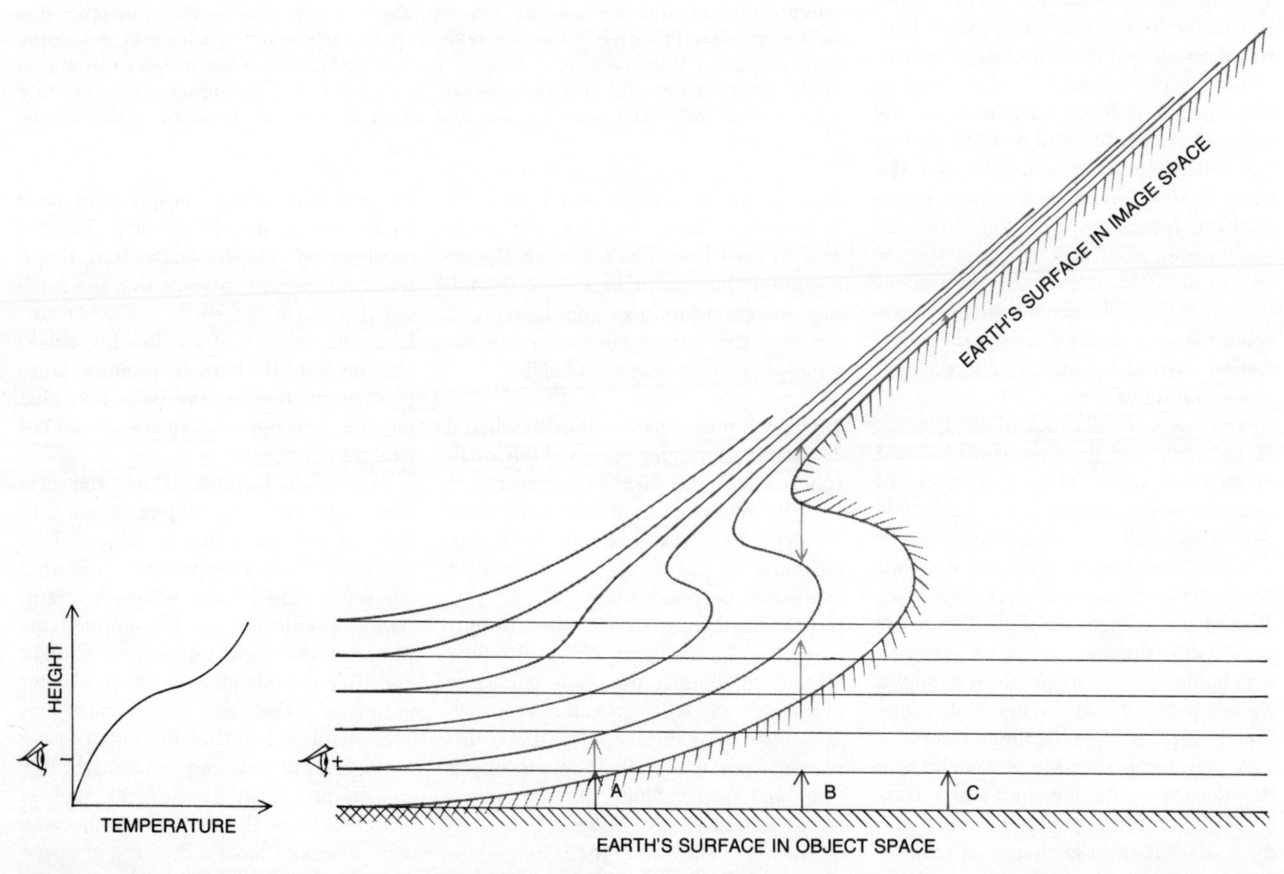

THREE-IMAGE MIRAGE can result from an inflection-point temperature profile. This profile often exists over an enclosed body of water on a sunny afternoon. The apparent shape of the water surface is a large, flattened *S*. An object *B* located where the surface folds is displaced up and seen as three images. An object *A* will appear as a single image, but if it extends through the depth of the diagram, the central part will be magnified and the top reduced. An object *C* would be seen as a greatly squashed single image.

convex upward. The impression is again of a bowl, but now the bowl is inverted and the observer is on the top of it with the surface curving down in all directions. As a result there is an optical horizon beyond which the surface vanishes as it curves out of the observer's sight. This phenomenon is sometimes called sinking.

It was an inferior mirage that made those two boys appear to be walking on the waters of Puget Sound. They were actually walking on a sandspit that had been uncovered at low tide, but their feet and the sandspit were beyond the optical horizon and could not be seen. The image of the rest of the boys' bodies was left suspended on the surface of the intervening water.

The temperature profile that gives rise to this effect is quite common over enclosed bodies of water in the early morning. The water retains its heat through the night but the surrounding land cools off. Cool air from the land flows out over the warmer water and is heated from the bottom, thus creating a temperature profile in which the temperature decreases with height.

From the viewpoint of a mathematician who wants to calculate the shape of the images in image space the difference between a superior mirage and an inferior one is slight. If he assumes that the temperature increases with height, resulting in a superior mirage, the surfaces bend up with distance and the observer is apparently in the bottom of the bowl. To obtain the corresponding inferior mirage the mathematician need only turn the diagram upside down. Now the temperature decreases with height and the observer is apparently sitting on the top of an upturned bowl. The shape of the bowl is the same in both cases, but there is a world of difference to the observer. With the superior mirage he sees the inside of the bowl, whereas with the inferior one he sees the outside of the bowl and so cannot see beyond its horizon. The meteorologist would likewise claim that a worthwhile distinction can be made between the superior mirage and the inferior one because the meteorological conditions that give rise to an increase of temperature with height differ from the conditions that produce the reverse effect.

Having examined the general characteristics of superior and inferior mirages, let us now look closer at their behavior. One question is whether the image has been magnified or reduced. If the temperature gradient is constant with height, there will be no magnification. The image will have been displaced from the position of the object, but everything at a given distance is displaced up or down by the same amount. For example, a person's head will be displaced by the same amount as his feet, so that he keeps the same proportions. Actually a constant temperature gradient is rarely encountered in the bottom few meters of the atmosphere.

The primary mechanism that determines the structure of the temperature profile near the surface of the earth is the transfer of heat between the surface and the atmosphere. Near the surface, in the bottom few centimeters of the at-

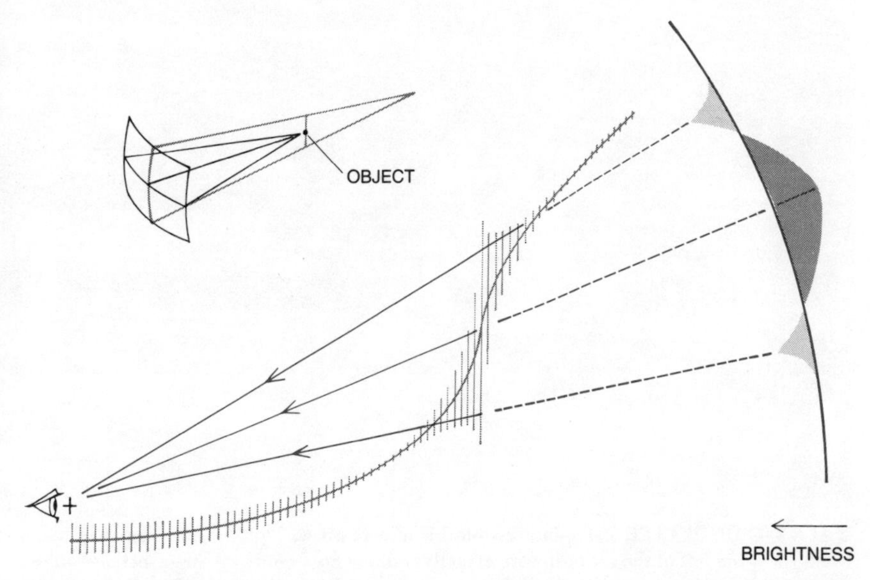

CONDITION FOR THE FATA MORGANA is also an inflection-point profile, but the temperature gradient near the inflection point is slightly smaller than it is for a three-image mirage. The apparent shape of the water surface does not fold over but instead rises up and forms a wall. The surface is blurred because of astigmatism; since the wave front (*upper left*) is not spherical, objects in the distance are out of focus as a blurred vertical line. Amount of blurring is indicated below by gray vertical lines. Astigmatic blurring also redistributes the brightness (*right*) so that center of wall appears dark and other parts bright.

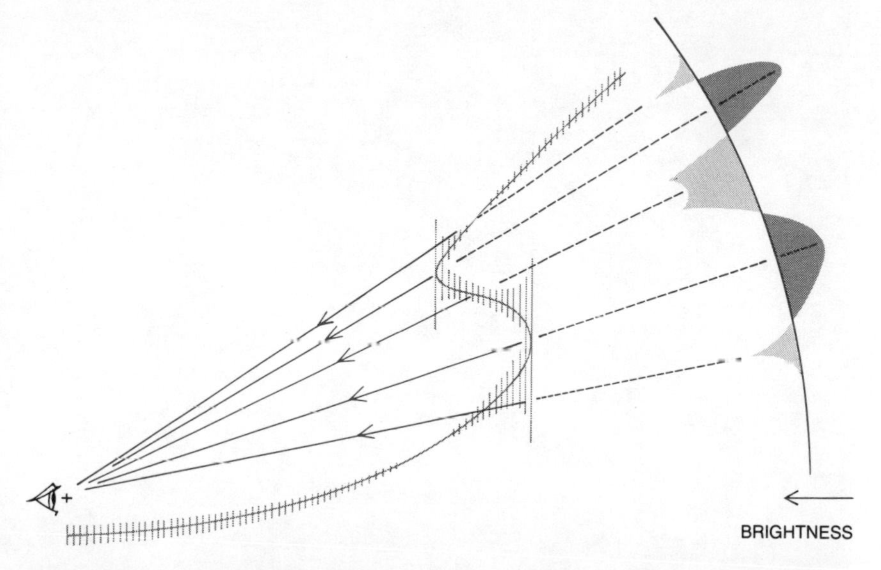

ASTIGMATIC THREE-IMAGE MIRAGE produces an overhanging wall so blurred that no detail can be seen on it. The brightness, however, is redistributed so that the center of the wall is bright. This strip of brightness appears to the eye as a bank of fog and has been called the Fata Bromosa. Internal gravity waves in the atmosphere can cause the image to oscillate back and forth between looking like a vertical dark wall and looking like an overhanging bright wall. The patchy distribution of brightness that results can make observers think they see a wide variety of phenomena, including mountains and elaborate buildings.

WALKING ON WATER is the impression this mirage gives. The two boys to the left of the sailboat were actually walking on a sandspit that had been uncovered at low tide in Puget Sound. In this inferior mirage with towering the sandspit and the feet of the boys have vanished below the optical horizon, so that the boys appear to be suspended on the water. Because the figures have also been magnified, they appear to be larger and therefore closer than the people in the boat, whereas they were actually much farther away.

FATA MORGANA MIRAGE appears as a whitish gray wall that breaks into pilasters and, toward the right side, arches. The wall is actually a greatly magnified portion of the sea. The atmospheric lens, however, is so badly out of focus that no detail of the original sea surface can be distinguished on the wall. The same defocusing can also redistribute the brightness of the surface into bright and dark patches that are easily mistaken for real eminences such as buildings and mountains. The mirage was in Puget Sound.

mosphere, heat is transferred mainly by molecular conduction and radiation; higher up free or forced convection usually dominates. The most efficient of these mechanisms is convection, and so it requires the smallest temperature gradient to transfer a given amount of heat. It is usual, therefore, for the temperature gradient to be at a maximum near the surface and to decrease with height. The resulting temperature profile has curvature, which changes the magnification of the image.

Stooping, the name applied to an image when it is reduced in size (has a magnification of less than unity), occurs if the temperature gradient decreases as the temperature increases. This condition frequently arises on sunny afternoons over lakes and sounds. Air that has been warmed over the surrounding land is carried out over the cooler water and so is cooled from below. The strongest temperature gradient and the lowest temperature are found at the surface of the water, so that with height the gradient decreases and the temperature increases. We are therefore dealing with a superior mirage: the image is displaced upward from the position of the object. Thus the bottom of the object, which is seen through a stronger gradient than the top is, will be lifted more than the top. The resulting image is squashed.

An observer on a beach can often watch a distant scene becoming increasingly stooped as the day progresses. The lateral dimensions do not change, because the horizontal gradient of temperature is negligible. The view of, say, a distant boathouse would therefore retain a fixed angular width while becoming increasingly small vertically.

Towering, the term applied to an image when it is magnified, occurs when the temperature gradient and the temperature increase together. Towering almost always accompanies an inferior mirage, since whenever the surface is warmer than the air above it, the air will be heated from below. The maximum temperature and the maximum temperature gradient are hence found at the bottom of the temperature profile, and both quantities decrease with height. Such conditions are to be found over enclosed bodies of water in the early morning, as we have mentioned, and also over sun-heated ground later in the day. Since the temperature increases near the surface, the image of a distant object will be displaced downward, but the bottom of the object will be displaced downward more than the top because the bottom is seen through a stronger temperature gradient. The photograph of the two boys apparently walking on water illustrates the phenomenon. They both appear to be as large as or even larger than the people in the sailboat in spite of the fact that they are considerably farther from the camera. They are towering.

We turn now to the two-image mirage, which is the phenomenon involved in the familiar example of the traveler in the desert who thinks he has sighted an oasis. The "water" in such a mirage is a second (inverted) image of the sky, seen below the horizon and therefore interpreted as if the sky light were reflected by a surface of water. This is a two-image inferior mirage.

The requirements for the existence of a two-image mirage are the same as for towering: the temperature and the temperature gradient must increase together. To give rise to two images instead of a single towering one, however, the temperature profile must have a somewhat greater curvature. In an inferior mirage the effect can be accomplished by an increase in the temperature gradient at the surface of the ground or of the water. A ray of light that travels through this region of strong temperature gradient becomes so strongly bent that it will no longer be able to join the eye with the bottom of some distant object but will instead join the eye with the top of the object to give a second, inverted image. The image is inverted because as the observer lifts his gaze slightly he is looking through a region of the atmosphere that has a weaker temperature gradient, so that the ray is less strongly curved. It will therefore join the eye to a point lower on the object rather than higher, as would usually be expected.

A diagram of the image space for a two-image inferior mirage [*see top illustration on page 84*] shows that the various surfaces do not extend indefinitely into the distance. The bottom surface terminates at the optical horizon, and the other surfaces terminate at a distance that increases with the height of the surface. They vanish at the tip of a noselike fold in the surface that causes a portion of the surface to appear in the diagram twice.

As an object at the height of one of the surfaces moves farther away from the eye it will appear first as a single image, then as a double image, and finally it will vanish from view. A person walking away from you on a desert would therefore slowly vanish from the

DESERT MIRAGE is another example of a two-image inferior mirage. In spite of the strong impression of water in the distance, nothing is there but dry desert and mountains. The bottom part of the mountains has vanished; upper part appears as an inverted image.

feet up. It would look as if he were wading into the sea, an impression strengthened by the seemingly reflected image of the upper portion of him that was still to be seen. He would ultimately "drown" as his entire body vanished.

UTILITY POLES appear to become submerged in water with increasing distance from the observer. The effect is caused by the two-image inferior mirage and can be understood by examining the image space shown in the top illustration on page 84. With increasing distance the poles seem to vanish from the bottom up. As they do so an inverted image of a portion of the poles gives the impression of being a reflection from a surface of water. Perfectly normal utility poles appearing thus distorted here are in Great Salt Lake Desert.

The distance to the edge of the "water" is the distance to the optical horizon. It is determined by the last ray that touches the surface of the ground, that is, the ray that is tangent to the surface. As the observer moves forward or backward, so does the optical horizon. The phenomenon is familiar to every driver who watches the "water on the road" recede up the road as he approaches it. Over the flat expanse of a desert the "water" surrounds the observer and moves as he moves. With the coming of evening the desert cools, the temperature gradient decreases, the curvature of the light rays decreases and the "waters" recede from the observer.

It is sometimes difficult for the observer to decide whether the scene before him is actually a reflection from a surface of water or whether it is a mirage, particularly when the two-image inferior mirage is seen over a real water surface (as in the photograph on the cover of this issue). If it is a mirage, what is seen depends strongly on the elevation of the observer's eye. When the observer kneels down, all the surfaces terminate much closer to him, so that distant objects vanish, only to reappear when he stands up. A reflection would not behave in this way.

To obtain a three-image mirage the temperature profile must satisfy the conditions for a two-image mirage (the temperature and the temperature gradient must increase together), and in addition the rate at which the temperature gradient increases must decrease. This is roughly equivalent to saying that the curvature of the profile decreases with increasing temperature.

One example is the "elbow" profile that appears next to a sun-baked wall. When the sun shines on a long, uniform wall, the temperature of the wall rises and the heat is transferred both into the building and out to the surrounding air. A temperature gradient builds up in the first few centimeters outward from the wall. Because the change of temperature is horizontal, the image is displaced sideways, yielding a lateral mirage. An object that is along the wall but close to the eye has a single image. If it is far from the eye, it has two images, and if it is in a narrow zone at an intermediate distance, it has three images.

Often on warm spring or summer afternoons a three-image mirage can be seen over an enclosed body of water such as a large lake, a bay or a sound. The images are caused by a temperature profile that has an inflection point. The temperature increases with height, because the warm air from the land has flowed out over the cooler water. As a result of turbulent mixing of the air as it flows over the water, the temperature gradient in the bottom few meters is small, but it increases with height at first and then decreases again. The curvature of the profile therefore decreases with height and indeed vanishes at the point where the temperature gradient is at its maximum: the inflection point.

The apparent shape of the surface of the water, as seen through this strange lens, is that of a large, flattened letter S [*see bottom illustration on page 84*]. An object located where the surface folds will be seen three times. A diagram of the image space reveals the remarkable way an image would change as the object moved farther away from the observer.

A temperature profile with an inflection point will not necessarily give rise to a three-image mirage. If the temperature gradient in the vicinity of the inflection point is just a little gentler, the apparent surface will not fold over but will instead rise and form a wall. An object located at the distance of the apparent wall will be seen greatly magnified. That is the simplest manifestation of the Fata Morgana.

If that were all there was to the Fata Morgana, it would have been incapable of generating the many strange images that have been credited to it, such as the castles and pilasters seen by Angelucci and the mountains of Crocker Land seen by Peary and MacMillan. An understanding of those phenomena requires a further discussion of the basic image-forming properties of the atmospheric lens.

Up to this point we have discussed the atmospheric lens in terms of its ability to produce multiple images of a distant object and to alter the magnification. Nothing has been said about how good the images are in terms of sharpness, although some of the photographs accompanying this article will reveal that the sharpness of the atmospheric lens can vary greatly. Some of the lack of sharpness can be ascribed to shimmering, which is caused by small irregularities in density and temperature that result from turbulence in the air. Also contributing to the lack of sharpness is the astigmatism of the atmospheric lens.

Astigmatism is apparent when the wave front of the light that reaches the eye is not spherical but instead is shaped like a small portion of an ellipsoid [*see upper illustration on page 86*]. A horizontal cross section through the wave front has a curvature different from that of a vertical cross section. One's ability to focus the light on the retina or on the plane of the film in a camera is determined by the curvature of the wave front, so that in this case one can focus to define the image sharply in either the horizontal or the vertical but not in both. In the atmosphere it is only possible to focus on, and thus to sharply define, the horizontal position of the image, because the curvature varies so rapidly in the vertical cross section. A discrete object will appear defocused as a fuzzy vertical line.

The astigmatic wave front gives rise to another curious effect: it can alter the brightness of the image. Consider a point source of light seen in the distance. The vertical defocusing of the image means that the light is spread into portions of the image that would otherwise have been dark. The light energy has been redistributed, leaving the apparent position of the light source darker than it would have been in the absence of astigmatism, whereas the regions above it and below it have become brighter. Because the amount of astigmatism varies greatly in different parts of the mirage, even a uniformly illuminated surface will appear to the eye as being nonuniformly bright.

That effect is the origin of the mirage phenomenon named the Fata Bromosa, or fairy fog. A perfectly flat and uniformly illuminated surface of the sea is deformed in image space to look like a slightly overhanging wall. The blurring effect of the astigmatism eliminates all detail in the "wall," and so the fact that it is an image of the ocean cannot be determined. The redistribution of brightness on the overhanging wall causes the wall to appear much whiter than its surroundings, so that it looks to the eye just like a fogbank out over the water.

The water surface that appears to have an overhanging wall is the result of an inflection-point profile that gives rise to a three-image mirage. If the temperature gradient at the inflection point is decreased a little, the apparent wall no longer overhangs but stands vertical. The astigmatism will still blur detail in the wall, but now it redistributes the brightness so that the wall appears to be darker than the surrounding regions. Only one more element remains to explain Angelucci's observation of the 10,000 pilasters, and that is the presence of gravity waves in the atmosphere.

The temperature profile we have been discussing is one that causes the atmosphere to be stably stratified. If the surfaces of constant temperature (and density) become tipped, they will oscillate back and forth around the horizontal position and give rise to waves. (It is rather like water sloshing back and forth in a swimming pool.) Gravity acts as the restoring force for these waves, which derive their energy from the wind.

For the purposes of this discussion all that matters is that the waves cause the strength of the temperature gradients in our temperature profile to vary slightly in a periodic manner. Looking through one portion of the wave, the apparent shape of the water will be slightly overhanging and thus bright; looking through another portion, the surface will be vertical and thus dark. Each wave will result in the appearance of another whitish gray pilaster. A slight increase in the strength of the mean temperature gradient makes possible the appearance of a white wall with periodic dark windows in it. In fact, slight variations in the shape of the temperature profile and the amplitude of the gravity waves are all that is required to produce any of the details of Angelucci's observations or of Peary's view of "Crocker Land."

Although the simple forms of the mirage only transmit and distort images of objects that actually exist, the Fata Morgana is capable of bringing about such a great transformation that the images seen bear no resemblance to the object that gave rise to them. The Fata Morgana can take a flat, evenly illuminated surface of water, snow or ground and transform it into a wall on which all prior information has been erased. As the brightness on the wall is redistributed, new pictures are created. It is not surprising that some mirages have been credited to fairies.

The Green Flash

10

by D. J. K. O'Connell, S. J.
January 1960

When the sun's disk is disappearing at sunset, or appearing at sunrise, its top sometimes momentarily turns a brilliant green. The flash results from dispersion of sunlight in air

Some clear evening as the sun is sinking below the horizon you may, if you are fortunate, witness one of nature's most unusual and beautiful displays. Just as the last of the solar disk is about to disappear, it may momentarily turn a brilliant green.

The green flash, as the phenomenon is called, is not easy to see from most places. People who have heard of it but looked for it in vain tend to dismiss it as a fantasy. Many who have seen it, including astronomers and physicists, have considered it an optical illusion. But the green flash is in fact purely objective and perfectly real. We shall see that it has a straightforward physical explanation.

Appropriately enough, widespread interest in the strange effect began not with scientific observation but with a work of science fiction. *Le Rayon vert,* a novel by Jules Verne published in 1882, describes a long search for the mysterious green ray of the title. (The word "ray" is used in both French and German to describe the effect, but "flash" is a better description.) It would be interesting to know what drew Verne's attention to the matter, because one can find hardly any mention of it before that time.

Since then many observers have seen the green flash and speculated about its cause. Most commonly they have described a thin green band, visible for a fraction of a second at or just above the top edge of the sun as it sinks out of sight. But the flash can take other forms. Sometimes it is blue, or turns from green to blue; sometimes it is even violet. On rare occasions it appears while the whole disk of the sun is still above the horizon, and then there may be a red flash at the bottom of the disk as well as a green or blue one at the top. The flash is about as common at sunrise as at sunset. Here too it may be blue or violet as well as green. If it changes color, it does so in the reverse order, passing from blue to green.

Normally the flash is extremely narrow. From top to bottom it covers only about 10 seconds of arc, which gives it the same apparent width as a one-inch ribbon at a distance of 570 yards.

To see the flash with the naked eye requires a sharp horizon and a sky free from haze—conditions most likely to be found in deserts or on mountains, or over water. The clear desert air of Egypt affords an exceptionally favorable setting, and the flash appears very frequently there. In fact, there is evidence that the ancient Egyptians were familiar with it. A stone pillar dating from about 2500 B.C. shows the rising (or setting) sun as a semicircle colored blue above and green below. Furthermore the people of the time seem to have believed that the sun is green during its nocturnal passage beneath the earth, an idea they might have got from seeing the green flash at sunset and sunrise.

Some modern attempts to account for the phenomenon are scarcely less naive. According to one theory the horizontal rays of the sun become green by passing through water waves. The fact that the flash can be seen over land disposes of that fancy. A medical man has suggested that biliousness of the observer is responsible!

A more plausible physiological explanation, which has had and still has many supporters, attributes the effect to retinal fatigue. It is a well-known fact that, after looking into a bright-colored light, our eyes become fatigued. Upon looking away we now see the complementary color. So, the theory says, in looking at the red sunset our eyes are dazzled and we see, immediately afterward, the complement of red, which is green. In other words, the green flash is all in the eye, and is purely subjective.

Although a number of distinguished workers have espoused this theory, it is patently wrong. The flash can also be seen at sunrise, *before* the sun's disk appears in the sky.

As long ago as 1899, in a letter to *Nature,* Lord Kelvin described a sunrise blue flash. He had seen the green flash at sunset some six years previously and wanted to see what happened at sunrise. A trip to the Alps, he wrote, "promised me an opportunity for which I had been waiting five or six years, to see the earliest instantaneous light through very clear air, and find whether it was perceptibly blue. I therefore resolved to watch an hour till sunrise, and was amply rewarded by all the splendours I saw. . . . In an instant I saw a blue light against the sky on the southern profile of Mont Blanc; which, in less than the one twentieth of a second became dazzlingly white."

One might think that this statement, from one of the greatest physicists of the time, published in a widely read journal, would have disposed of the subjective theory once and for all. But it still keeps cropping up. Paradoxically the advent of color photography has given it a new lease of life. Attempts to photograph the flash on color film with an ordinary camera are always futile, even when it is quite apparent to the eye. This has led unsuccessful photographers to conclude that the color they saw was merely subjective. The real trouble is simply that the focal length of their camera lens is too short to form a visible image of the

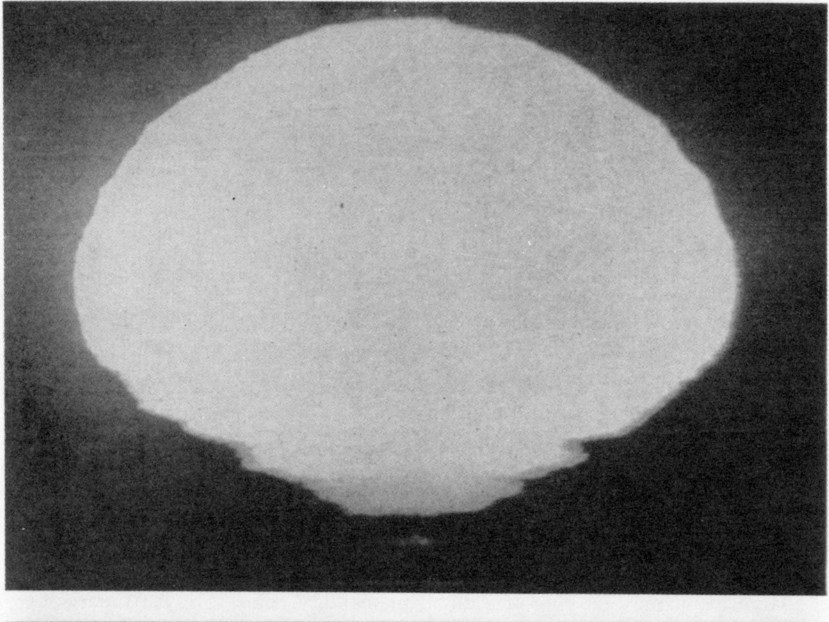

DISTORTION in the sun's disk is caused by discontinuities in the atmosphere. In top photograph the color print shows a green rim at the top of the disk and a red rim at the bottom, together with a red flash below. At bottom the green flash remains after sun has set.

narrow band of color. For example, in a 35-millimeter camera with a lens five centimeters in focal length the image of a normal green flash would be only .005 millimeter wide, much too narrow to be recorded on the film.

For years I had looked in vain for the green flash—at sea, on mountains, in many parts of the world. Then, a few years ago, I saw it very clearly at sunset over the Mediterranean from the Vatican Observatory at Castel Gandolfo. (The Observatory, at an altitude of 1,500 feet, provides an unobstructed view across the Roman Campagna to the sea horizon, 50 miles away.) I soon began to wonder whether it might be possible to photograph the flash on color film with one of our telescopes. With a focal length of several feet, they should magnify the image sufficiently to make it visible. At any rate it seemed worthwhile to try. If we succeeded, we would be able to administer the quietus to the subjective theory of the flash.

In 1954, after a long series of experiments, C. Treusch, S.J., instrument-maker and photographer of the Observatory, took our first color photographs of the green flash. Developing a successful technique was no easy task. For any such fleeting phenomenon it is very difficult to find the exposure that will give a faithful reproduction on film of what is seen. Moreover, the intensity of the light changes very rapidly during the last moments of sunset, and the exposure must be continually adjusted to compensate. Many trials were required before we learned how to do it. A further complication is that, while the sun is still above the horizon, the light from the disk is much brighter than that from a green flash. Thus if the exposure is right for the flash, the rest of the sun will be overexposed. With the help of the Kodak Research Laboratories we made extensive studies of the effects of over- and under-exposure on the colors in film.

All these difficulties were compounded in the beginning by the necessity of sending the Kodachrome film we were using to France or to the U. S. for development. This meant waiting several weeks to see the results of each test run. As soon as Kodak Ektachrome film became available, we began to use it almost exclusively. It can be developed right at the Observatory.

Our earlier photographs were taken with a 24-inch Zeiss reflector, the focal length of which is 95 inches. We used a reflector rather than a refractor

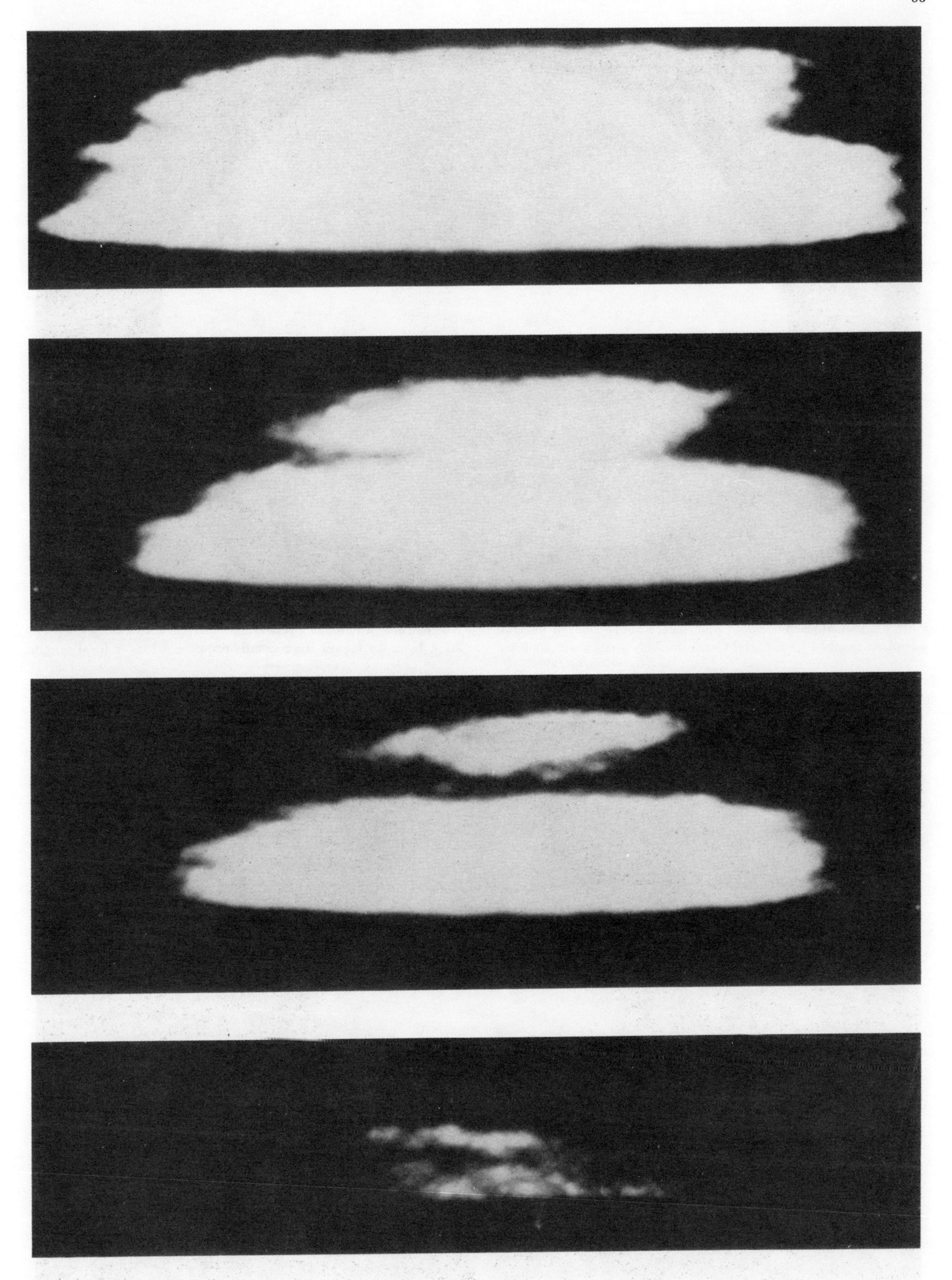

DEVELOPMENT OF A GREEN FLASH is seen in these black-and-white reproductions of color photographs. At top the sun has almost set, and the upper rim of the segment is green. In the next two pictures a green flash is splitting off from the solar disk. At bottom sun has disappeared and a green flash remains suspended above horizon. Time between top and bottom views was 8.4 seconds.

RED FLASH appearing about one degree above the sea horizon was photographed from Castel Gandolfo. This picture was made on Kodachrome film using a reflecting telescope with a focal length of eight feet. To record finer details requires a longer focal length.

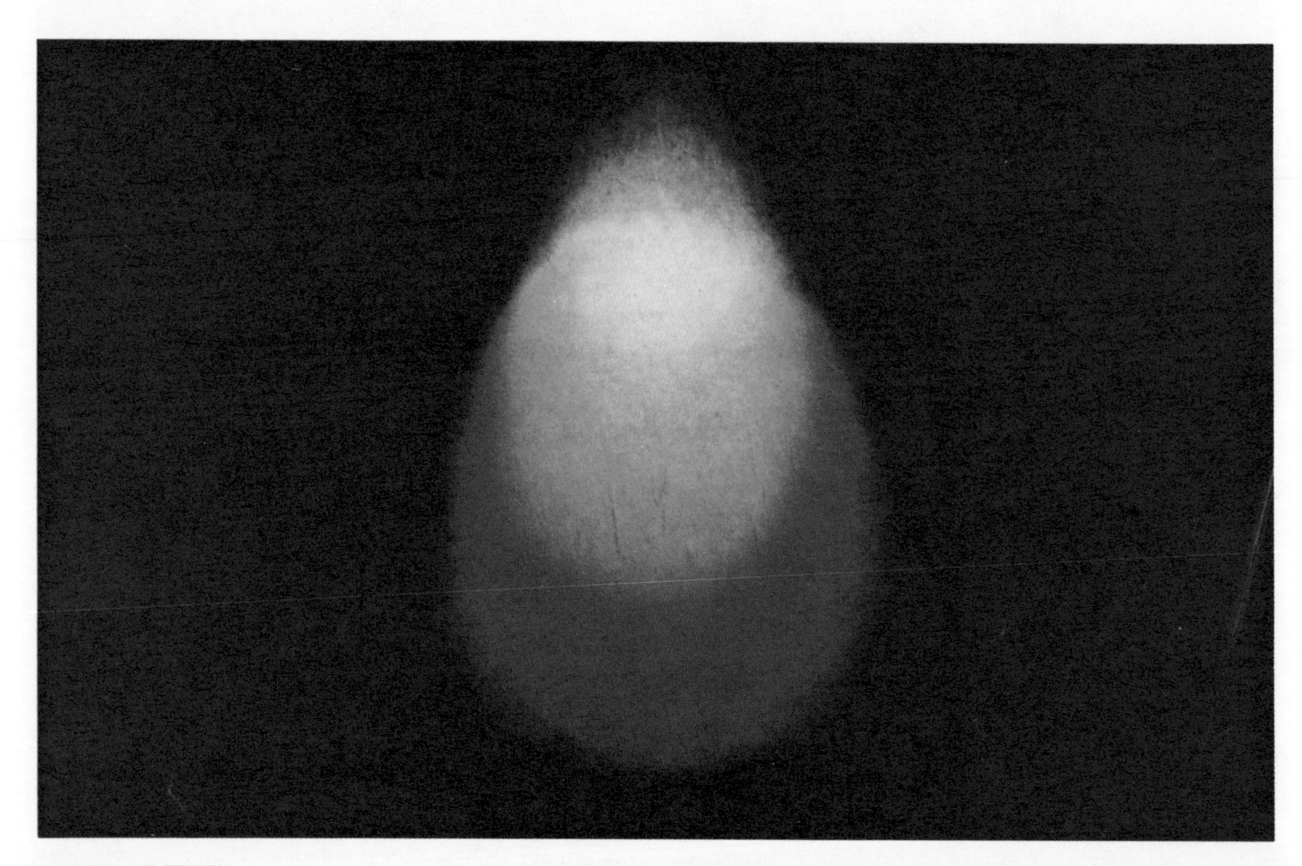

SETTING OF VENUS was photographed on Ektachrome daylight film with a refracting telescope 20 feet in focal length. The planet appears as a series of separate but overlapping images ranging from blue-green at the top to red at the bottom. The blue image as seen visually was at least twice as elongated as the image recorded on the film. The colors are the result of atmospheric dispersion.

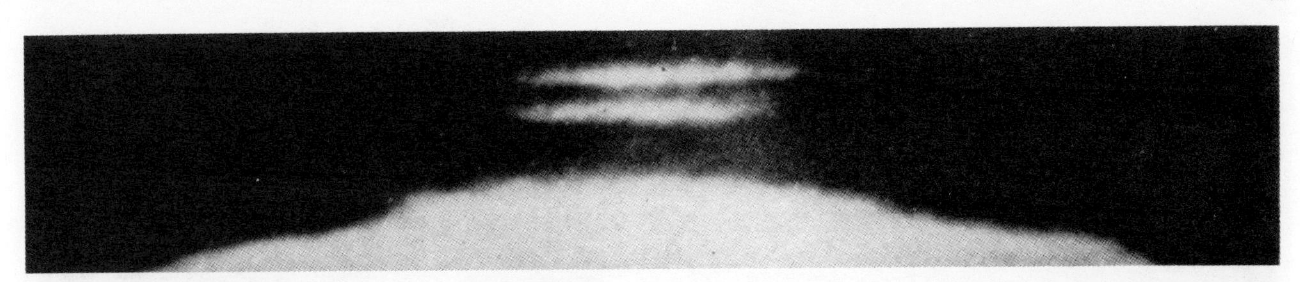

MULTIPLE FLASHES are sometimes seen. Those pictured here were a brilliant blue-green in color and floated for about two seconds above the last tip of the setting sun. There was also a third strip below the two shown here, but it was too faint to be recorded.

to ensure that the instrument introduced no spurious color effects, as a lens might have done. Later on we tried our Zeiss 16-inch refractor (focal length 20 feet). Its lens is an exceptionally good one, very well corrected in the visual region. It turned out to give no spurious color effects and its pictures are in no wise inferior to those obtained with the reflector. The longer focal length makes it possible to photograph very fine details that are lost with the shorter focus. Most of the later photographs were taken with this telescope.

Thanks to Treusch's skill and patience we now have pictures showing the diverse forms the green flash can take, the ways in which it develops, the influence of various atmospheric conditions and the strange forms that the setting sun can assume as it is seen through different layers of the atmosphere. There are also photographs of the red flash which sometimes appears below the sun, as well as of various phenomena seen at the setting of the moon and of Venus.

The green flash is even more difficult to photograph at sunrise than at sunset. One has to judge the exact point of the horizon where the sun will rise and the precise moment when a flash may appear, as well as the correct exposure. At Castel Gandolfo conditions are particularly bad because the eastern horizon, running along the crest of the Alban Hills, is mostly covered with forest. Only at one point is the horizon clear of trees, but through this gap we have made some color photographs of the green flash at sunrise.

A much wider selection of the Castel Gandolfo photographs than can be included in this article are reproduced in a recent book by the writer: *The Green Flash and Other Low Sun Phenomena* [see "Bibliography"]. The ones shown here and on the cover, however, are quite enough to demonstrate the reality

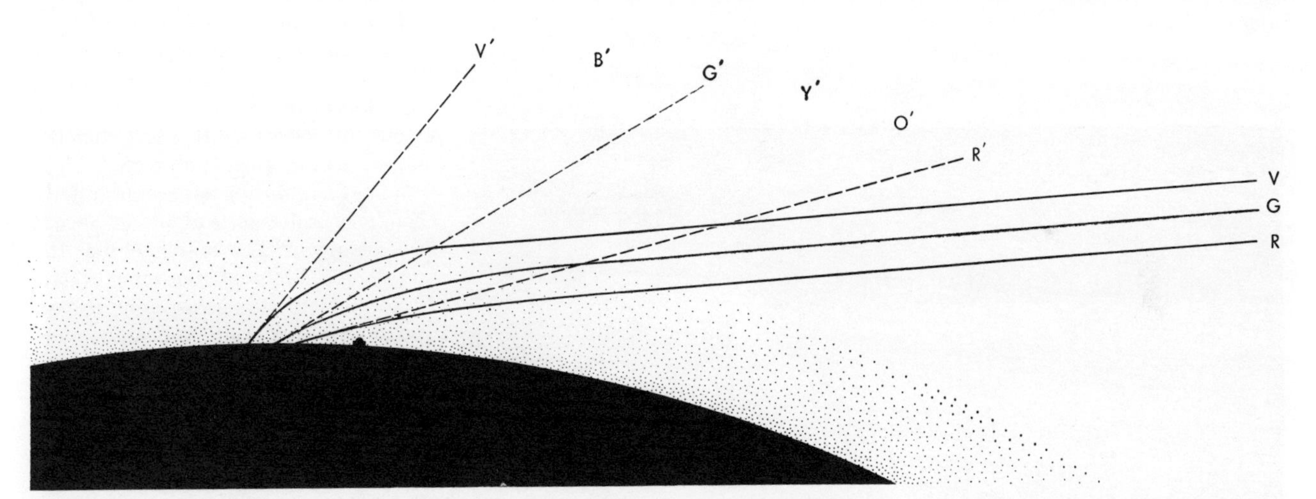

CAUSES OF THE GREEN FLASH are dispersion, scattering and absorption of sunlight by the earth's atmosphere (*black-dotted area*). Solid lines show the paths of three spectral colors, violet (V), green (G) and red (R), in the light coming in at a small angle to the earth's surface from a low-lying sun. The broken lines show the directions from which the rays appear to be coming toward observers on the ground as a result of bending of their paths in the air. They have been dispersed into a spectrum. Between the three rays shown in the drawing fall the other colors of the spectrum, blue (B), yellow (Y) and orange (O), in normal order.

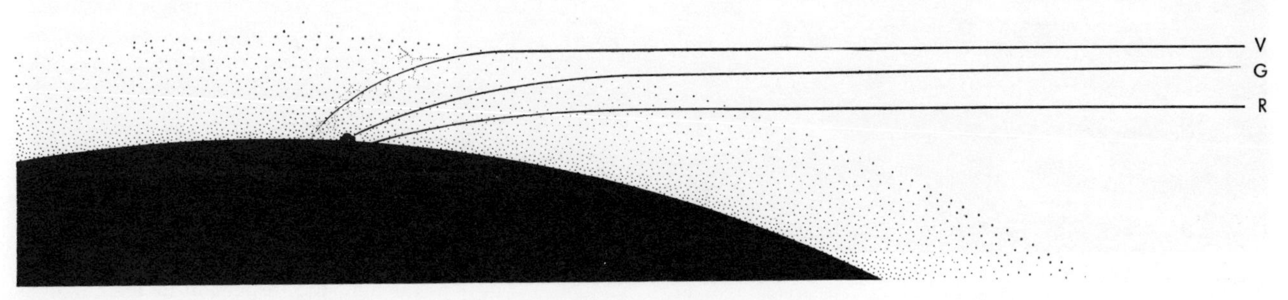

GREEN FLASH APPEARS ABRUPTLY when the lower, red rays have sunk below the observer's horizon. Orange and yellow are largely absorbed by the atmosphere. Blue and violet rays are scattered by particles in the air, leaving green as the color seen.

RISING SUN is seen from Castel Gandolfo behind castle at Rocca Priora, 6.5 miles away. The green flash appears at sunrise, but is harder to photograph at that time.

CASTEL GANDOLFO, the site of the Vatican Observatory, is seen from the west, looking across Lake Albano toward the eastern horizon. Two observatory domes are at left center.

of the green flash. What does produce the phenomenon? The answer can be found in the laws of physics.

When light from the sun enters the earth's atmosphere it is slowed down, and therefore bent or refracted, just as when it passes through a glass prism. The amount of refraction depends on the wavelength of the light, the shorter waves being bent more than the longer ones. Thus the white sunlight is spread out or dispersed into a spectrum, with the longest (red) wavelengths at one end and the shortest (violet) at the other. The lower the sun, the greater the thickness of air through which its light must pass before reaching the eye of an observer. Hence the dispersion is greatest at sunset and sunrise.

Because the short waves are bent more sharply, they strike the eye at a steeper angle and appear to be coming from a point higher in the sky than the longer ones [*see middle illustration on preceding page*]. Thus the spectrum extends from violet at the top to red at the bottom. As long as a fair portion of the sun's disk is visible, light rays from its various parts overlap, and one cannot see the spectrum. (Sometimes a green or blue rim does appear at the top of the disk and a red rim at the bottom.) When the sun sinks below the horizon, the colors of its spectrum disappear one by one, the lowest red rays first, then the orange, yellow, green and so on.

Why, then, do we not see an orderly change of color instead of an abrupt flash of green? The reason is that the atmosphere filters out the other colors. In addition to dispersing light the air also absorbs and scatters it. Absorption, due mainly to water vapor, oxygen and ozone, affects chiefly the orange and yellow light. Scattering is stronger for short wavelengths. (Preferential scattering of blue light accounts for the color of the sky.) Thus when the red rays have sunk below the horizon, the orange and yellow are attenuated by the thick layer of air through which they are traveling toward our eyes. The blue and violet light is largely scattered away. The color least affected is green, and this is what reaches us. At high altitudes, and when the air is very clear, the shorter waves may still come through, and the flash can be blue or violet.

The same processes that operate on sunlight affect the light from stars or planets. Since the images of these objects are so much smaller than that of the sun, the dispersion of their light is apparent when they are near the horizon. Because of absorption and scattering, the spectra are not continuous. Instead we see a set of overlapping images in different colors. The brighter the object, the clearer the effect; the setting of Venus provides a vivid example [*see bottom photograph on page 94*].

While dispersion, selective absorption and scattering are chiefly responsible for the green flash, other factors favor or hinder its appearance. Dust or haze in the air absorbs light and makes it less likely that a flash will be seen. On the other hand, abrupt variations in the density and temperature of different layers of the atmosphere may act to increase the intensity and duration of the effect. The English astronomer J. Evershed, while traveling by sea from Australia to Java, saw a brilliant green flash at sunset every evening and noted that on each occasion there was also a mirage. This was caused by a thin inversion layer—a region where the temperature of the air increases with height—near the surface of the ocean. When the sun was near the horizon, two images of it could be seen—the direct one and a mirage reflected upward from the inversion layer. As the sun sank further, the reflected image moved up to meet the direct one, and the green flash appeared when they coalesced. Evershed observed a similar effect at the setting of Venus.

The duration of the flash depends on the rate at which the sun is sinking below the horizon, and at any one place this rate varies with the time of year. Moreover, the sun sinks more slowly, and the green flash lasts longer, as one moves from the Equator toward the poles. For example, at Hammerfest in Norway (latitude 79 degrees north) the flash at midsummer may last as long as 14 minutes, seven minutes during the slow sunset and another seven at sunrise, which follows immediately. The longest display on record was reported by Admiral Byrd's expedition to Little America (78 degrees south) in 1929. The sun grazed the irregular horizon of the barrier ice and the green flash was seen, on and off, for about 35 minutes.

Refraction, in addition to producing the flash, extends its duration. The downward bending of light rays makes the sun appear higher than it really is, and keeps it visible for an appreciable time after it has actually sunk below the horizon. At Castel Gandolfo, under normal conditions, refraction raises the sun about 1.3 degrees, or about two and a half times its diameter. Depending on atmospheric conditions the effect may sometimes be much greater, delaying sunset still more and lengthening the life of the flash proportionately. Some remarkable instances of abnormally great refraction have been recorded. In 1597 some Dutch sailors in the Arctic Circle saw the sun for 14 days when it should have been entirely below the horizon. It had been raised no less than four degrees by abnormal refraction.

With so many factors involved it is small wonder that the appearance of the green flash is quite capricious. As has been mentioned, the chances of seeing it, at least with the naked eye, are much better in high mountains, tropical seas or deserts. I should add that it is much easier to see if you already have a good idea of what to look for. People who have often tried and failed to find the green flash have been able to see it without difficulty after studying our color photographs.

If you decide to make a serious attempt to look for the green flash, you will find the search much facilitated by a modest optical aid such as low-power binoculars or a small telescope. It is not light-gathering power that is needed, but simply magnification of the extremely narrow band of color. Of course you must be very careful to avoid damaging your eyes. Binoculars should be turned on the sun only at the last moment, when the disk has almost disappeared. With a telescope you might use a neutral filter or a solar eyepiece to reduce the intensity without altering the color of the light.

Even if you succeed in seeing the green flash, you will not be able to photograph it in color with an ordinary camera of short focal length. At the Vatican Observatory we have obtained good pictures using a focal length of eight feet, but the finest details can be recorded only with the refractor of 20-foot focal length. Amateurs should not be discouraged, however; there are many interesting sunset phenomena that can be photographed with much more modest equipment. It is not so difficult, for instance, to record the changing, and often very curious, shape of the sun's disk as it nears the horizon.

Beyond its interest as an odd and often beautiful sight the green flash may help increase our understanding of the earth's environment. As we have seen, it is a wholly atmospheric phenomenon, and its appearance is influenced not only by the lower layers of air, but also by conditions at very high levels. It is possible that studies of the green flash and related effects will augment the information about the upper atmosphere now being gathered by rockets and artificial satellites.

11 The Mechanism Of Lightning

by Leonard B. Loeb
February 1949

The dramatic phenomenon of nature is closely studied in the laboratory and in the field. It is found to be an intricate series of physical events

LIKE many other commonplace natural phenomena, lightning has been uncommonly difficult to explain. Although it is two centuries since Benjamin Franklin made a start by showing that lightning was an electrical discharge, not until the present decade have we begun to understand the extraordinarily intricate mechanisms involved in the origin and development of a lightning flash. Recent researches, however, appear to have removed much of the mystery from this awesome spectacle.

Let us begin by considering the atmospheric situation that leads to a lightning storm. Whenever weather conditions produce rapid updrafts of warm, moisture-laden air that rise well above the freezing level in the atmosphere, the region involved becomes a huge generator of static electricity. The water droplets of this thundercloud, in which the updrafts may reach a velocity of 160 miles an hour or more, become electrified. Just how the charges are generated is still a matter of conjecture. It may be that the wind currents tear at the surface of the droplets, producing a fine, negatively-charged spray and leaving the larger droplets positively charged. Another possibility is that electrical fields already existing in the clouds induce charges on falling droplets. Still another is that ice crystals at the upper levels are electrified by friction or by some process that accompanies freezing.

In any case, large masses of charged water droplets and ice crystals become segregated in positively- and negatively-charged groups and collect at different localities within the thunderhead (*see diagram on opposite page*). Between these huge groups of opposite charge very high potential differences and electrical fields develop. It is these highly charged regions of the cloud that account for the split-second electrical discharges called lightning.

As the charged, wind-driven thunderhead approaches a given ground area, electrical fields gradually build up between the earth and the cloud. Near the earth the fields rarely exceed 2,700 volts per centimeter of their length, but even at such a field a vertical conducting rod from the earth only 10 feet long (about 305 centimeters) would have a potential difference of more than 800,000 volts ($305 \times 2{,}700$) with the uncharged surrounding atmosphere. Such a field would make the hair of a person seated on the ground literally stand on end. It accounts for the corona discharge, popularly called St. Elmo's Fire, which is sometimes seen issuing from a church steeple or from the wingtips of an airplane during a storm.

In a cloud that produces lightning, the area of the charged region generally has linear dimensions of some 1,000 feet. At nearby points the charged region develops fields of 30,000 volts per centimeter, so the field of the cloud is vastly greater than the one at the ground. Thus lightning discharges usually originate at the cloud and work downward. On the other hand, a very tall grounded conductor such as the Empire State Building may develop a potential high enough to initiate a lightning discharge from the earth. Be this as it may, however, the distribution of charges in a thundercloud is such that there are many more discharges within the cloud than between the cloud and the ground. These discharges, largely concealed by the cloud, account for the so-called sheet lightning sometimes seen in distant thunderclouds on dark nights.

Lightning strokes vary in length from 500 feet to two miles or more. Calculations based on the length of these paths and on the fields at the earth indicate that the electrical potential between a thundercloud and the earth is of the order of hundreds of millions to billions of volts. If the space between the cloud and the earth were a vacuum, these huge potentials would accelerate electrons and ions to a speed sufficient to smash the nuclei of atoms. In the air, the accelerated particles cannot attain any such energies, for they are slowed by countless impacts with the air molecules. Nevertheless the power of a lightning stroke remains impressive. The brief currents it generates will vaporize No. 12 copper wire; the magnetic fields produced by a relatively short stroke down a copper tube one centimeter in diameter with walls one and a half millimeters thick will collapse the tube. A lightning stroke which travels down the moist interior of a tree in 30 feet liberates enough heat and steam literally to blow the trunk open.

THE visible flash of lightning is produced by the heating of air molecules in the path of the stroke, which may reach a temperature of 30,000 degrees Centigrade. Camera studies show that the channel of the stroke remains luminous for some 100 millionths of a second after the stroke itself. Owing to the enormous power of the stroke, the channel expands explosively, and this accounts for thunder, which comes from the shock waves produced by the channel's expansion. The explosion of segments of the channel near an observer is heard as a sharp crack; rumbling thunder comes from more distant segments, from repeated strokes and from echoes. These rumbles, it may be noted, cannot be heard as far as those from major gunfire, which indicates that lightning discharges, impressive as they are, do not yield as much power as ordinary man-made explosives, to say nothing of an atomic bomb. The noise of big guns can be heard for a distance of some 15 miles; the audible limit for thunder is only about seven miles.

Analysis of the lightning stroke shows that it is very similar to the long sparks that jump the gap between two widely separated condensers of high potential. Like these sparks, a lightning stroke follows a crooked path and develops branches or forks that advance in the direction of the stroke, so one can always deduce from its branches the direction in which a stroke is traveling. Unlike condenser sparks, lightning does not oscillate back and forth. Damping due to the high resistance of the electrical feeders from which the stroke originates permits it to discharge in only one direction. A lightning stroke may come from a positive or negative cloud, but most strokes, except in mountain storms, are from negative clouds.

The speed of lightning is no idle metaphor: a lightning stroke travels at a velocity of approximately one billion centimeters per second. It lasts no longer than five to 500 microseconds (millionths of a second), the median being some 30 microseconds. The quantities of electricity involved, however, are huge: a single discharge may transfer 200 coulombs of electricity (a coulomb being the quantity of electricity transferred by one ampere of current in one second); in terms of current a stroke may carry as much as 500,000 amperes. A stroke of 200 coulombs and one billion volts which lasts 200 microseconds produces a thousand billion kilowatts of power.

The spark channel down which this huge packet of energy travels at first is tiny—an inch or less in diameter—but the power of the completed stroke expands the channel at the explosive rate of 50,000 inches per second. Thus it is difficult to define the diameter of a lightning stroke, either as it appears to the eye or in a photograph. The lightning channel loses much of its luminosity after it expands to a diameter of a foot or more, so that it appears reasonable to estimate that the visible lightning stroke ranges from an inch to a foot in width.

THESE observations serve to describe the phenomenon, but they do not explain the mechanics of the lightning stroke itself. The basis for an explanation of that fundamental question was laid by several independent studies made just before the recent war. One group of studies, conducted by the writer and J. M. Meek in the U. S. and independently by H. Raether in Germany, analyzed the mechanism of ordinary electrical sparks in air. Other basic information was provided by photographic analyses of the progress of electrical strokes on a microsecond time scale. These were made in South Africa by B. J. F. Schonland and his associates, who used a camera with a rapidly revolving lens to photograph actual lightning strokes, and in England by T. E. Allibone and J. M.

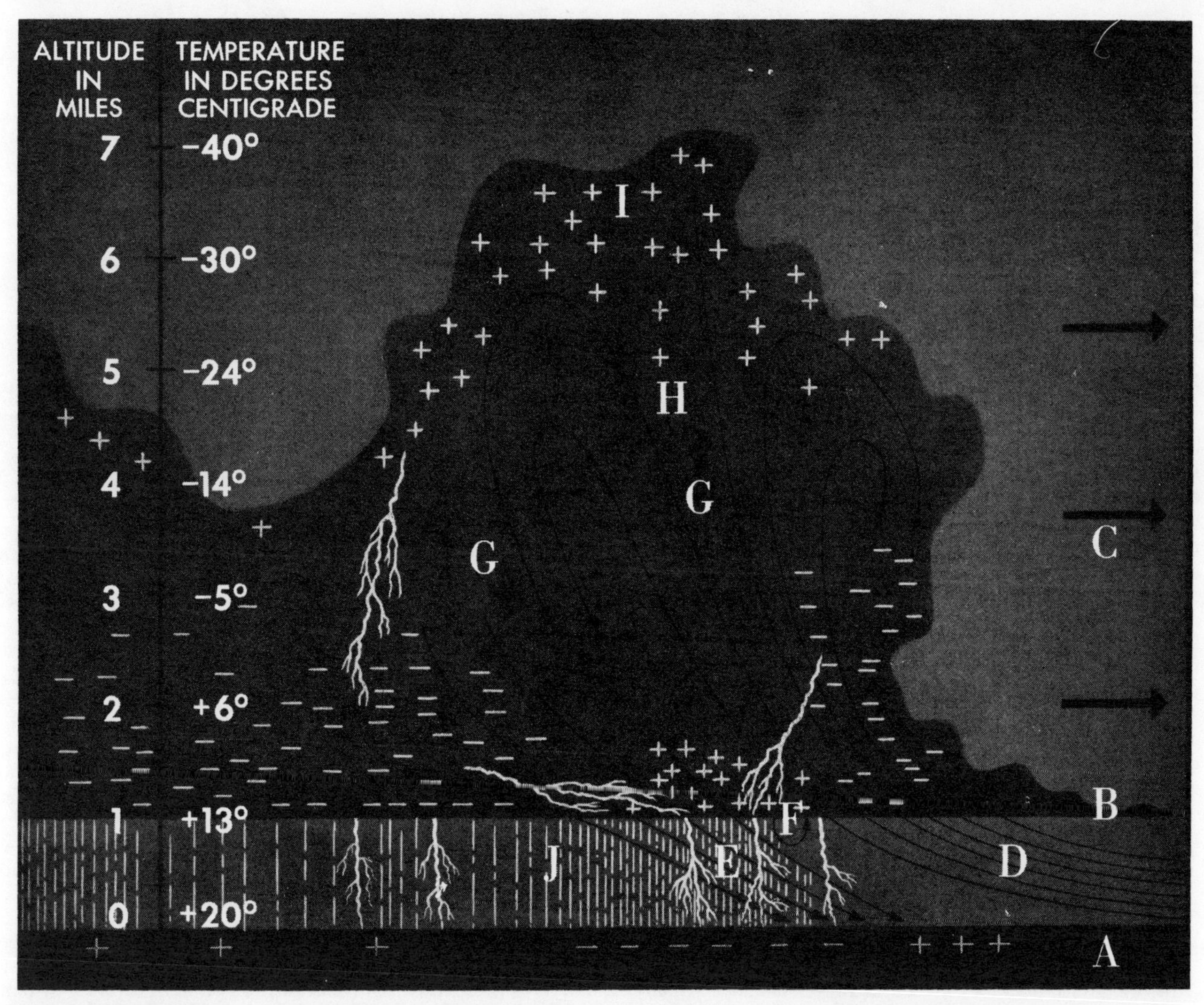

THUNDERCLOUD is a mighty generator of static electricity. The lightning flashes in this drawing are massive discharges between regions of differing potential. Some flashes are within the cloud. Others play between the cloud and the ground. The latter is indicated by A. B is the base of the cloud. C is the wind that drives the cloud. D is the ascending, moisture-laden air current. E is the descending air current. F is the roll or scud cloud. G indicates up and down drafts. H is the region where hail is generated. I is the highest region of ice formation. J is rain falling from the cloud. A scale of height in miles and temperature in degrees Centigrade is at the left.

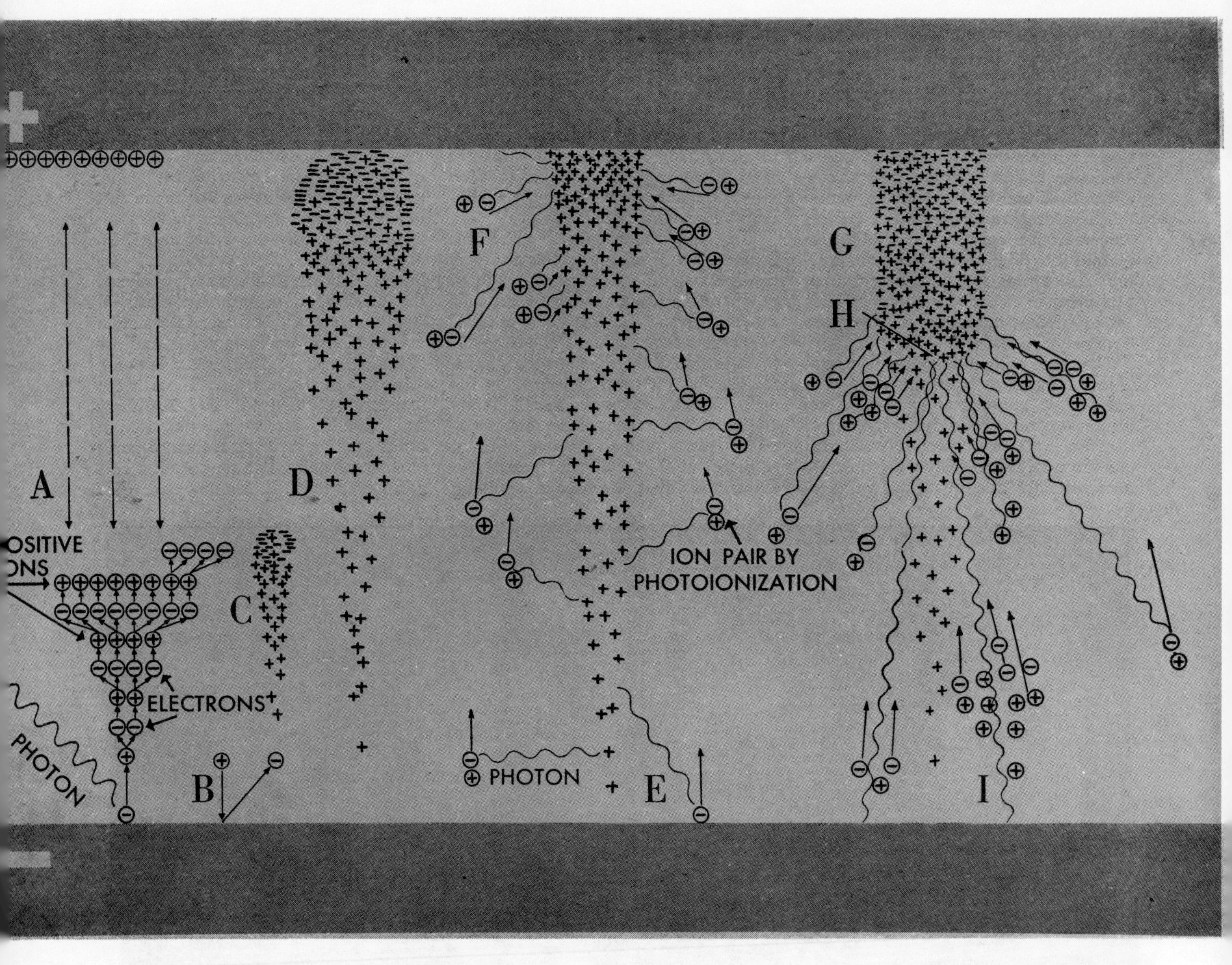

SEQUENCE OF EVENTS in a flash of lightning is outlined by analyzing a discharge between two parallel plates. The upper plate is positively charged; the lower plate, negatively. The field between them (A) has a strength of 30,000 volts per centimeter. At the lower left a random photon knocks a single electron from an atom. Moving towards the positive plate, the electron collides with other atoms to liberate an avalanche of electrons. In the wake of the electrons remain positively-charged atoms, or ions (C and D). These ions reinforce the charge of the positive plate, thus attracting new electrons (F) that have been liberated by radiation (E)

Meek, who made similar photographs of long sparks with a revolving film camera.

The investigation of ordinary sparks showed that the path they follow in the air is created by a so-called "streamer" mechanism. This process begins when air at atmospheric pressure is placed in a field of about 30,000 volts per centimeter, or as little as 10,000 volts if water droplets are present. A single electron starts the process, and there are always stray electrons, liberated by cosmic rays or radioactivity in the air, on hand to start it. The electron, advancing from the negative end of the field towards the positive end, gains energy from the field despite its billions of collisions per second with gas molecules. When it gains enough energy, it begins to knock electrons out of some of the molecules it strikes. These in turn repeat the process, so that the liberated electrons soon become an avalanche. After a run of one centimeter in a field of 30,000 volts per centimeter, the single initial electron produces an avalanche of 10 million free electrons. Raether has photographed such avalanches in a Wilson cloud chamber by condensing water drops on the ions left in the avalanche tracks.

In their wake the freed electrons of course leave large numbers of positively-charged ions, for each molecule from which an electron escapes is ionized. The ions create a positive electrical charge throughout the space in the path of the electrons. When the ions left by the electron avalanche are deserted by the electrons at the positive end of the field, they add to its positive potential. The augmented field soon becomes strong enough to draw photoelectrons, created as an accompaniment to the avalanche process, from the negative side of the field. The photoelectrons, feeding into the ion space-charge left by the initial avalanche, produce more ionization and enlarge the space-charge so that it expands backward towards the negative end of the field. The situation can be pictured by imagining a magnet that by its field draws iron filings from a distant pile in such a way that the filings, building on to one another from the magnet backward, form a path back to the pile.

THE entire chain of events, illustrated in the series of diagrams above, takes place in a matter of microseconds; the avalanche of electrons, for example, ad-

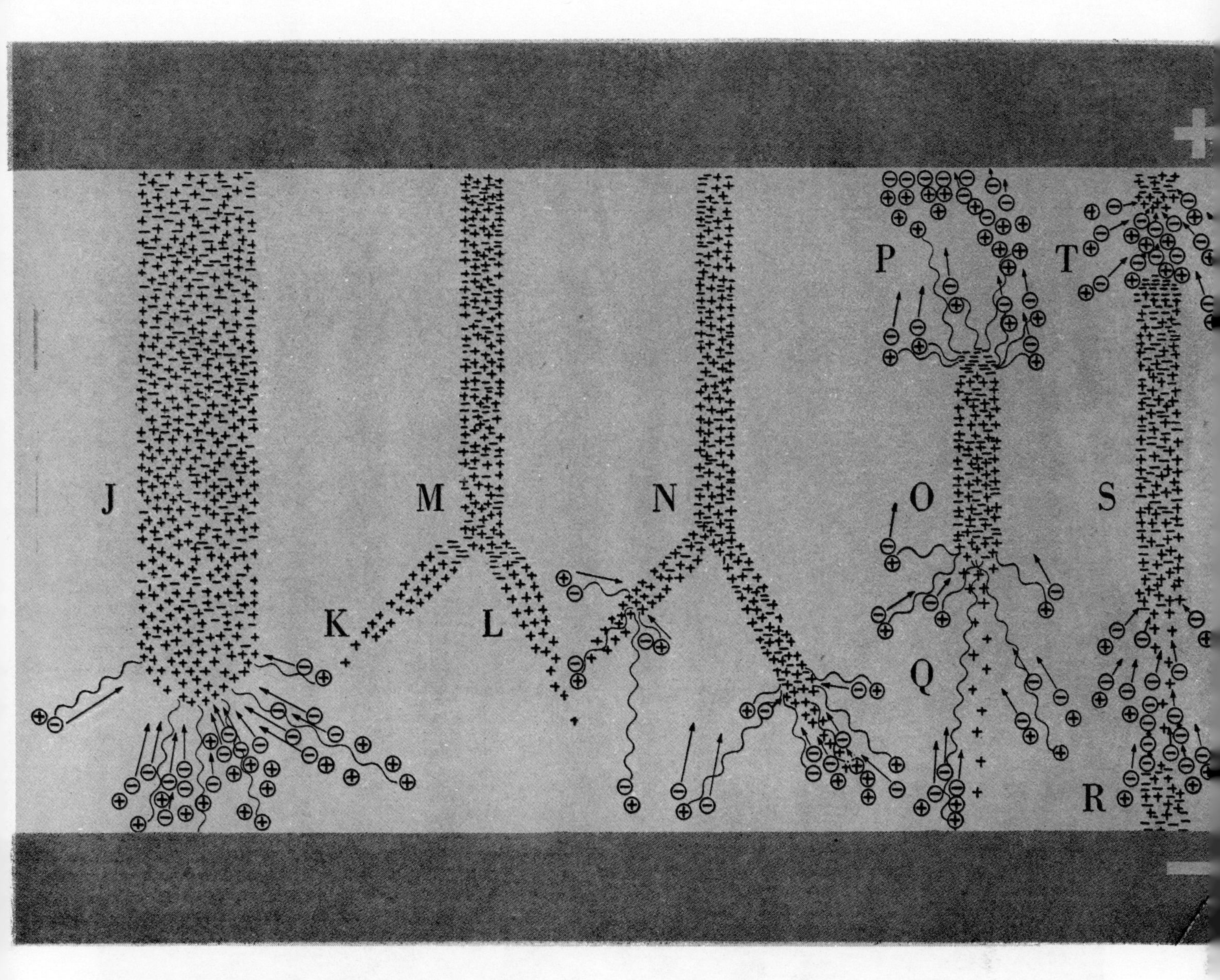

during the previous events. The electrons, in turn, ionize more atoms so that a heavily ionized region (G) begins to extend towards the negative plate. This process continues until there is a bridge of ions (J), called a streamer, between the two plates. It is this streamer that provides the channel for a spark or for lightning. The next drawing illustrates the process by which streamers form branches. A streamer (M) attracts two avalanches (K and L). The avalanches are then reversed to form a branched streamer. The remaining example shows how a streamer may begin before an avalanche has reached positive plate. Streamer then works towards both plates.

vances at the rate of 20 million centimeters per second. The ionized path that is formed as the end result of the process is called a streamer, and it is this streamer that provides a channel for the spark or actual lightning stroke.

In a thunderstorm, such a streamer bridges the gap between the charged cloud and the ground. The bridge is a conducting filament of ions with electrons streaming over it, and it acts as a kind of tear in the electrical field, accentuating the electrical stress at its ends. The instant the bridge is completed, it releases a cataclysmic burst of electrons from the negative terminal—in this case the ground. The burst of electrons sends a potential wave up the streamer channel to the cloud. This literally tears electrons out of most of the molecules in the centimeter-wide channel. The cloud's charge and energy are then drained away down the conducting channel for some 10 microseconds, making the channel brilliantly luminous. The speed of the potential wave, called the return stroke, is from one to ten billion centimeters per second—one thirtieth to one third of the speed of light. This brilliant flash constitutes the phenomenon we call lightning.

Once the spark channel has been established, there may be repeated strokes from the cloud down the same channel. The discharge of the section of the cloud from which the stroke comes changes its potential with respect to other sections of the cloud, and strokes within the cloud then recharge this section, causing new strokes to the ground. As many as 40 successive strokes down a single channel have been observed—the legend that lightning never strikes twice in the same place is wrong in more ways than one. The repeated strokes follow one another very rapidly, at intervals ranging from tenths to hundred-thousandths of a second.

The process that has been described is that for a stroke from a positive cloud to negative ground. For a stroke from a negative cloud the mechanism is similar, except that the streamer is built up by steps.

The foregoing description represents the fundamentals of the process, but in an actual lightning storm the sequence is a bit more complex because of the length of the strokes. When Schonland photographed the lightning discharge with his

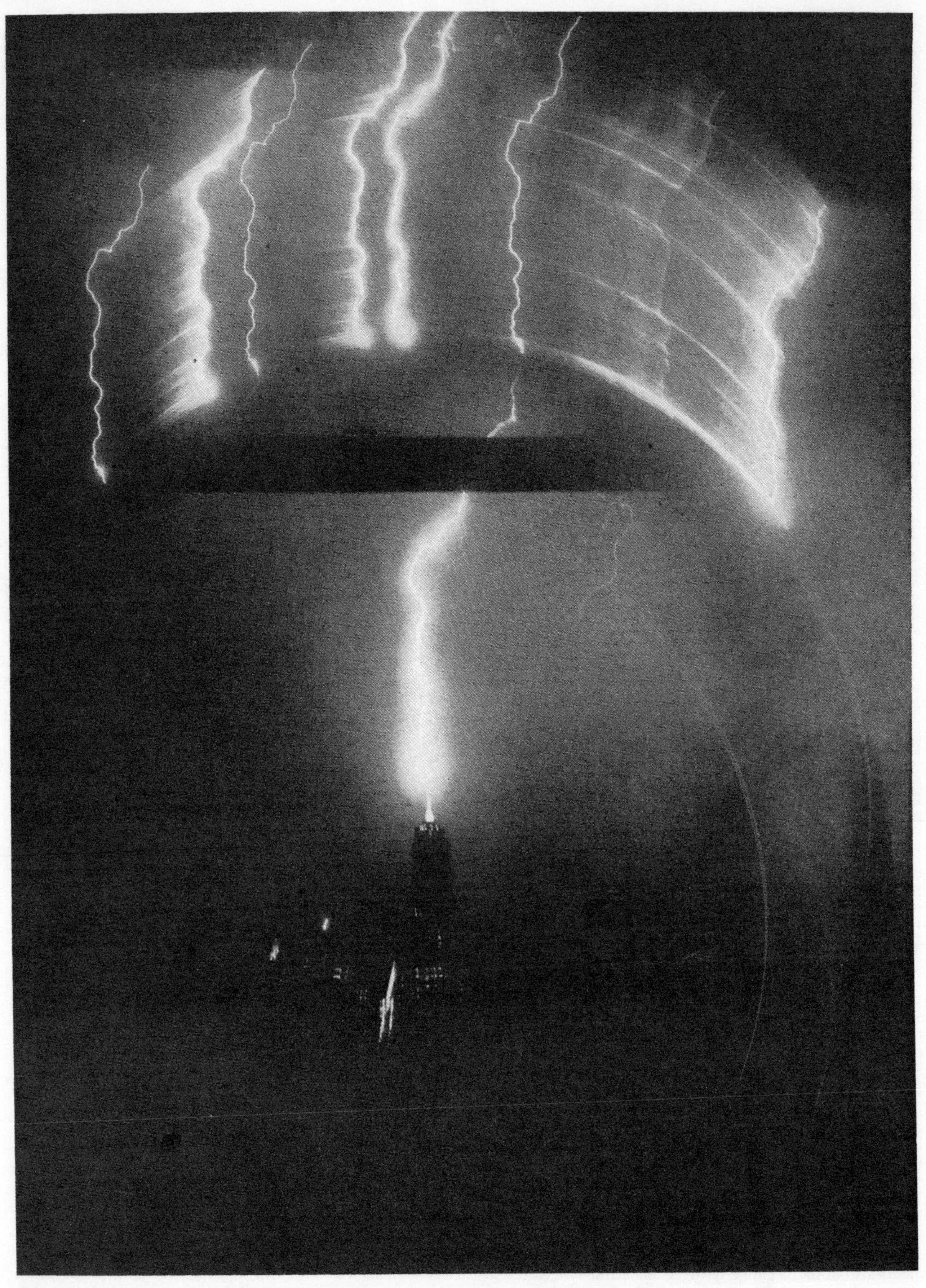

COMPLEX STRUCTURE of lightning is illustrated by a photograph of a single bolt striking the Empire State Building. This photograph was made by researchers of the General Electric Company with the revolving-lens Boys camera. At the top the stroke is dissected into one long discharge (*at right*) and six subsequent ones.

AVALANCHE of electrons, depicted in drawings on pages 100 and 101, is photographed in a cloud chamber.

ADVANCE of avalanche within the chamber is apparent in a photograph made tiny fraction of a second later.

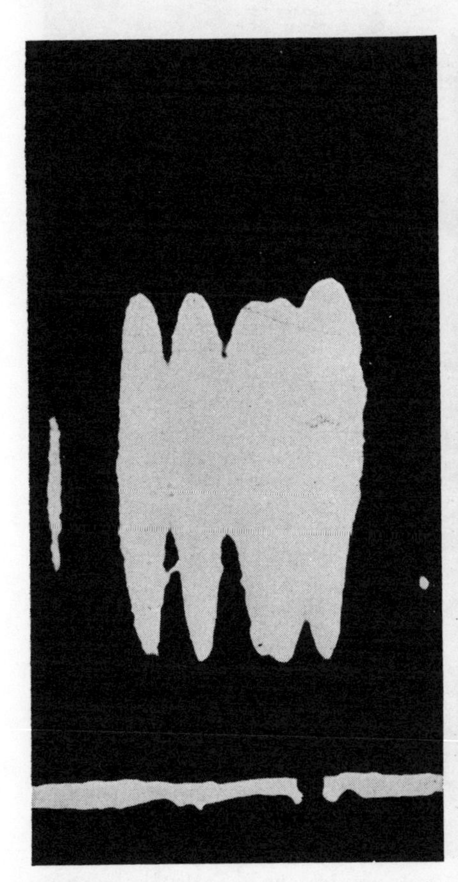

GROWTH of avalanche is shown in third photograph. Electrons ionize atoms so droplets condense on them.

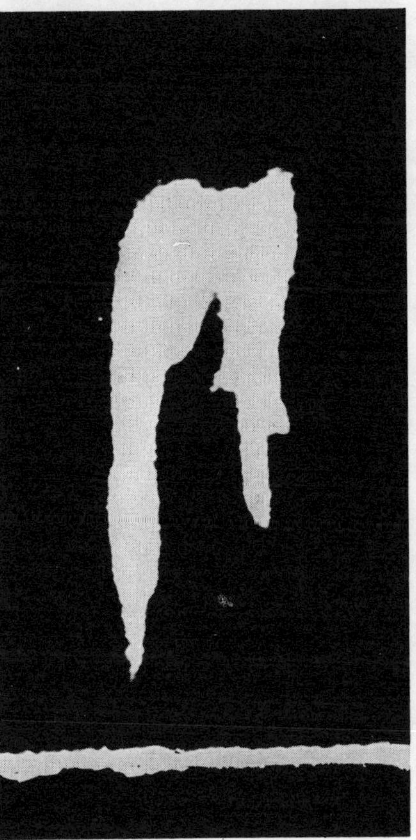

STREAMER develops from another avalanche. These experiments were done in the laboratory of H. Raether.

revolving-lens camera, the pictures showed that the stroke advanced in a series of jumps. The mechanism for this process is deduced to be as follows: When a streamer is initiated from the cloud it starts towards the ground as a "pilot" streamer. After some 30 to 90 microseconds, during which the tip of the streamer has advanced 10 to 30 feet, the ionization in the streamer's upper or cloud end decays as the result of the recombination of ions. This builds up a high resistance, and in consequence a high potential, at the upper part of the channel. When the potential reaches a critical limit, the stress is great enough to re-ionize the part of the channel on the earth side of the region of stress. A rapid pulse of ionization then sweeps down the channel, increasing in speed and intensity as it approaches the tip of the streamer. When it reaches the tip, the latter is given a boost in energy. It lights up brilliantly, and often produces branches at this point. The pilot streamer then advances for another 40 to 90 microseconds, whereupon decay again sets in at the cloud end and the process is repeated. Thus the pilot streamer forges ahead, the mechanism being known as the "stepped leader" process.

As the streamer approaches the ground, the field distortion, particularly near conducting elements in the ground, increases. The streamer speeds up and heads in at a nearly vertical angle to the ground. If the stroke is from a negative cloud, in the last microsecond before the streamer is completed a positive streamer may advance from the ground to meet the pilot streamer, usually 15 to 40 feet above the ground. In any case, as soon as the pilot streamer makes contact with the ground or with a positive ground streamer, an enormously steep potential wave sweeps up the channel to the cloud. This return stroke, which ionizes from one to 30 per cent of the gas molecules in the channel, is the lightning flash that we see. After this stroke discharges the section of the cloud to which the channel leads, the cloud is recharged as we have already indicated, and it then yields a new discharge towards the ground. Because the channel is now fully ionized, the discharge proceeds not by steps but directly to the ground. This wave, called a "dart leader," accounts for the repeated strokes of lightning. When the dart leader reaches the ground it calls forth a new bright return stroke. Dart leaders and return strokes repeat themselves in rapid sequence until the cloud element is drained of its high potential.

PULSATING BALL LIGHTNING probably caused the luminous streak at the top of this color photograph. The image may, however, have been caused by bead lightning, a form related to ball lightning. The photograph was taken during a severe thunderstorm at Los Alamos, N.M., in August, 1961, by B. T. Matthias and S. J. Buchsbaum of the Bell Telephone Laboratories. The exposure was made by aiming the camera at an approaching thundercloud about a mile away and leaving shutter open for five minutes.

Ball Lightning

12

by Harold W. Lewis
March 1963

This rare form of lightning may represent a stable configuration of a plasma of charged particles. If it does, it may help to solve the problem of confining a plasma within a thermonuclear reactor

A minor legend of the American West relates that forest-fire watchtowers are equipped with tall stools to provide a refuge for the fire warden who is attacked by ball lightning. The idea is that the lightning ball will snap around the legs of the stool and then, frustrated, depart through a door or a window. This is undoubtedly an anthropomorphic exaggeration of the behavior of ball lightning, but the fact is that there are reasonably believable accounts of instances in which balls of lightning have hung suspended in the air or floated eerily through it, have climbed through windows, dived down chimneys, popped out of ovens, glided along fences and telephone wires and performed other improbable maneuvers.

Any normal, cynical scientist, on hearing of ball lightning for the first time, almost instinctively places it in the category of folklore, along with flying saucers and ectoplasm. A brief survey of reported events, however, quickly convinces the skeptic that enough reputable observers have seen and possibly even photographed ball lightning to leave no doubt that the phenomenon is real, although it is rare and as yet unexplained. The fact that lightning balls, in contrast to lightning bolts, have been observed to persist for considerable periods of time has long piqued the curiosity of physicists. In the past five years there has been a surprising resurgence of interest in ball lightning, stimulated by a controversial hypothesis put forward by the noted Soviet physicist Peter L. Kapitza. His suggestion has in turn led to other theories, to renewed but unsuccessful efforts to reproduce the phenomenon in the laboratory and to some rather unlikely conjectures about the possible applications of the ball-lightning "technique" in the control of thermonuclear reactions for the production of power.

Any explanation of ball lightning must take into account the curious properties attributed to it by those who have seen it. The most recent survey of sightings was conducted in 1960 at the Oak Ridge National Laboratory, where J. Rand McNally, Jr., asked 15,923 employees if they had ever seen ball lightning. The 515 who gave an affirmative response were asked in some detail about the size, duration, color and other physical properties of the ball. These replies, which form the basis of the description that follows, developed a picture substantially like that obtained in previous surveys.

It would appear that the typical lightning ball—or, to use its German name, *kugelblitz*—is a luminous sphere perhaps as bright as a strong household fluorescent lamp. The sphere may range in diameter from a few inches up to a few feet, most often from six inches to a foot. It usually materializes immediately after an ordinary lightning stroke. The ball can be almost any color, although green and violet are rare. Most seem to shine steadily, but some pulsate. Normally the ball moves about, sometimes along a conductor or an insulator and sometimes directly through the air. It can last from a second or less up to several minutes; the median, if one may judge from the estimates of startled observers, is a few seconds. Some balls fade out; others disappear abruptly, occasionally with an explosive report. Lightning balls seldom damage anything badly, although they sometimes leave physical evidence of their occurrence. They have scorched wood and burned through wires. A lightning ball in the U.S.S.R. was reported by a number of observers to have partly melted the tip of the propeller of an airplane that encountered it at an altitude of 10,000 feet. The matter of damage is important because some estimate of the amount of energy stored in the ball is essential to an evaluation of the various theories. Since lightning balls have not been produced in the laboratory, estimates of their energy content are based entirely on the subjective reports and on inferences from their effects.

It is easy to set an upper limit to the energy stored in a lightning ball by assuming that the air within the ball is at most singly ionized. This means that each atom or molecule in the air has lost one electron and that there is an equal number of free electrons. A gas in this state is called a fully ionized plasma. The search for a way to control thermonuclear reactions involves the creation and maintenance of a plasma within some sort of magnetic "bottle" [see "Fusion Power," by Richard F. Post; SCIENTIFIC AMERICAN Offprint 236]. In ball lightning the energy that went to ionize the air is stored in the plasma until the charged particles recombine, releasing the energy in the form of light, heat and sound. At sea-level atmospheric pressure, single ionization of the air would produce an energy density of about 100 joules per cubic centimeter. (The joule is a unit of work equal to 10^7 ergs, or to the work done in one second by a current of one ampere flowing through a resistance of one ohm.) An average lightning ball 25 centimeters (10 inches) in diameter and singly ionized at normal atmospheric density would contain about one megajoule (million joules) of energy. I am indebted to M. L. Goldberger of Princeton University for pointing out to me that a megajoule can be visualized as the amount of energy that would be released by the chemical combustion of a large jelly doughnut.

In order to determine the energy content of an object the investigator normally employs calorimetry. He places the object in a known quantity of water and measures the extent to which the water is heated. Obviously catching a lightning ball in a calorimeter, or in any container of water, would be a considerable achievement. The amazing thing is that just such an experiment was accidentally performed by a lightning ball in the presence of an apparently sober and reliable resident of the London area. According to *The Daily Mail* for October 3, 1936, the observer reported that the glowing ball came out of the sky, cut a telephone wire, scorched a window frame as it entered a room and finally dived into a butt (a small barrel) containing four gallons of water. The water boiled "for some minutes," indicating that the lightning ball must have persisted most of this time. In support of his claim that the water actually boiled, the observer testified that he could not keep his hand immersed in it 20 minutes later. To boil so much water the lightning ball must have had an energy of not one megajoule but at least four megajoules and perhaps as much as 10 megajoules. The concentration of energy in the ball must have been considerably higher than that in our 25-centimeter, one-megajoule model: the ball was said to be the size of a large orange. The calculated upper limit of one megajoule for a 25-centimeter ball, although it is less than the energy of the ball in the

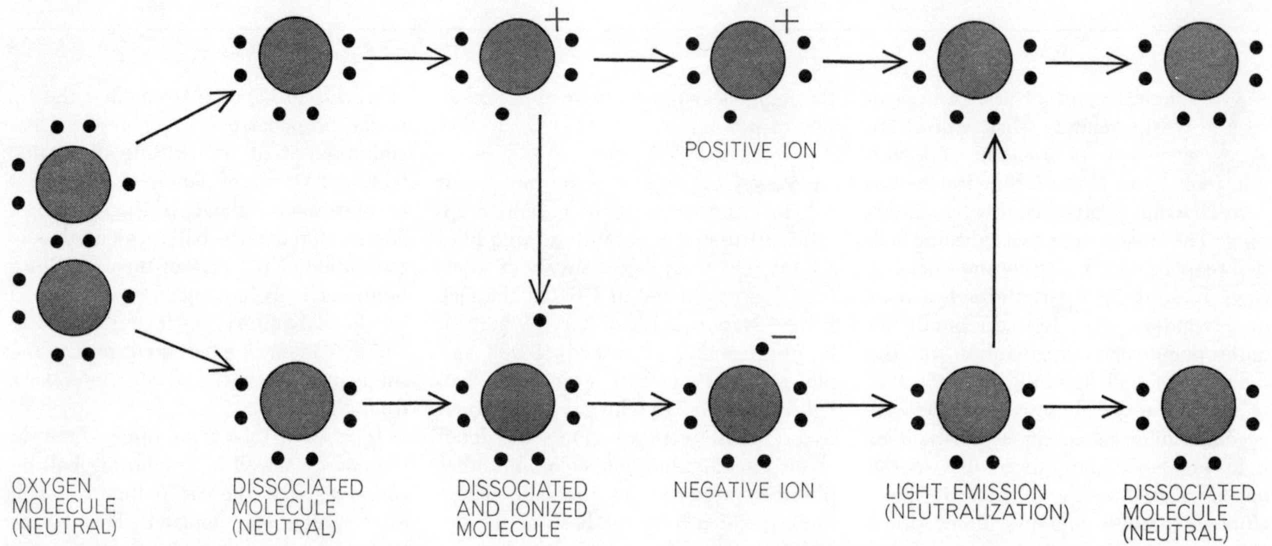

ONE TYPE OF IONIZATION that can occur in air involves first the dissociation of oxygen molecules by radiation. The radiation acts further to knock electrons off some of the atoms, which acquire a positive charge. Other atoms may take up these electrons, thereby becoming negatively charged. The process of recombination of a positive atom with an electron is highly oversimplified here.

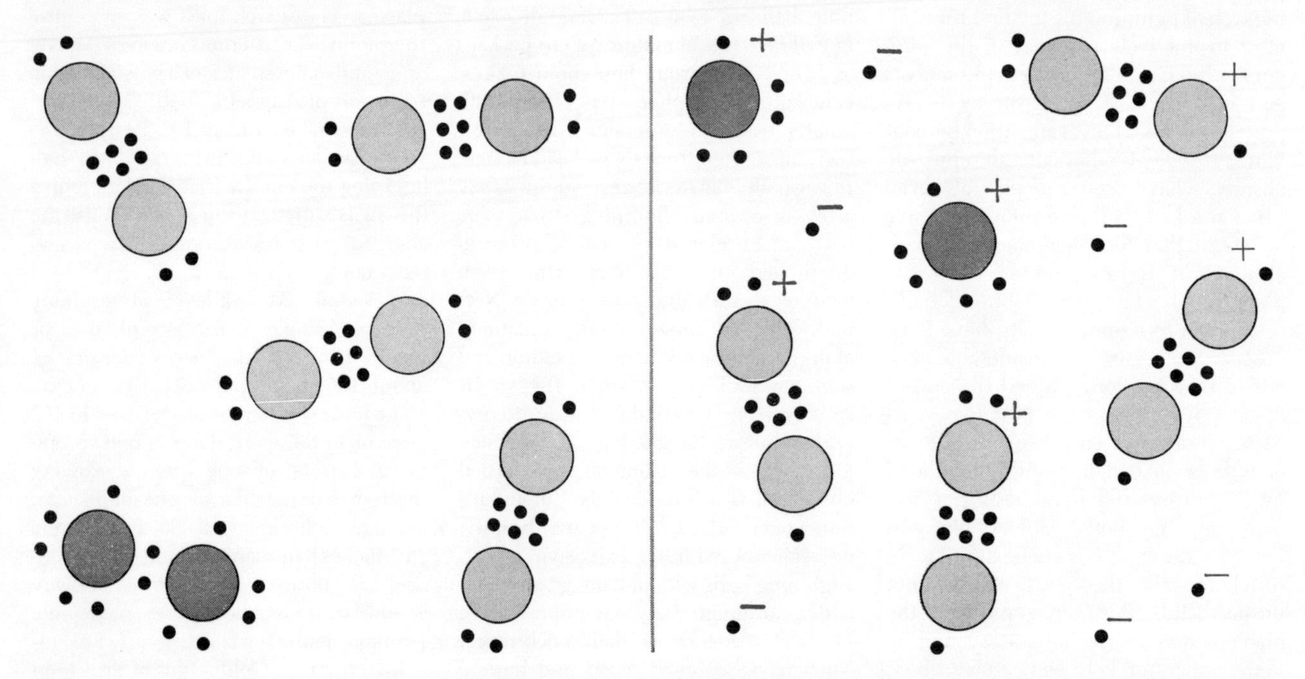

NITROGEN IONIZATION occurs in air along with the oxygen process. Here four molecules of nitrogen and one of oxygen are shown (air is approximately 78 per cent nitrogen and 20 per cent oxygen). Radiation can cause the dissociation of oxygen molecules and their ionization (*right*); nitrogen molecules usually ionize before dissociating. Recombinations (*not shown*) are highly complex.

water-butt experiment, is at least not too far from the truth.

Now, the only way to store so much energy would be in the ionization of practically all the atoms and molecules in the lightning ball. This constitutes a real dilemma, because it is difficult to reconcile ionization with the length of time the ball lasts, Ionized air is in a highly unstable state. Several mechanisms cause the free electrons to combine either with the positive ions or with electrically uncharged but chemically active atoms, particularly those of oxygen. Ionic capture of an electron, a process known as recombination, neutralizes the electric charge of the ion. If an uncharged atom picks up an electron, an event known as electron attachment, the atom becomes a negative ion. Such an ion regains its electrical neutrality by combining with a positive ion. The pressure, temperature and other characteristics of the gas determine whether neutralization occurs directly or through the intermediate formation of a negative ion. In the lightning ball the latter process is probably dominant. Even though the process is slightly roundabout, calculations based on well-known physical laws show that neutralization of the air in a lightning ball will take only a tiny fraction of a second, far too short a time to account for the relatively long life of the ball. Since not even the wondrous kugelblitz is permitted to amend the laws of nature, one must conclude either that the energy for ionization is being supplied continuously by some outside agency or that there are no free electrons in a lightning ball.

Kapitza employed the former idea in his theory, which he published in 1955. He suggested that the highly electric environment of a thunderstorm could create electromagnetic standing waves. Such a wave arises at the meeting place of two or more wave fronts of the same frequency traveling in different directions. The region where the waves reinforce each other is called the antinode and the region where they cancel each other the node. Kapitza said that at the antinode the waves might be intense enough to separate electrons from atoms in the air, thereby producing a small ionized region. It had been known before Kapitza published his paper that an ionized gas will resonate to and absorb electromagnetic waves of the appropriate frequency; the absorption of energy in this manner will finally result in a cascade of ionization. In the air the cascade would form a lightning ball. The ultimate size of the ball would be directly re-

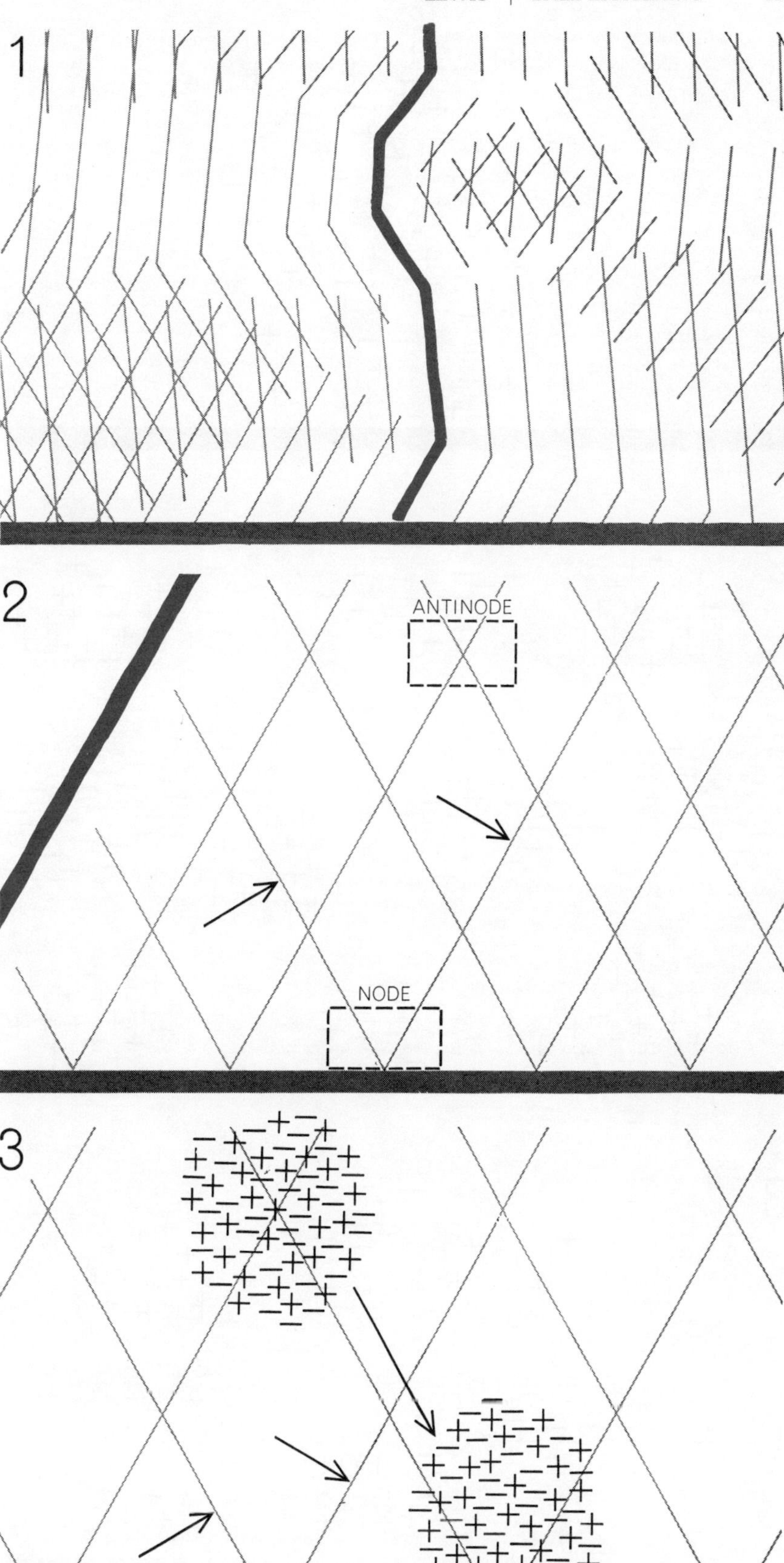

KAPITZA THEORY of the formation of lightning balls postulates the creation (*1*) of electromagnetic waves (*gray*) by a bolt of lightning (*color*). These waves are reflected by conducting surfaces (*2*), including the earth, and set up standing waves. Short arrows show direction of motion of wave fronts. The node is the region of weakest electric field; the antinode, of greatest field strength. Energy from the waves ionizes a region of air at an antinode, creating luminous ball that moves to node (*long arrow*), where radiation pressure holds it (*3*). Energy is fed in continuously. This representation of theory is highly schematic.

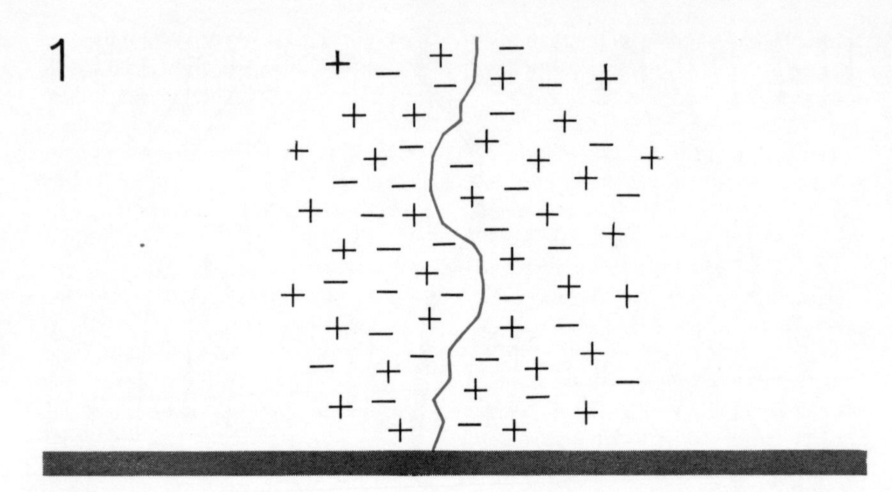

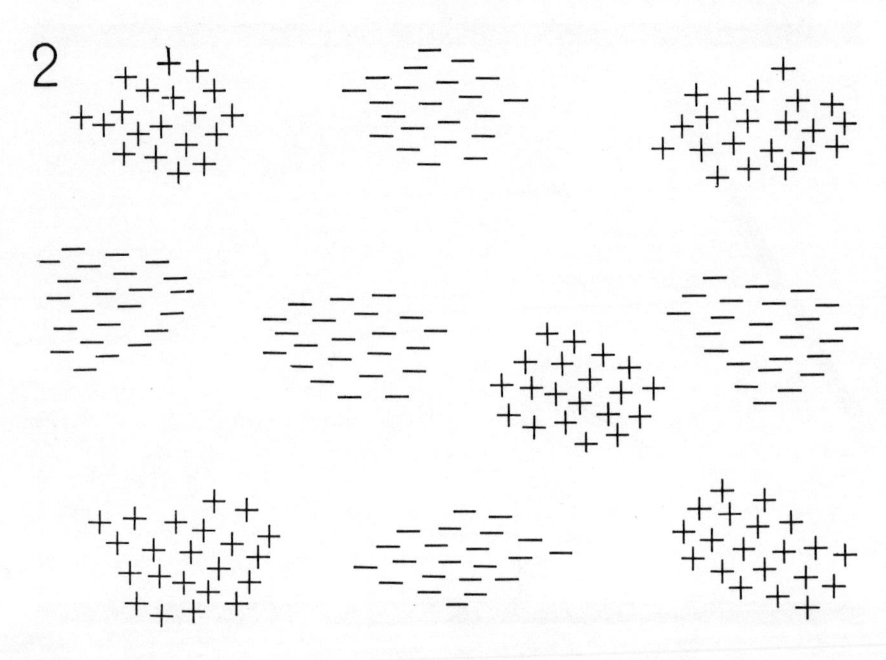

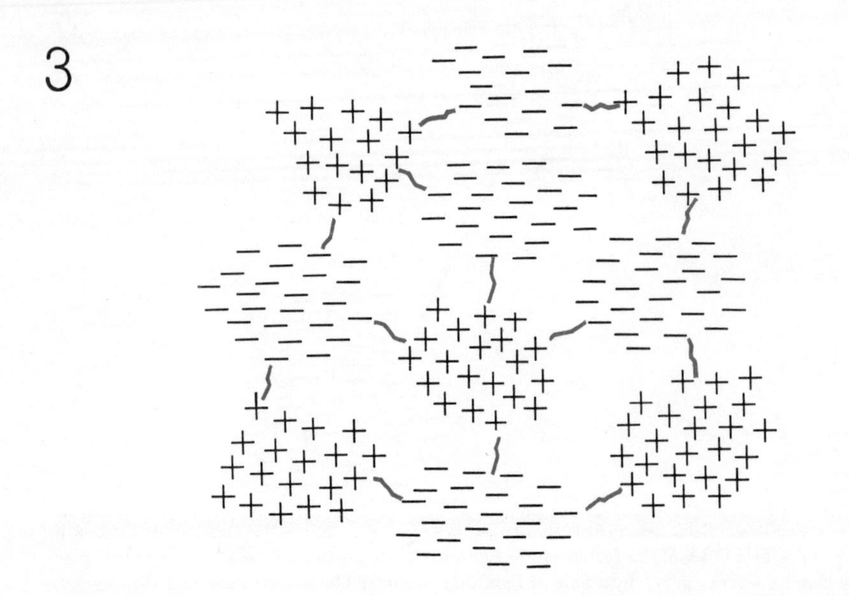

HILL THEORY also starts with lightning stroke that brings on partial ionization in air (*1*). Somehow the charges become separated so that the situation resembles a miniature thundercloud (*2*). Then tiny discharges between parts of the cloud give the ball its appearance of luminosity (*3*). This model stores all the energy inside and has no continuing supply.

lated to the frequency of the radiation furnishing the energy. Kapitza pointed out that the reported size of lightning balls indicates radiation in the range of several hundred megacycles, which is part of the radio-communication band. He simply made the assumption that during a thunderstorm there is a great deal of such radiation around. Noting that lightning balls frequently come through doors and windows, and particularly down chimneys, Kapitza wrote that these apertures were of an appropriate size to act as wave guides for the radiation. Bead lightning, in which small balls appear to be strung together, might arise, he added, from the formation of several small balls at adjacent antinodes.

A number of investigators have studied the problem of the stability of a Kapitza lightning ball and have agreed that the ball would form at an antinode (the point of greatest electric field strength) and then move to a node (the point of least strength), where it would tend to stay. If it should start to move away from the node, radiation on both sides would push it back. Thus the ball would be held loosely in place and would follow the vagaries of the radiation as it moved about. This could account in a reasonably natural way for the apparently capricious motions of some lightning balls.

There are two major troubles with the Kapitza theory. In the first place, a large amount of ultrahigh-frequency radiation has never been detected during a thunderstorm. In fairness to the theory it should be said that such radiation need not arise very often, since ball lightning is so rare. The fact remains that the communication bands in this range of frequencies, which are used by aircraft, are known to be relatively free of static during a thunderstorm. This suggests, of course, that there is little natural radiation at these frequencies.

The other trouble is that the Kapitza theory completely fails to account for the water-butt observation. A ball operating according to the Kapitza mechanism would have had enough energy to burn through the telephone wire and to scorch the window frame, but when it fell into the water butt it would have been completely cut off from its energy supply. Even salt water will not conduct enough electromagnetic radiation to supply a kugelblitz, much less a kugelblitz energetic enough to bring four gallons of water to a boil and keep it there for many seconds and perhaps minutes. Even if the Kapitza model involved energy storage, which it does not, the ball would not contain enough energy to raise the water to a high temperature. In the Kapitza model recombination occurs constantly, converting radio energy into light. The energy required to illuminate a sphere 25 centimeters in diameter until its surface glows with the brightness of a fluorescent lamp is 250 watts (allowing for the 10 per cent efficiency of the typical lamp). In fact, although the lamp is a pretty bright object, in four seconds it will use up only a kilojoule (a thousand joules) instead of the megajoule or more that a completely ionized ball of air 25 centimeters across could contain. Hence if one is to accept the Kapitza theory, one must completely discount the water-butt report. Since there is no way to verify the report, the credibility of the Kapitza theory rests in the end on the frequency with which large quantities of radiation accompany thunderstorms, and this remains to be determined.

Two alternative explanations of ball lightning involve the storage and expenditure of energy at a rate suitable for the phenomenon. The first explanation is suggested by the jelly doughnut and its stored megajoule. Chemical energy is in fact very efficient. One ounce of heating oil also contains a megajoule. Could not a flaming mass of pitch, ejected from a tree that has been hit by lightning, account for some observations of ball lightning? Unfortunately there is too much variety in kugelblitze for them all to be explained in this way. The air at 10,000 feet, where the lightning ball damaged the propeller of the Soviet airplane, would not contain lumps of such combustible material.

Edward L. Hill of the University of Minnesota has offered another interesting suggestion. He discounts the plasma aspects of the phenomenon largely because plasmas recombine too quickly. Instead he suggests that the lightning stroke preceding the ball somehow induces a separation of positive and negative charges carried by atoms, molecules, clumps of molecules, bits of dust and other minute objects in the air. It is known that charges carried in this way and not by free electrons and ions do not recombine so quickly as the charged particles in a plasma. Hill's model is virtually a miniature thundercloud. In the full-sized cloud, however, concentrations of positive and negative charge are separated by considerable distances. Hill proposes that in the case of the lightning ball the positive and negative particles stay more or less in their separate locations until the turbulent motion of the air creates a situation in which the strength of the electric field exceeds the minimum necessary to produce an electric discharge, i.e., lightning. In the usual kind of lightning this occurs over distances so large that well-defined lightning strokes appear; in Hill's model everything happens on such a small scale that the startled observer might see the discharge as a uniformly illuminated ball. Hill's idea may explain the appearance of the ball, but it leaves open the question of how the charges are separated in the first place.

Hill credits the water-butt experiment sufficiently to propose that his model contains at least a megajoule of stored energy. The model would work just as well if it contained only a kilojoule. The theory does not stand or fall according to the amount of energy the ball might contain, as long as the energy is enough to account for the luminosity. Hill's theory is primarily an effort to account for the long life of the kugelblitz.

Several other hypotheses to explain ball lightning have been offered, but they leave us as far from any real understanding of the phenomenon as the Kapitza and Hill theories do. No theory accounts for all the observations. Meanwhile the theories are inspiring efforts to produce lightning balls in the laboratory. The Bendix Research Laboratories has succeeded in creating small localized regions of ionized luminous gas by focusing microwaves in a small volume. Confinement of the luminous region occurs simply because the microwave generators are adjusted to achieve this. The glowing region cannot move about like a kugelblitz. Similar experiments have been performed elsewhere in the U.S., the U.S.S.R. and other countries, but none have made lightning balls.

If it is ever discovered that the lightning ball is a stable configuration for a plasma, the finding will have considerable importance for the thermonuclear power program. This, of course, is why some workers are studying the matter. The Oak Ridge survey was inspired by this possibility; in fact, I became interested in the subject during a summer spent working on the power problem at the Los Alamos Scientific Laboratory. As far as I know, no one working in the field thinks the kugelblitz can serve as a model for a thermonuclear reactor, but it may hold important lessons. Moreover, as long as we do not really know what a kugelblitz is, it holds out the hope that there is such a thing as a stable plasma configuration.

Thunder

13

by Arthur A. Few
July 1975

It is the acoustic signal generated by a rapidly expanding channel of heated air. From the information in this signal it is possible to deduce the location, shape and orientation of a lightning flash

It is obvious today that the ultimate cause of thunder is lightning; it is less obvious just how an electrical discharge in the atmosphere produces the variety of powerful noises heard in a thunderstorm. A lightning flash dissipates a prodigious amount of energy, but how is that energy (or a portion of it) transformed into sound waves? Moreover, a lightning stroke is essentially instantaneous; how does it generate the often protracted sequence of rumbles, claps, booms and other sounds heard in a peal of thunder?

Considering that thunder has been a topic of speculation since antiquity, one might expect that these questions would have been answered long ago. Actually even the most general principles of thunder production were not established until this century, and a theory of thunder that is both comprehensive and detailed has been approached only in the past 10 or 15 years. A number of important questions remain unresolved today.

With my colleagues at Rice University and in cooperation with workers elsewhere I have investigated thunder by recording its "signature": the pattern of sound waves received at a particular location from a single lightning flash. The study of such signatures has revealed much about the nature of thunder itself, and it has led to the formulation of a theory of how features of the lightning discharge are expressed in the acoustic signal. The theory is successful enough for us now to employ thunder as a tool in searching for the source of electricity in clouds.

The relation of lightning to thunder was well established by the end of the 19th century. It remained to be determined what effects of the lightning discharge are responsible for producing the sound. Four principal theories were proposed, and in the first decade of the 20th century they were vigorously debated. (In 1903 *Scientific American* published four contributions on the nature and cause of thunder.)

The first of the theories held that the lightning stroke creates a vacuum and that thunder is produced when the vacuum collapses. Another maintained that water drops in the path of the lightning flash are turned into steam and that the rapid expansion of the steam is accompanied by a loud report. A third proposal suggested that the electrical discharge decomposes water molecules by electrolysis and that the hydrogen and oxygen thereby produced subsequently recombine explosively. Finally, the simplest explanation ascribed thunder to the sudden heating of the air in the path of the lightning flash. Because air has electrical resistance, it is heated by the passage of a current just as a wire is; the expansion of the heated air was considered sufficient to explain the ensuing thunder.

We now know that the last explanation is the correct one. Each surge of current in the lightning flash heats the air in its path, creating a channel of gases at high temperature and pressure. The gases expand into the surrounding air as a shock wave, which after traveling a short distance decays into an acoustic wave.

It is interesting to note that the other theories were not entirely without foundation. For example, a region of diminished air pressure is briefly created in the aftermath of lightning; this partial vacuum is an effect of thunder, however, rather than a cause. Moreover, water drops are certainly evaporated in the lightning channel, and water molecules are decomposed. There is excellent evidence for the decomposition in the optical spectrum of lightning: one of the most prominent features of the spectrum is an emission line of hydrogen. Both of these phenomena, however, are merely collateral effects of lightning; they make no significant contribution to thunder.

The interest in the nature of thunder that culminated in the theories of the early 20th century declined soon thereafter, and the study of thunder was largely ignored for 50 years. Isolated experiments and observations were made during that period, but they tended more to confuse than to enhance understanding. For example, from measurements of the duration of thunder one can estimate the length of the lightning channel; in many cases such measurements were found to yield lengths much greater than that of the observed lightning flash and in some cases the lengths were greater than the height of the cloud itself. This discrepancy has only recently been explained.

Since 1960 scientific interest in thunder has revived and research into the nature and source of thunder has been renewed. The investigation is greatly assisted by a technology that was unavailable to earlier workers. The fundamental principles are no longer in question; the challenge to research today is to account for the detailed features of the thunder signature.

What is heard in a peal of thunder depends in large measure on the characteristics of the particular lightning flash that produced it. Both the temporal sequence of events in the discharge and their arrangement in space must be considered; these two complex factors determine not only the frequency and the amplitude of the radiated acoustic waves but also the order in which the waves

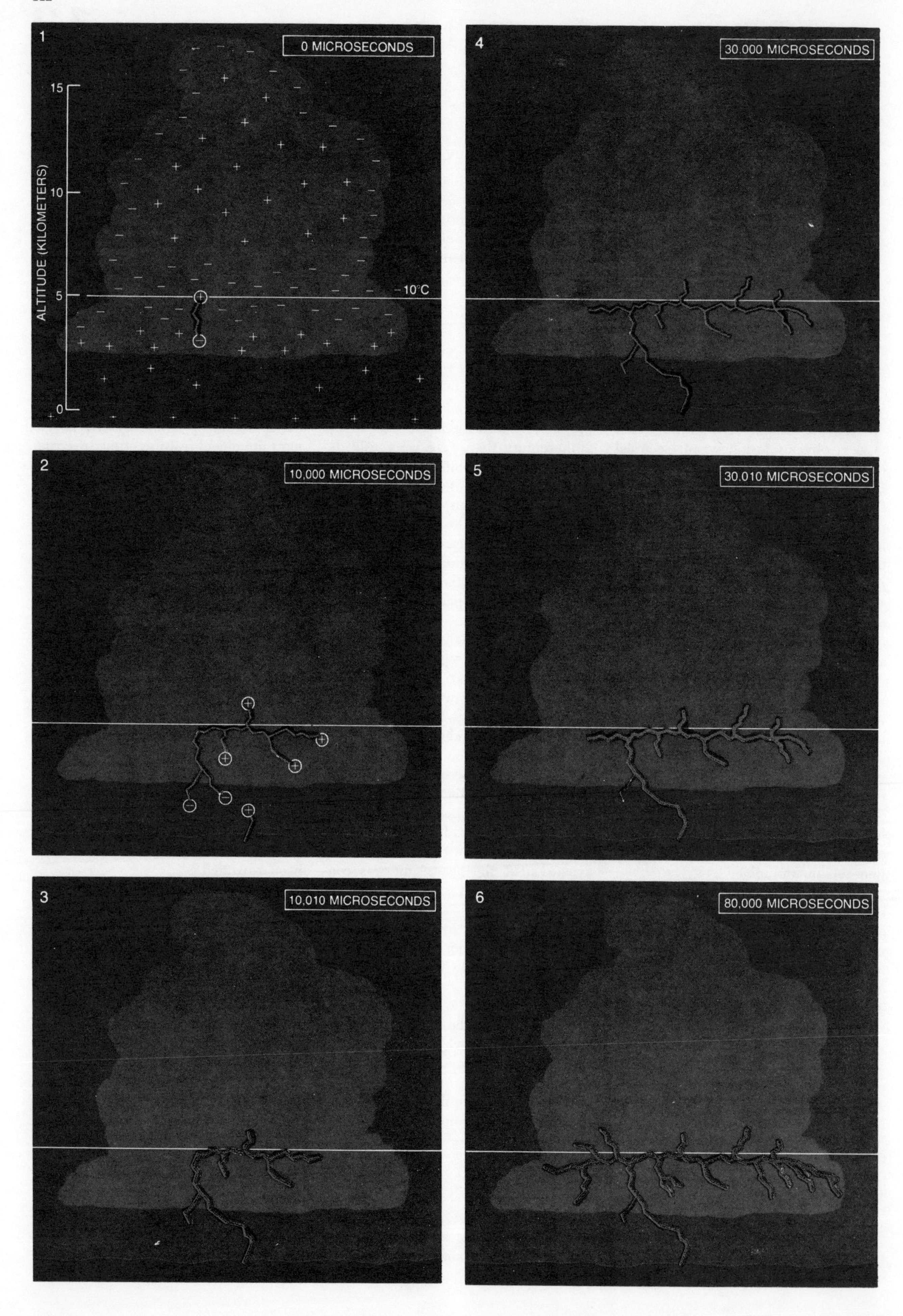
1
0 MICROSECONDS
ALTITUDE (KILOMETERS)
15
10
5
0
−10°C
2
10,000 MICROSECONDS
3
10,010 MICROSECONDS
4
30,000 MICROSECONDS
5
30,010 MICROSECONDS
6
80,000 MICROSECONDS

are received at a given observation site.

A lightning flash in most instances begins near the base of a cloud in a region of dense negative charge. This stratum is typically at an altitude of about five kilometers, where the temperature is approximately −10 degrees Celsius. That is the region of the cloud where water droplets freeze, a process that may be connected with the generation of electric charge.

The negatively charged zone of the cloud can be at a potential of as much as 300 million volts with respect to the ground, but even that enormous potential is insufficient to support a spontaneous arc across five kilometers of air. The main discharge can begin only after the channel has been traced by a preliminary low-current discharge called the stepped leader. The stepped leader begins to form when electrons emitted by droplets in the cloud are accelerated by the intense electric field; the electrons collide with molecules in the air, freeing many more electrons and leaving a conductive path of partially ionized air. Such a cascade of accelerated electrons typically progresses only 50 to 100 meters, but with each step a portion of the cloud's charge is transferred downward, and the next step can begin from the tip of the advancing leader.

The course of the stepped leader is highly irregular and in its progress toward the ground it forms numerous branches. Each step is accomplished in less than a microsecond, but there is a pause of about 50 microseconds between steps. As the leader approaches the ground the potential gradient—the voltage per meter—increases, and sparks are emitted from objects and structures on the ground, usually from the highest points first. When one of these sparks meets the downward-propagating leader, a conducting path between the cloud and the ground is completed. Since the potential difference across the path is a few hundred million volts a surge of current immediately follows; this large current is called the first return stroke.

The stepped leader may require 20 milliseconds to create the channel to the ground, but the return stroke is completed in a few tens of microseconds. In some cases that is the end of the lightning flash; more commonly, however, the leader-and-stroke process is repeated in the same channel at intervals of tens to hundreds of milliseconds. The subsequent leaders, called dart leaders, progress faster and more smoothly than the stepped leader because the electrical resistance of the path they follow is lower than that of the surrounding air. As the dart leader progresses toward the ground, intracloud processes extend the channel within the cloud so that additional areas are discharged. The subsequent return strokes, however, are usually less energetic than the first one is. A typical lightning flash has three or four leaders, each followed by a return stroke; one flash has been photographed that had 26 strokes.

Each surge of current in the flash, including the steps in the stepped leader, the dart leaders and the return strokes, heats the gases in the lightning channel and thereby generates an acoustic signal. The amplitude and duration of the signal produced by each current surge depend on the magnitude of the current. The complete thunder signal therefore reflects a complex sequence of events in the lightning flash. Ordinarily it is not possible to distinguish in a recorded thunder signature the acoustic pulses generated by individual leaders and strokes, but the sequence of strokes and the currents they carry nevertheless determine what sound is produced. In a few exceptional recordings the correlation of thunder pulses with lightning strokes is evident.

The spatial arrangement of the lightning flash has perhaps a greater influence on the resulting thunder than the temporal organization. The lightning channel is said to be tortuous: it consists of straight segments separated by sharp bends. The structural elements are classified in three size ranges. The large-scale features of the channel, called the macrotortuous elements, are straight segments at least 100 meters long. Straight segments between five and 100 meters long are classified as mesotortuous, and those shorter than five meters are called microtortuous. The mesotortuous and macrotortuous elements are not, of course, actually straight lines; they are made up of many smaller segments and seem to be straight only when considered at the proper scale. Moreover, the classification of a feature as mesotortuous or microtortuous is not determined by a fixed rule but depends on the energy of the discharge. Five meters is an appropriate minimum length for mesotortuous features in strokes carrying large currents, but smaller elements can be considered mesotortuous when the current is comparatively small, as it is in stepped leaders and dart leaders.

The mesotortuous segments are the primary radiators of the acoustic pulses of thunder. The entire channel can be considered a "string of pearls," each pearl a mesotortuous segment radiating a series of pulses determined by the sequence of pulses in the lightning flash. The macrotortuous segments determine the spatial organization of the individual acoustic radiators and hence have a profound influence on what is heard by the observer.

A pulse of thunder begins in a channel of hot gases at high pressure. Spectroscopic evidence indicates that the temperature in the channel can reach 30,000 degrees C., and the pressure can exceed atmospheric pressure by from 10 to 100 atmospheres. Initially the high-pressure core expands as a shock wave. A shock wave is distinguished from an acoustic wave in that it compresses and heats the medium in which it propagates and thereby increases the speed of sound. Because the speed of sound increases as the temperature rises, the shock wave travels faster than sound does in the same medium. The extent of the compression and heating, and therefore the magnitude of the increase in velocity, depend on the amplitude of the wave. Behind the shock wave the air continues to move outward, and a region of low pressure forms.

The expanding shock wave dissipates its energy in performing work on the surrounding air. When all the energy imparted to the shock wave by the lightning stroke has been expended, the wave "relaxes" and the pressure in the vicinity of the channel returns to normal. (The core of the channel itself, however, re-

LIGHTNING CHANNEL develops from a region of concentrated negative charge near the base of a cloud, an area associated with the freezing of water droplets. The flash begins with the formation of a stepped leader (*1*), which moves downward in steps from 50 to 100 meters long and simultaneously extends streamers horizontally through the charged region. As the leader nears the ground, sparks propagate upward to meet it (*2*); when the pathway is completed, the large current of the first return stroke flows (*3*). Subsequent leaders, called dart leaders (*4*), progress much faster and extend the channel to other parts of the cloud; each is followed by a return stroke (*5*). Much of the length of the completed channel is horizontal and only a small portion of it is visible below the cloud (*6*). Each leader and stroke heats the gases in the lightning channel and contributes to the signal ultimately perceived as thunder.

FOUR LIGHTNING FLASHES were photographed near Tucson, Ariz. The prominent kinks and bends are the large-scale, or macrotortuous, elements of the lightning channels; segments between major bends are on the order of 100 meters long. The macrotortuous features determine the overall pattern of claps and rumbles in thunder. The four flashes were not simultaneous but were recorded in a single photograph by making a time exposure lasting for about two minutes. The photograph was made by Henry B. Garrett of Rice University. The forked lightning channel at left was produced by a flash consisting of multiple strokes that followed divergent paths.

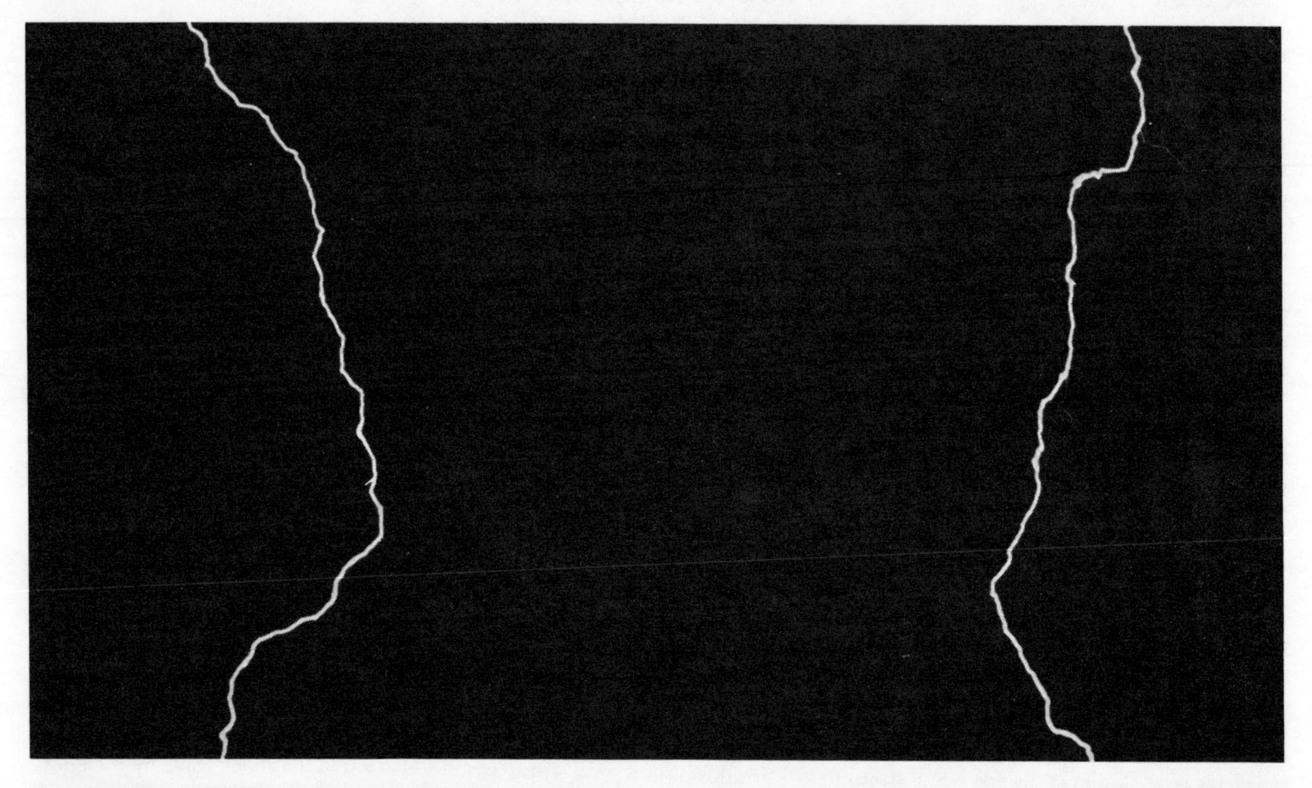

SMALLER FEATURES of a lightning channel are discernible in a photograph made with a telephoto lens. The area seen is the lower portion of the forked channel in the photograph at the top of the page. The many small straight segments are on the order of 10 meters long; elements of the channel in this size range are classified as mesotortuous and can be considered point sources of thunder. The microtortuous features (those shorter than about five meters) can rarely be resolved in photographs and have little effect on thunder.

mains a region of high temperature and low density.) From the ambient pressure and the energy per unit length of the lightning stroke one can calculate the radius at which the shock wave relaxes. The relaxation radius is in turn related to the wavelength of the thunder. Thus the wavelength, or pitch, of thunder is determined by the energy of the lightning stroke and the ambient air pressure in the region where the thunder is generated. The more powerful the stroke or the lower the air pressure, the lower the pitch of the resulting thunder. A typical value is 60 hertz [*see illustration below*].

The relaxation radius serves as the actual measure distinguishing microtortuous elements from mesotortuous ones. Features of the channel that are smaller than the relaxation radius are blurred in the expansion of the shock waves. They have little influence on the form of the resulting acoustic waves and therefore cannot be resolved in data derived from thunder. Microtortuous features are thus defined as those that are too small to be detected in the thunder signature.

The shock wave is not an efficient source of acoustic radiation: less than 1 percent of its energy is transmitted to the acoustic wave. The remaining 99 percent is dissipated in heating the air in the vicinity of the lightning channel. The total energy of the shock wave is very large, however, and even the small fraction of it converted into sound generates an acoustic wave of large amplitude. The result is one of the loudest sounds in nature.

Sound produced by the mesotortuous features of the channel is not radiated with equal power in all directions. The study of large sparks produced in the laboratory has shown that more than 80 percent of the acoustic energy is confined to a zone 30 degrees above and below a plane that bisects the spark perpendicularly. A microphone placed end on to the spark will receive much less sound than one placed broadside to it. This directed quality of the radiated sound is one of the most important determinants of what is heard bv the listener on the ground.

From studies of photographs of lightning it has been found that the average change in direction or orientation between adjacent mesotortuous segments is about 16 degrees. Because this angle is substantially smaller than the zone 60 degrees wide into which each segment radiates most of its energy, a number of mesotortuous segments strung along a single macrotortuous segment will radiate most of their energy in roughly the same direction [*see top illustration on following page*]. It is this property that is responsible for the sudden claps and prolonged rumbles of thunder.

Acoustic pulses from all parts of a lightning stroke are, of course, emitted almost simultaneously. The entire acoustic output of a lightning flash is produced in less than a second, in the time required for the complete sequence of leaders and strokes. The thunder persists much longer than that because the lightning channel is long (at least five kilometers long and often considerably more) and signals from its nearest segments reach the listener sooner than those from the farthest extensions. The orientation of the channel, and in particular the orientation of the macrotortuous segments, determine the character of the sounds heard.

If a macrotortuous segment is oriented end on to the observer, the sound received will be of comparatively low amplitude, since most of the acoustic energy is radiated perpendicularly to the segment. Moreover, the wave fronts from each mesotortuous element will reach the listener in succession, beginning with signals emitted by the nearest part of the segment. The result is a protracted rumble or roll of thunder.

If the macrotortuous segment is broadside to the listener, a much larger portion of the radiated energy will be received. Equally important, the wave fronts from all the mesotortuous elements will arrive almost simultaneously. As a result a brief but intense clap of thunder will be heard [*see bottom illustration on following page*].

Thunder produced by a single flash of lightning is commonly perceived as a combination of claps and rumbles because various elements of the lightning channel are oriented differently with respect to the listener. Thunder from the same lightning flash will be perceived differently at different locations, since each location has a unique position and orientation with respect to the lightning channel.

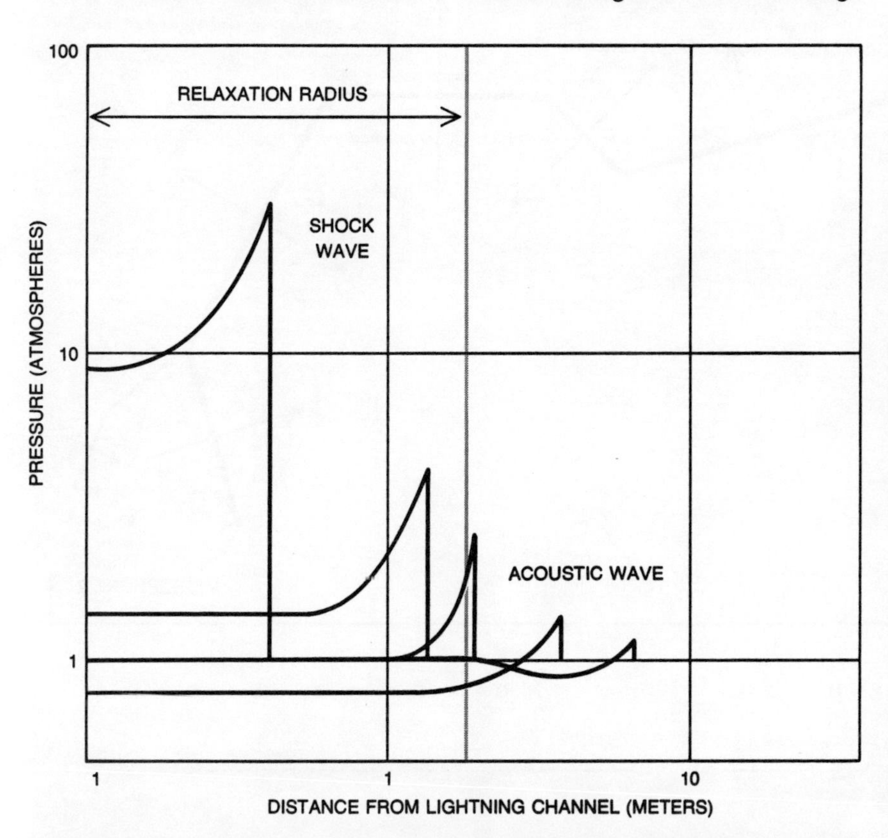

EXPANDING CHANNEL of hot gases produced by a lightning stroke propagates as a shock wave and then as an acoustic pulse of thunder. Because the initial shock wave compresses and heats the air, it quickly dissipates its energy. A few meters from the lightning channel it "relaxes" to yield an acoustic wave of lower amplitude, and the pressure in the region behind the wave is briefly reduced. Only about 1 percent of the energy of the shock wave is transferred to the acoustic wave; the rest is expended in heating the air near the channel.

Between the lightning channel and the listener the thunder signal is altered by the medium in which it travels. The atmosphere attenuates, scatters and refracts the signals; in addition they are subject to nonlinear propagation effects

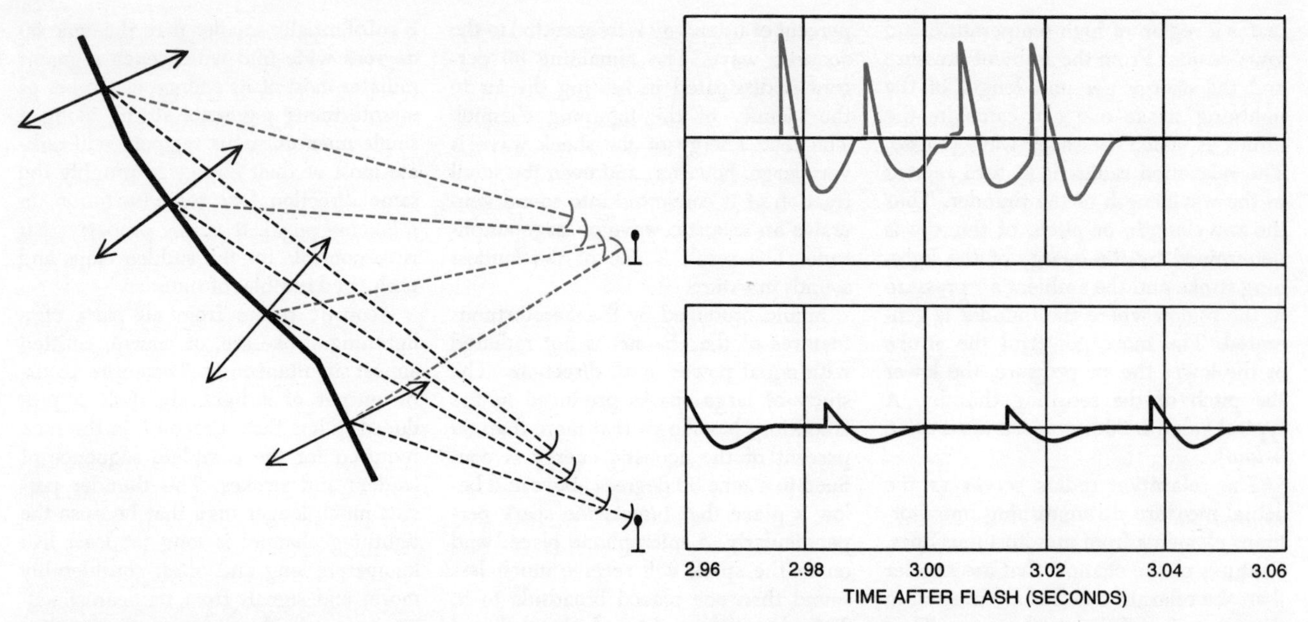

MESOTORTUOUS SEGMENTS of the lightning channel are the primary radiators of thunder. Each of the four segments shown can be considered an independent point source of sound. Acoustic pulses are emitted from the segments almost simultaneously, but the pattern of sounds perceived on the ground depends on how the channel is oriented with respect to the observer. Sounds emitted perpendicularly to a segment (*color*) are more powerful than sounds emitted parallel or nearly parallel to it (*black*). Since each segment differs in orientation from the adjoining segments by only a small angle, signals from several segments reach an observer whose position is perpendicular to the channel almost simultaneously; the result is a brief but loud clap of thunder. An observer looking up the length of the channel, on the other hand, receives wave fronts at greater intervals and hears a prolonged rumble.

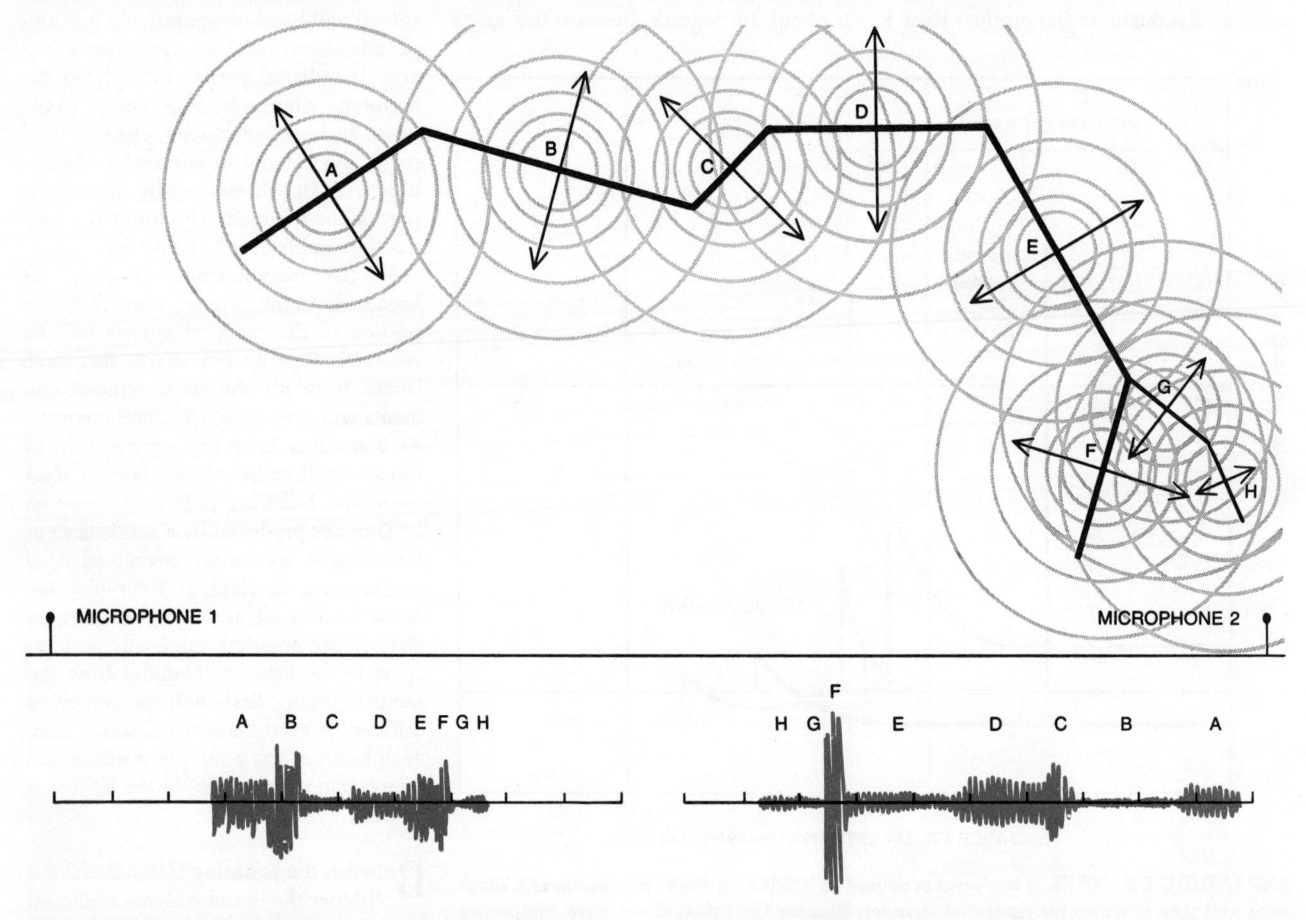

MACROTORTUOUS ELEMENTS of the lightning channel determine the overall pattern of claps and rumbles in the thunder signature. The amplitude and duration of the signal produced by each segment are determined by the orientation of the segment and its distance from the observer. Microphones placed at different positions will record different signatures from the same lightning channel. For this hypothetical lightning flash the segments have been labeled along with the sounds they would produce at two locations.

and, once they reach the ground, to reflection.

"Nonlinear propagation" refers to processes that affect one part of a wave more than others or some frequencies of sound more than others. Nonlinear propagation therefore alters the waveform of the signal (the shape of the individual waves) or its spectrum. One such effect tends to lengthen each pulse as it propagates; sound waves of large amplitude are the most severely affected, so that the process is most significant close to the lightning channel.

The attenuation of thunder in the atmosphere is caused by two independent processes. "Classical" attenuation results from the viscosity of air, that is, from the fact that air is not a perfectly elastic medium. Classical attenuation is well understood, and its magnitude can be predicted. "Molecular" attenuation results from a complex interaction of sound waves with water molecules and oxygen molecules in which acoustic energy excites internal vibrations of the molecules. It can be evaluated only if the temperature and humidity are known for all points along the path traveled by the thunder signal. Molecular attenuation is usually the more important of the two effects. Attenuation increases as the square of the frequency of the signal; thus a 20-hertz wave will travel four times as far as a 40-hertz wave before it is attenuated to the same degree.

The scattering of the thunder signal is even more difficult to predict than its attenuation. The principal agents of scattering are turbulent eddies in the atmosphere, which range in size from microscopic disturbances a few microns in diameter to the thunderstorm cell itself. For the scattering of thunder the most significant eddies are those that are approximately the size of the wavelength of the thunder (less than 50 meters). Again, the losses are severer for higher frequencies.

Because of scattering and attenuation, when a thunder signal travels several kilometers through a turbulent medium, only the lowest-frequency components of the original spectrum survive without major modification. As a result a lightning flash of low energy, which produces little low-frequency sound, will not give rise to audible thunder except at close range.

The refraction of thunder is a large-scale phenomenon caused by variations in the speed of sound in the atmosphere; a refracted "ray" of sound is bent away from the straight-line path between the source and the observer. The laws governing refraction are well understood, and given sufficient information on atmospheric conditions, the curved path followed by the acoustic ray can be calculated. The most important variables are temperature and wind.

In the lower atmosphere temperature ordinarily decreases with altitude, typically at a rate of about 6.5 degrees C. per kilometer. Below a thundercloud the temperature gradient is generally steeper, reaching a maximum of 9.8 degrees per kilometer. Because sound travels faster in warm air than it does in cool air, the temperature gradient tends to curve the acoustic rays upward [*see illustration below*]. For this reason thunder generated by the lowest parts of a lightning channel cannot be heard at a distance, and there is a distance beyond which thunder cannot be heard at all, since all the sound passes over the head of the observer.

Wind has two effects on the propagation of sound waves. First, the actual velocity of a wave front is the sum of the speed of sound in air and the wind velocity. Sound therefore propagates downwind faster than it does upwind, and in a precise analysis of thunder signals this difference must be taken into account. Second, wind shear, the variation of wind velocity with altitude, imparts a further bend to the acoustic rays. Wind velocity generally increases with altitude. In the upwind direction the effects of wind shear add to those of temperature refraction; in the downwind direction they subtract from them.

In combination with attenuation and scattering, the temperature gradient and wind shear impose an ultimate limit on the range over which thunder can be heard. The maximum range can be as little as 10 kilometers, but it is sometimes much greater, depending mainly on the altitude of the lightning channel and the wind velocity.

Reflection adds a final modification to the thunder signal before it reaches the listener. For low-frequency sound the amplitude of a reflected wave is roughly proportional to the angular size of the

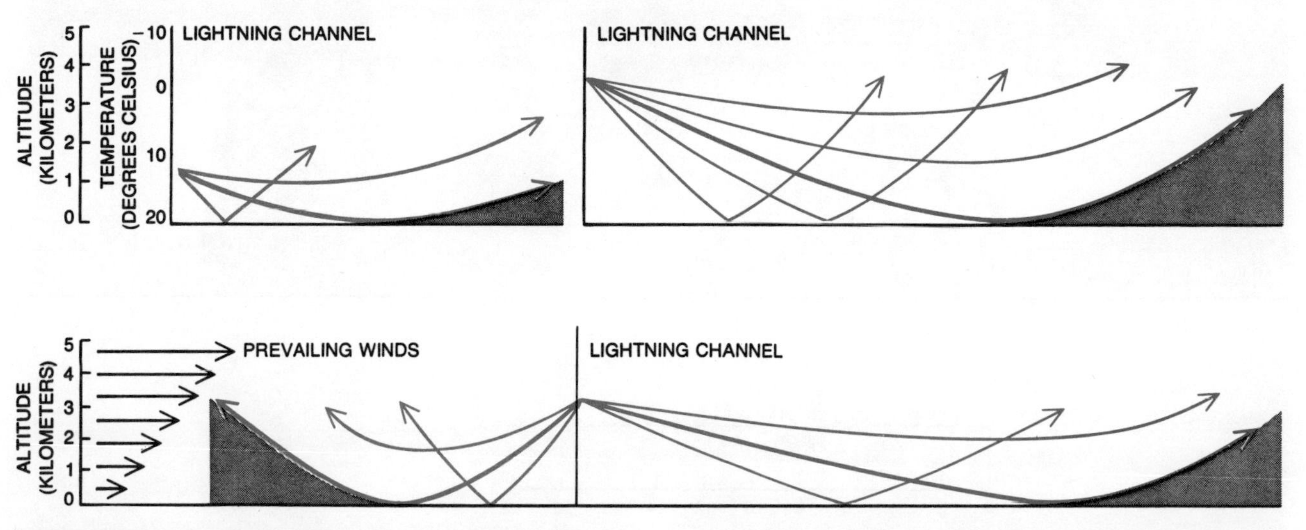

REFLECTION AND REFRACTION alter the thunder signal and limit its range. The temperature gradient in the atmosphere bends sound waves upward, so that thunder emitted from the lower portions of the lightning channel can be heard only nearby (*top left*); there is a maximum range beyond which thunder cannot be heard (*top right*). The distribution of sound produced by temperature refraction is altered by the wind, which usually has progressively higher velocity at higher altitude. Upwind of the channel the refraction is increased; downwind it is diminished (*bottom*). Low-frequency sound waves striking the ground are efficiently reflected.

reflecting surface as it is viewed by the observer. In most cases there is only one reflecting surface large enough to produce amplitudes comparable to those of signals received directly: the ground. Since flat ground completely fills the field of view when one is looking downward, the amplitude of the reflected wave will be equal to that of the directly received wave if no sound is absorbed at the surface, which for low frequencies is probably a correct assumption. The perceived sound is thus the sum of the direct and the reflected waves.

Depending on one's height above the ground and the angle of the incident signal, the direct and the reflected waves will add constructively for some wavelengths and destructively for others. In recording thunder signatures we avoid destructive interference by placing our microphones at a height equal to a fraction of the shortest wavelength to be measured.

The thunder perceived at any given position is unique to that point of observation. A microphone placed a few meters above an identical microphone at ground level will detect a slightly different signal. If the two microphones are both near the ground but are separated by a horizontal distance of 20 to 30 meters, the recorded thunder signatures will be similar, but minor differences will be distinguishable because the two positions give a significantly different perspective on the mesotortuous elements of the lightning channel. If the microphones are placed 100 meters apart, few details of the recorded signatures will correspond, but the underlying pattern of claps and rumbles will be preserved; the spacing is now comparable to the size of the macrotortuous elements. Microphones separated by more than a few kilometers will record thunder signatures that may have only one or two features in common.

A thunder signature contains a great deal of information about the lightning channel that produced it and about the atmosphere between the lightning and the observation site. The signature is a complex waveform, but it is possible to analyze it and recover much of the information. When that is done, one can reconstruct the lightning channel, a particularly valuable capability when the channel runs inside a cloud and photography and other optical methods of study are not possible.

The thunder signature is recorded by

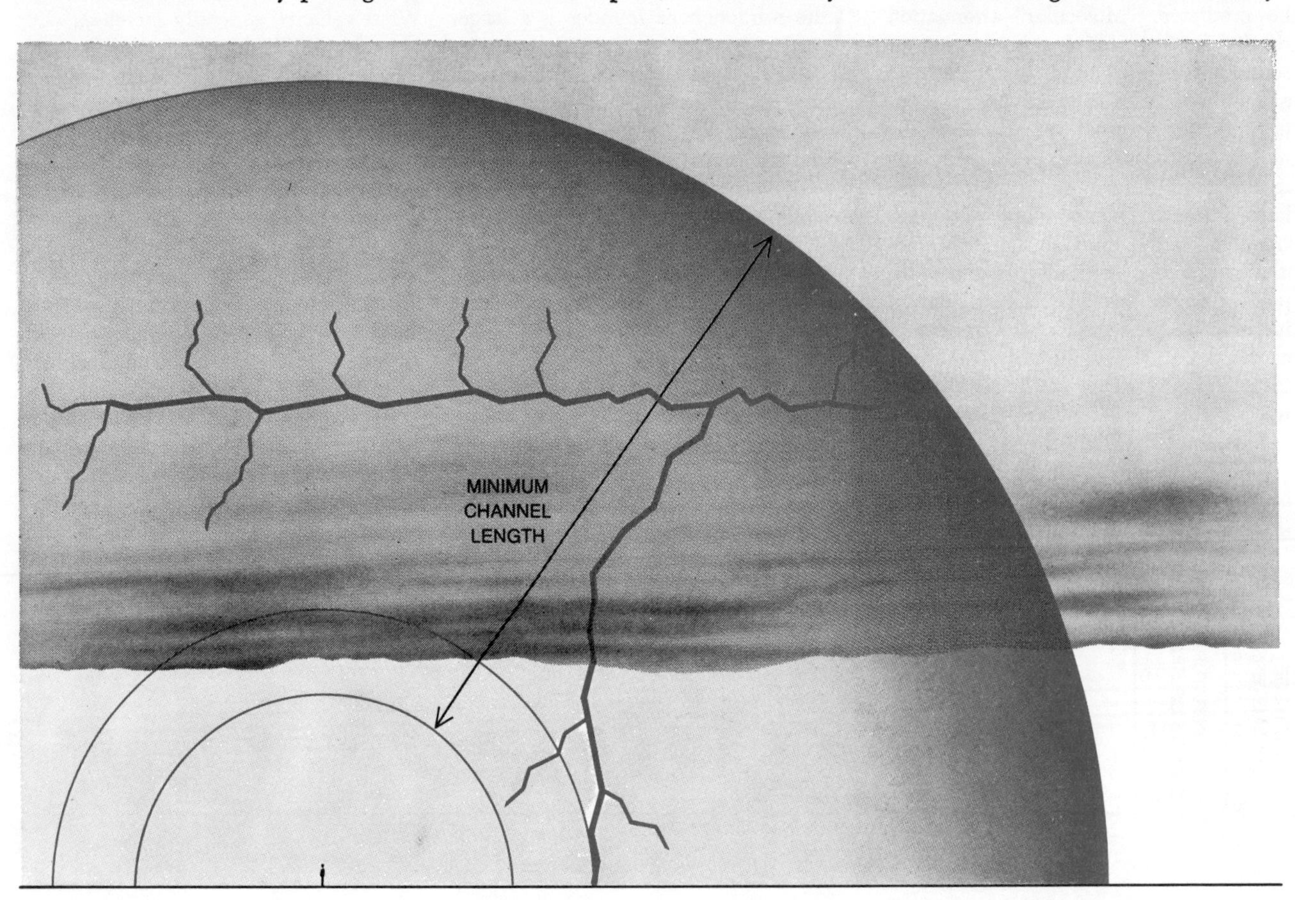

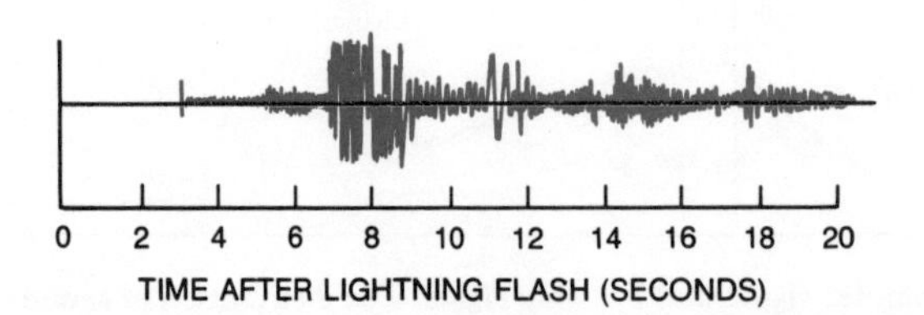

AMATEUR OBSERVATIONS of thunder require only a wristwatch. By measuring the time elapsed from the lightning flash to the first thunder heard, the loudest clap of thunder and the last rumble, one can determine the distance to the nearest branch of the lightning channel, the main channel and the farthest feature. The approximate distance in kilometers is given by the time in seconds divided by 3. By the same method the minimum length of the channel can be calculated from the total duration of the thunder signal.

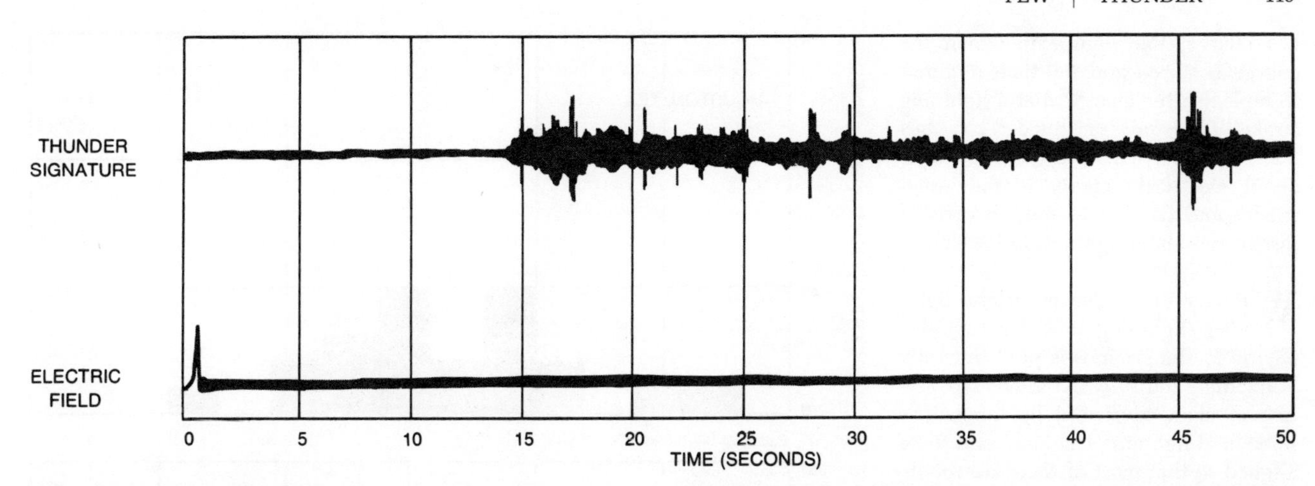

THUNDER SIGNATURE recorded by one microphone in an array of several microphones contains detailed information about the lightning channel that produced it. The time of the lightning flash is recorded by the momentary change in the electric field at the beginning of the graph. From the time of arrival of each feature of the thunder signature the distance to the corresponding segment of the channel can be calculated; with information from additional microphones the direction can also be determined and hence the location of the channel. Corrections must also be made for effects introduced by the wind and by temperature variations in the atmosphere.

an array of microphones on flat ground. For the purposes of analysis the thunder signal is broken up into short sections, each one-fourth of a second to half a second long. The microphones measure the direction of arrival of each section, and in addition the time of the lightning flash and the time of arrival of each section of the signature are known. With the aid of a digital computer and a mathematical model of the atmosphere it is then possible to determine the origin of each section, that is, to deduce its position at the time of the lightning flash. In this way the entire channel can be reconstructed in three-dimensional form. The technique is sensitive enough to locate the main channel and some of the larger branches, but most small branches are unrecoverable because the thunder produced by them has too small an amplitude.

By studying the acoustic record of an entire thunderstorm we have been able to compile a graphic history of the large-scale electrical discharges during the lifetime of a storm. From the same data we have derived information on the process by which the cloud acquires elec-

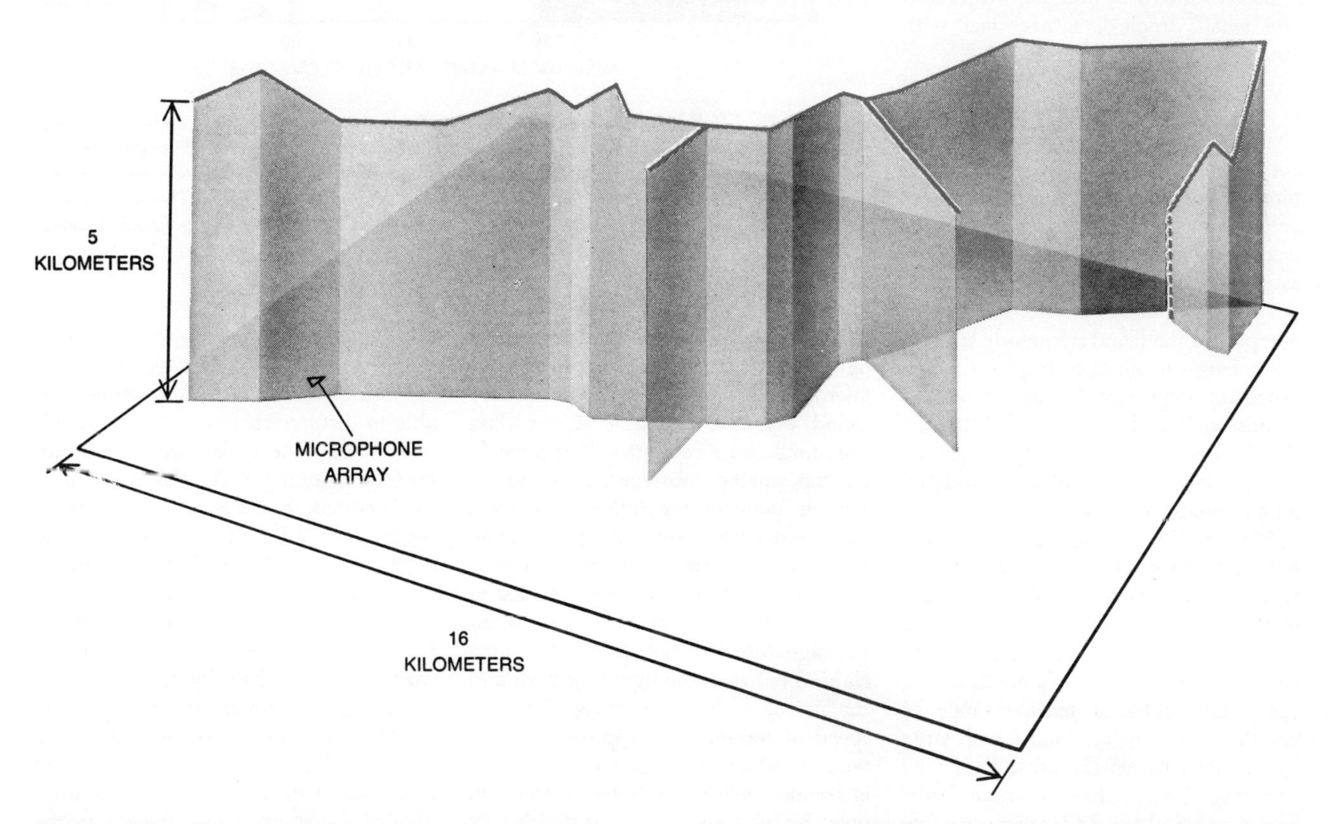

RECONSTRUCTION of a lightning channel was derived from the thunder signature at the top of the page. Most of the channel is horizontal and is hidden from visual observation by cloud. The macrotortuous segments of the main channel and the larger branches are recorded, but smaller features cannot be resolved and small branches cannot be reconstructed because the thunder they produce is too feeble. Thunder from lowest portion of the channel (*broken line*) could not be detected at the microphone array because of refraction.

tric charge, the volume in which the charge is stored and the time required to replenish the charge after a lightning stroke. The study of thunder has thus made possible tentative generalizations about electrical activity in the atmosphere and has led to some surprising discoveries about lightning inside clouds.

We have found that intracloud lightning (lightning that does not discharge to the ground) is predominantly horizontal, and so is the intracloud portion of cloud-to-ground lightning. The horizontal lightning channels tend to be aligned so that most of them are roughly parallel.

Our results also show that the negative charge center in the lower part of the cloud is commonly disk-shaped, about two kilometers thick and 10 kilometers in diameter. The positive charge, on the other hand, appears to be dispersed throughout the upper part of the cloud. The negatively charged region is consistently located near the −10 degree isotherm, an indication that the development of the charge is related to the freezing of droplets or to the coexistence of ice and droplets in the same part of the cloud. The development of charge also seems to be correlated with regions of inflow or updraft, which contain small droplets, rather than with areas where raindrops are found.

Lightning early in the life of a storm is confined to the lower, negatively charged region; the upper zone becomes active late in the cycle. Successive lightning flashes draw charge from different volumes of the cloud, but the channels frequently intersect at some point. It is as if one flash took up where the previous one left off. Moreover, lightning in one part of the cloud frequently triggers a discharge in another region. Certain processes important in the physics of clouds, such as the growth of cloud particles, appear to be strongly influenced by the electric field and are correlated with lightning activity.

Finally, the average length of the lightning channel seems to vary with the type of storm. Small, local storms have relatively short lightning discharges (typically about five kilometers long), and all the lightning channels in the storm fall within a narrow range of lengths. In storms associated with large frontal systems, on the other hand, the lightning channels have a broad distribution of lengths and a large mean value (about 15 kilometers). In the larger storms much of the channel length is horizontal.

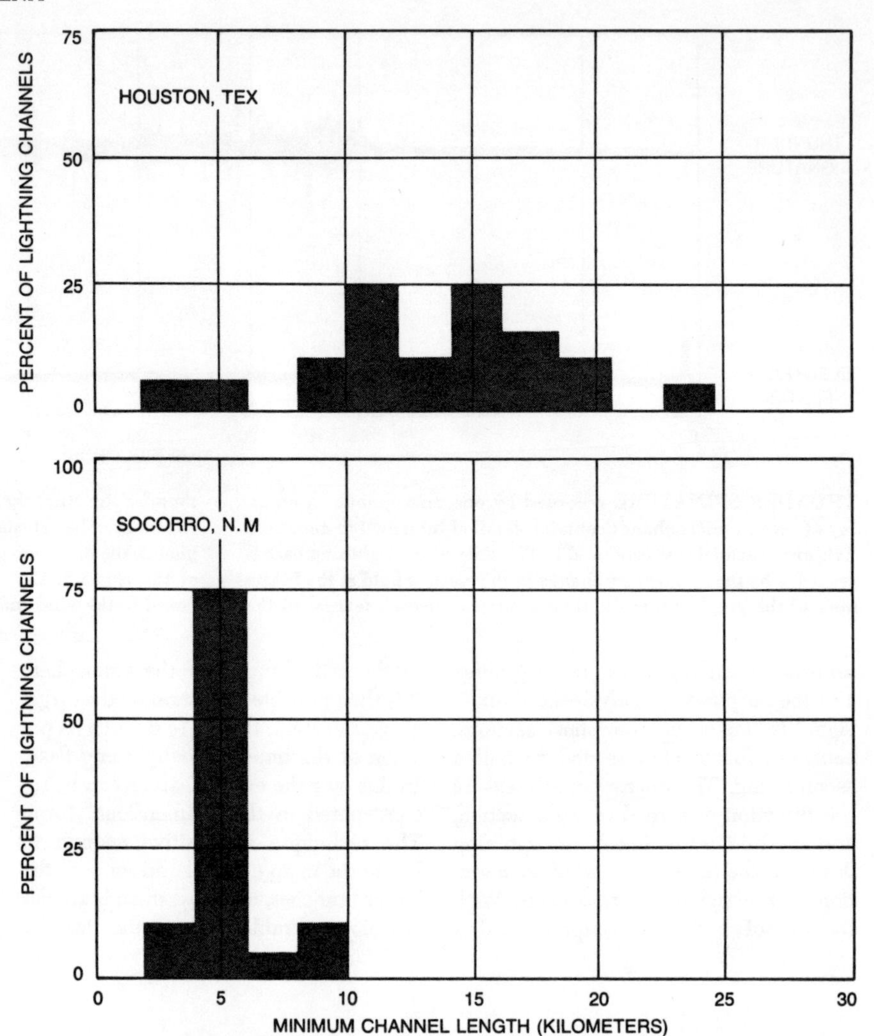

LENGTH OF LIGHTNING CHANNELS varies from storm to storm. A thunderstorm near Houston, Tex., associated with a frontal system, had a broad distribution of lengths, including many quite long channels. In a local storm at Socorro, N.M., however, there were no very long channels and the great majority were clustered around a single value. The lengths measured were the minimum possible lengths, as determined by duration of the thunder signal.

The complete reconstruction of a lightning channel requires arrays of microphones sensitive to low-frequency sound, equipment capable of recording the sound and a computer. It is possible for the amateur observer, however, to recover some of the information from the thunder signature with apparatus no more elaborate than a wristwatch. You can obtain the approximate distance to the source of a feature in the thunder signature by multiplying the time elapsed between the lightning flash and the arrival of the acoustic signal by the speed of sound. (The approximate distance in kilometers is given by the time in seconds divided by 3, the distance in miles by the time in seconds divided by 5.) Similarly, you can estimate the minimum length of the channel from the total duration of the thunder, again multiplying the elapsed time by the speed of sound [*see illustration on page 118*].

With careful observation you may be able to distinguish some of the individual events that make up a cloud-to-ground lightning flash: the creation of the tortuous, branched channel forged by the stepped leader, the brightening of the channel with the first return stroke and a flickering produced by multiple return strokes. Occasionally a lightning flash forks when a dart leader deviates from the path of the original channel.

During a thunderstorm at night you might also want to try photographing lightning. Set up the camera on a tripod and point it toward the most active region of the storm. Close the iris to the smallest aperture possible (that is, set the lens to the largest *f* number), focus at infinity and make time exposures of from 20 to 30 seconds. Typically one in every three or four frames will contain a lightning photograph, although the number depends on the activity of the storm.

More information about the lightning channel can be obtained through careful attention to the thunder signature. Measure the delay between the flash and the first thunder heard, the loudest clap and the final rumble. From these times you can estimate respectively the distance to the branch nearest you, to the main channel and to the farthest branch. Also note the total duration of the thunder in order to calculate the minimum length of the channel.

A discharge that has struck the ground nearby gives rise to a loud crack, sometimes preceded by a brief rumble or a ripping noise that probably originates in a small branch extending from the main channel toward the observer. When a nearby flash is made up of several strokes, it is sometimes possible to distinguish the acoustic pulses produced by each stroke. The sound is somewhat like a short burst of machine-gun fire.

When thunder powerful enough to shake windows is heard as a boom rather than as a clap or a crash, it is usually the product of an energetic flash but a distant or high-altitude one. At a distance of a few kilometers the high frequencies are attenuated with respect to the low frequencies, and the resulting thunder is felt as much as it is heard. In some cases the pitch of the thunder grows progressively lower as sounds arrive from higher or more distant portions of the channel.

Finally, a somewhat rare thunder signal is a ripping noise that can be imagined as the tearing of some cosmic cloth. It is usually attributed to a stepped leader that fails to reach the ground. When the leader does complete a path to the ground, the sound it generates is overwhelmed by that of the subsequent return stroke.

Obviously caution is in order when you are observing a thunderstorm. Do not expose yourself to the risk of personal participation in a lightning discharge by standing in open country or near trees, power lines, fences or other objects likely to be struck. The safest observation post is a closed space, such as a building or an automobile, provided that you avoid touching exterior walls and conducting surfaces.

14 Atmospheric Halos

by David K. Lynch
April 1978

Rings around the sun and moon and related apparitions in the sky are caused by myriad crystals of ice. Precisely how they are formed is still a challenge to modern physics

Anyone who spends a fair amount of time outdoors and keeps an eye on the sky is likely to see occasionally a misty ring or halo around the sun or the moon. The phenomenon is well established in folklore as a sign that a storm is coming. Actually the halo is only one of a number of optical effects that arise from the same cause, which is the reflection and refraction of light by crystals of ice in the air. Whenever cirrus clouds or ice fogs form, arcs of light appear overhead, woven into the veil of cirrus in a splendid variety of circles, arcs and dots.

The effects are best seen when the clouds are thick enough to fill the air with ice crystals but not so thick as to hide the sun. The commonest effect is the 22-degree halo, so named because its radius subtends an angle of 22 degrees from the eye of the observer. The halo appears as a thin ring of light (about 1.5 degrees wide) centered on the sun; sometimes it is pale white and sometimes it is brightly colored, with red on the inside and blue on the outside. The colors are clearest when the clouds form a uniform, featureless haze. If the clouds are patchy, the halo may be incomplete.

Frequently the 22-degree halo appears in company with two "sun dogs," which are bright and sometimes colored patches of light on each side of the halo, either on it or just outside of it. The formal name for them is parhelia, from the Greek for "with the sun." Occasionally visible is a larger, fainter halo with a radius of about 46 degrees.

If the cloud cover is uniform, one can sometimes see a ring of light encircling the sky parallel to the horizon. It is the parhelic circle. It passes through the sun and the parhelia and, if they are visible, through the anthelion (a whitish patch opposite the sun) and the paranthelia (which are like the anthelion but are located at azimuths of plus and minus 120 degrees from the sun).

Other phenomena commonly observed are the circumscribed halo and the circumzenith arc. The circumscribed halo surrounds the 22-degree halo and is bilaterally symmetrical with it. At the top and bottom the two are tangent. The circumzenith arc appears as an inverted rainbow centered on the zenith, facing the sun.

Many other phenomena of this kind have been identified, and I shall describe a number of them. As the optical effects of atmospheric ice crystals are enumerated, however, a point is reached where their existence and properties become questionable. Rare, one-of-a-kind observations haunt the published material, and quantitative measurements are almost unknown. Was the halo real? Could it have been a known halo mistaken for a new one? Was it described accurately? Theoretical work by Robert G. Greenler of the University of Wisconsin at Milwaukee and his colleagues predicts certain arcs that have not been seen; have they been overlooked or is the theory incomplete?

Even though the theory of halos is primarily encompassed by classical optics, the principles of which have been known for centuries, the present understanding of these wonderful arcs is imperfect. Hence the study of halos is as fascinating now as it was 100 years ago.

The belief that a halo signifies the onset of bad weather has a basis in fact. A falling barometer is usually caused by an advancing low-pressure system. Violent convection carries moist surface air to altitudes of from 9,000 to 15,000 meters (30,000 to 50,000 feet), where the temperature is well below freezing. The air becomes supersaturated with water vapor, which condenses out and forms cirrus clouds. High-velocity winds above the system carry the wispy cirrus ahead of it, which is why halos can be seen in these clouds as the lovely first harbingers of foul weather.

Since the ice in the clouds is the source of the optical effects, one is led to consider its structure. The delicate snowflakes of winter show that ice is a hexagonal crystal. Such a crystal has four axes of symmetry: three *a* axes, which are of equal length and intersect at an angle of 120 degrees, and a *c* axis, which is of a different length and is perpendicular to the plane of the *a* axes.

Although many forms of ice can occur, only about four are important in meteorological optics. The others are either too rare or do not have smooth, regular optical faces. The important forms are the plate, which resembles a hexagonal bathroom tile, the column, the capped column and the bullet (a column with one pyramidal end).

In each of these forms, except for the single pyramid, the two end faces are parallel and lie perpendicular to the *c* axis. They are termed the basal faces. The angles between the crystal faces are always the same: 120 degrees for adjacent prism (side) faces, 60 degrees for alternate prism faces and 90 degrees for the junctions between ends and sides. The 60-degree and 90-degree combinations are responsible for nearly all of the halo phenomena.

Atmospheric ice crystals form by direct sublimation from air that is supersaturated with water. The type of crystal depends primarily on the air temperature, although the degree of saturation relative to ice can be a factor when the saturation is less than 108 percent. Then only plates and columns can form. Most of the optically interesting crystals form when the saturation is between 100 and 140 percent and the temperature is between minus 5 and minus 25 degrees Celsius. When the saturation is higher than 140 percent, the growth of crystals is so rapid that rime (an amorphous deposit of frozen droplets) grows on the crystals and destroys their optical faces.

The relations of temperature and saturation can be represented in a diagram in which different regions are designated by Roman numerals [*see illustration on page 129*]. Composite crystals are formed when the growing crystal moves from one region to another. For example, a capped column begins as a simple column. During its period of formation it passes from Region VII to Region II, perhaps because it is descending through a stratified cloud. Once it is in the plate-forming region the columnar growth stops but the basal faces continue to develop outward from the *c* axis. If the column initially has two flat ends, it grows a cap at each end and comes to resemble a spool. A plate will not form on a column that terminates in a pyramid. It is important to realize that a capped column and all other mixed forms originate as single crystals and go through two separate periods of growth; they are not separate crystals that came together after they were formed.

The orientations of the ice crystals as they fall through the air are responsible for the wide variety of halos. Exceedingly small crystals (less than about 20 micrometers in diameter) are subject to Brownian motion, the random movement resulting from the impact on the crystals of the molecules of the air. The random collisions with air molecules cause the crystals to tumble constantly, so that all orientations are present.

When the crystals reach a size of from 50 to 500 micrometers, aerodynamic lift dominates the Brownian motion and forces the crystals into certain positions relative to the direction of their fall. If all of the ice particles are of one kind, they become aligned with one another. (Imagine what such a cloud looks like on a microscopic scale: billions of sparkling prisms lined up uniformly, glinting in the sunlight, each one producing its own family of tiny halos. Most halos are formed in this way.) When the ice crystals reach a size of from .5 millimeter to three millimeters, they tend to spin as they drift downward. These whirling crystals produce yet another class of halos, to which I shall return.

The 22-degree halo is formed by sunlight that passes through alternate side faces of randomly oriented crystals. All the crystals are less than 20 micrometers in diameter and all have 60-degree faces. Since the sunlight strikes the ice crystals at every possible angle, it may seem strange that a cloud composed of countless independent crystals should direct light chiefly at an angle of 22 degrees.

The principle underlying this effect is called the principle of minimum deviation. It is a cornerstone of classical optics and finds application in many areas of meteorological phenomena, including rainbows. Since the crystal faces are inclined to one another by 60 degrees, the problem of the 22-degree halo becomes one of understanding the passage of light through an ordinary spectroscope prism in a plane perpendicular to the *c* axis.

The angle between the incident ray and the emergent ray is termed the deviation; it is the angle by which the light changes direction in the crystal. As the angle of incidence increases from zero degrees (perpendicular to the face) the deviation decreases steadily, reaches a broad minimum and then increases again. In the vicinity of the minimum a

COMMON HALO, known as the 22-degree halo because its radius subtends an angle of 22 degrees from the eye of the observer, surrounds the sun. The photograph was made in Pasadena, which is why one sees a border of palm fronds. The halo is the result of randomly oriented ice crystals in cirrus clouds. This photograph and the one below were made with a wide-angle lens.

HALO COMPLEX was photographed near the South Pole. Visible are the 22-degree and 46-degree halos, the parhelic circle and its parhelia, the upper Parry arc and the circumzenith arc. The variety shows that the cirrus clouds had various shapes and orientations of ice crystals.

SUN AND PARHELION were photographed from White Mountain in California. The parhelion, or "sun dog," is the bright spot at the right. The formal name is derived from the Greek for "with the sun." Frequently two parhelia are visible on opposite sides of the sun.

THREE CROSSED ARCS meet at the antisolar point, which is always below the horizon when the sun is up. The arcs result from multiple reflections off the side faces and one end face of column-shaped crystals. The optics are much the same as in an ordinary kaleidoscope.

change in the angle of incidence produces no change in the deviation. Light therefore accumulates at the angle of minimum deviation. The 22-degree halo is this concentration of light, and it is circular because all orientations are present. Since the deviation of light can be more than 22 degrees but not less, the halo is actually doughnut-shaped, with a bright and sharp inner edge due to minimum deviation and a diffuse outer region resulting from the rays that traverse the crystal at other angles.

Two quantities determine the angle of minimum deviation: (1) the angle between the faces and (2) the index of refraction. The mean index of refraction for ice is about 1.31; as in all solids, however, it varies slightly with color, that is, with wavelength. This property is termed dispersion. It causes white light to be split up so that each component color travels in a slightly different direction. Hence the angle of minimum deviation is a bit different for each color, being smallest for red light. Thus the halo is in fact composed of a continuum of superposed halos, each of a slightly different color and size.

Without dispersion the overlapping halos would combine and would appear pure white, like the light from which they originated. Since the reddish halos are smaller than the others, however, they are seen at the inner edge of the composite halo. Being slightly separated from the rest, their colors are washed out the least. Other colors are considerably smeared because red rays near the minimum deviation can fall on them at the minimum deviation, whereas the opposite effect is impossible.

The 46-degree halo is formed in exactly the same way except that the crystal faces that refract the light are a basal face and a side face that share an edge. Such faces always intersect at 90 degrees rather than 60 degrees. Other halos formed by minimum deviation in randomly oriented crystals are rarely observed, but they can be explained in terms of the prisms with pyramidal terminations. They include six halos ranging in size from eight to 32 degrees.

The commonest optical effects caused by oriented crystals are the parhelia. They are at least as common as the 22-degree halo and are much easier to see because they are brighter. The crystals responsible for these "mock suns" are capped columns, bullets of moderate size and plates, all with vertical *c* axes. Aerodynamic lift forces the crystals to descend in this position.

As before, the light passes through alternate side faces. Since the faces are vertical and the sun is above the horizon, however, sunlight enters the crystal obliquely, and the plane on which the light travels is not perpendicular to the *c* axis. Thus a true minimum deviation does not occur. A "quasi-minimum" deviation does take place, and it concentrates light; the angle is always more than 22 degrees, however, so that the parhelia are formed outside the 22-degree halo. Only when the sun is on the horizon are the conditions for true minimum deviation fulfilled, and then the parhelia do lie on the 22-degree halo.

The colors of the parhelia, which are often dazzling, result from refraction, as within the 22-degree halo. Like most oriented crystals, the plates and capped columns tend to wobble around their mean orientation; the movement smears out the optical effect. The brilliance of a parhelion can be affected by the degree of alignment of the available crystals. Parhelia often show a bluish-white tail that extends horizontally away from the sun. It is caused by the rays that traverse the crystal near but not at the quasi-minimum deviation. The tail is most evident when the sun is low.

Analogous to the parhelia of the 22-degree halo are the parhelia of the 46-degree halo. They are rarely reported and indeed may not exist at all. If the phenomena reported as parhelia of the 46-degree halo are really that, they are formed by a quasi-minimum deviation in a 90-degree prism of a crystal that

VARIETY OF OPTICAL EFFECTS appear in another photograph made near the South Pole. The sun, which is obscured by the flag, is surrounded by parts of the 22- and 46-degree halos and the colorful circumzenith arc, which is the concave arc near the top of the photograph.

TWO EFFECTS, a column and an upper tangent arc, are visible in this photograph. The column rises upward from the sun, and the arc is the bright spot above it in the sky. Such an arc is sometimes seen tangent to a 22-degree halo; from the air one sometimes sees the lower arc.

has its *c* axis horizontal and its refracting edge vertical.

One of the loveliest members of the halo family is the elusive circumzenith arc. Although it is formed in the same crystals (capped columns and bullets) as the parhelia are, it is observed far less often because it can occur only when the sun is below 32.2 degrees of elevation. (Moreover, people seldom look straight up, which is the direction of the arc.)

In forming such an arc light enters the upper horizontal face of a crystal and emerges through a vertical side face. At elevations larger than 32.2 degrees the light is totally reflected internally. At 32.2 degrees the emerging light travels straight down. Hence the circumzenith arc appears as a bright spot at the zenith. As the sun drops below this elevation the spot opens up into a splendid arc of color centered on the zenith and facing the sun. Although the circumzenith arc is not a mimimum-deviation phenomenon, it does achieve its maximum brightness when the sunlight passes through the crystal at the minimum deviation. That happens when the solar elevation is 22.1 degrees, at which point the 46-degree halo and the circumzenith arc are tangent.

Complementing the circumzenith arc is the colorful circumhorizon arc. It is formed in the same crystal but by light that enters through a vertical side face and leaves through the bottom horizontal face. On the basis of symmetry one can readily infer that the circumhorizon arc cannot appear when the sun is below an elevation of 57.8 degrees (90 degrees minus 32.2 degrees). The arc starts out as a ring of color on the horizon. As the sun rises so does the encircling circumhorizon arc. The maximum brightness again occurs at the minimum deviation, when the elevation of the sun is 67.9 degrees (90 degrees minus 22.1 degrees). Because this arc is a high-sun phenomenon it is one of the few halos that cannot be seen from any place on the earth. It is visible only in the Temperate Zone latitudes from 55.7 degrees north to 55.7 degrees south, where the sun can get high enough. Since the sun's elevation is highest in the summer, when cirrus clouds are less likely to form, the circumhorizon arc will probably always remain a rare sight.

Analogous to the circumzenith and circumhorizon arcs for the 60-degree faces are the upper and lower Parry arcs, named for an arc phenomenon described by the British explorer Sir William Parry in 1821. The arcs are formed just above and below the 22-degree halo. They change shape dramatically with changes in the elevation of the sun. This relation has undoubtedly caused confusion in identification, since the arcs can masquerade in many forms. The associated crystals are columns that are oriented with their *c* axes horizontal and with two side faces also horizontal, one on the top and one on the bottom.

Two halos, the parhelic circle and the solar pillar, deserve special attention because they are formed primarily by external reflection from oriented crystals. They are therefore colorless. The parhelic circle is formed by reflection from the vertical side faces of capped columns and plates (*c* axes vertical) and from the end faces of horizontal columns. Since there is no preferred azimuthal orientation, the side faces scatter light in all horizontal directions while preserving the vertical component. Thus light appears to come to the observer from every point of the compass but from a single altitude. The parhelic circle is seen as a horizontal ring of light running through the sun and encircling the sky parallel to the horizon. It is seldom seen in its entirety because the clouds usually do not cover the sky uniformly.

The solar pillar, a commoner phenomenon, is a vertical shaft of light extending upward from the sun. It is most often observed above the rising or setting sun. Occasionally it is tilted or seen below the sun. The pillar is caused by reflection from the basal faces of plates and capped columns. As the crystals descend (with their *a* axes horizontal, like a leaf) they wobble around the mean orientation and smear the reflected solar image out vertically. Pillars therefore provide strong evidence of oriented crystals and also show that the crystals oscillate. Although pillars produce no color of their own, they take on the color of the sun and so often appear to be orange or red.

When the pillar, the parhelic circle and the 22-degree halo appear together, they often form crosses in the sky. This effect has undoubtedly led some people to interpret halos as signs from heaven. The most famous account of a well-timed cross resulted from an incident in the Swiss Alps in the summer of 1865. Edward Whymper and his companions were returning from the first ascent of the Matterhorn when four of them fell and were killed. Some hours later Whymper saw a circle with three crosses in the clouds, "a strange and awesome sight, unique to me and indescribably imposing at such a moment."

Most of the halos I have described are visible in the general direction of the sun. Looking the other way one finds several interesting phenomena. The colorless anthelion (counter-sun) and the two paranthelia (with the counter-sun) are often visible, looking like beads strung out along the parhelic circle. They also can appear when the parhelic circle is absent. Both are at the elevation of the sun. The anthelion is at an azimuth of 180 degrees relative to the sun, and the parhelia are at plus and minus 120 degrees.

Another class of halos arises from spinning crystals. Nine different arcs, attendant on either the 22-degree or the 46-degree halo, can be identified in this group. Because they are refractive phenomena they can be brightly colored.

The mechanism begins with crystals of moderate size that are drifting down through the air. They quickly become oriented and reach a terminal velocity of about 20 centimeters per second. At that point the force of gravity is balanced by lift and viscous drag. Air flows smoothly around the crystal and remains relatively undisturbed after its passage. As the crystal grows, the grace-

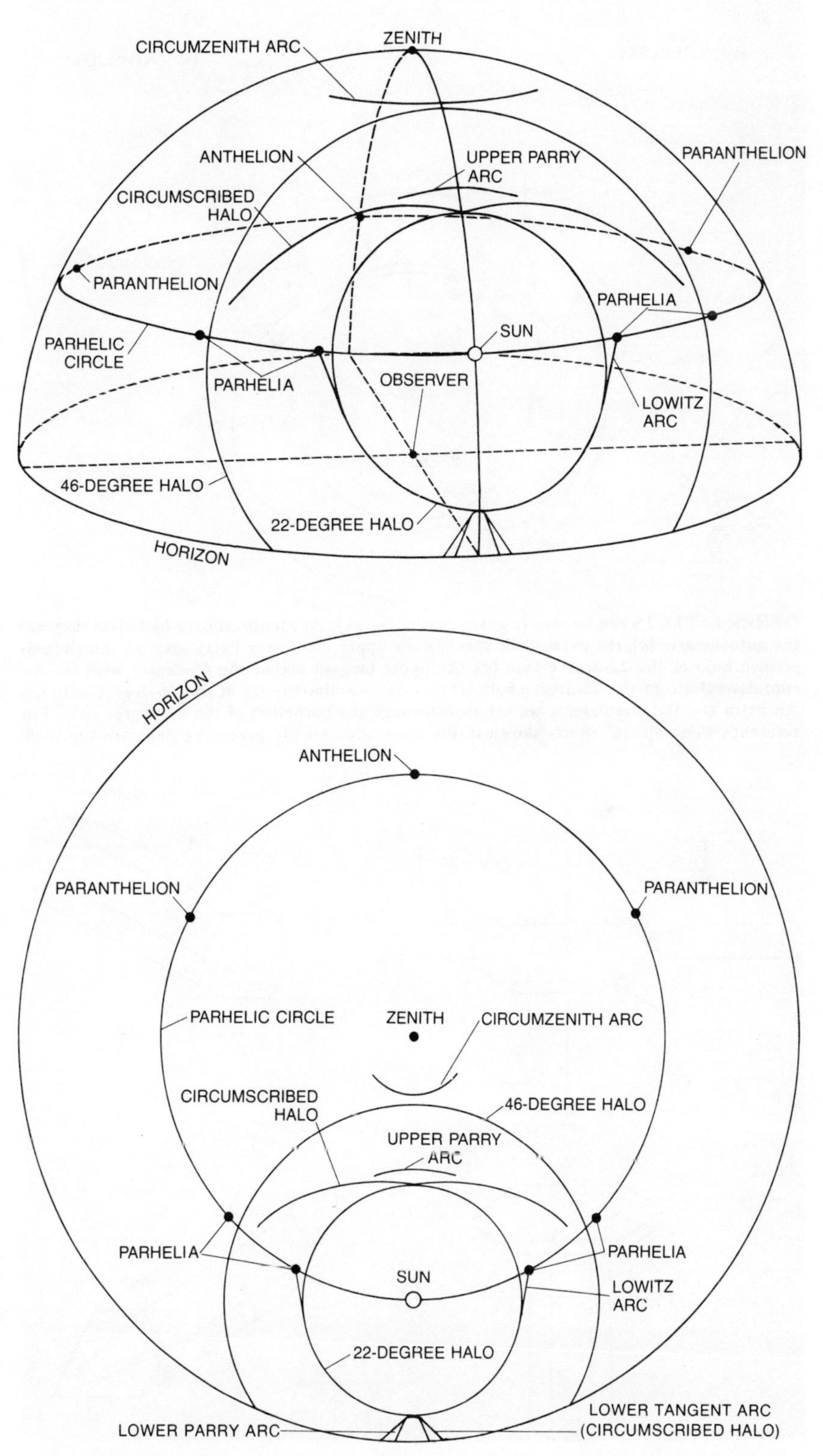

COMMONEST HALOS occur in the general direction of the sun and are portrayed here in two ways: in a perspective (*top*) from outside the hemisphere and in a view (*bottom*) straight upward to the observer's zenith. The same atmospheric optical effects appear in both views.

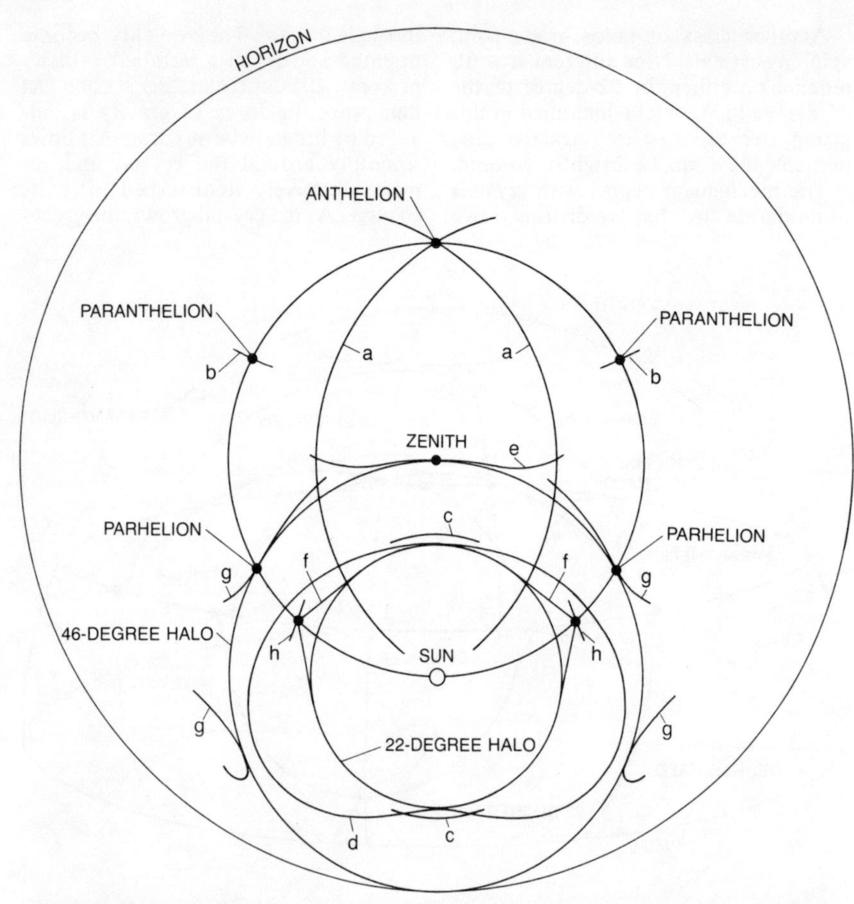

OTHER EFFECTS can be seen in every part of the sky. As identified here by letters they are the anthelic arcs (*a*), the paranthelic arcs (*b*), the upper and lower Parry arcs (*c*), the circumscribed halo of the 22-degree halo (*d*), the upper tangent arc of the 46-degree halo (*e*), the supralateral arc of the 22-degree halo (*f*) and the supralateral arc of the 46-degree halo (*g*). An extra arc, the mesolateral arc (*h*), runs through the parhelion of the 22-degree halo. For reference some optical effects shown in the illustration on the preceding page are repeated.

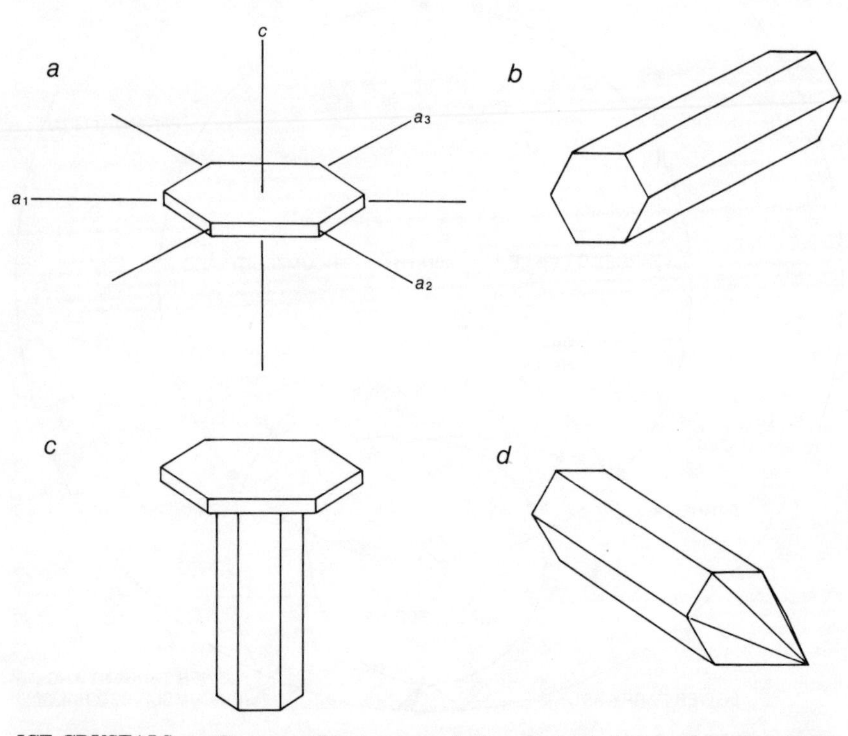

ICE CRYSTALS usually responsible for atmospheric optical effects have these four forms. They are the plate (*a*), with its four axes indicated, the column (*b*), the capped column (*c*) and the bullet (*d*). Although the crystals are drawn to the same scale, they occur in a variety of sizes.

ful streaming becomes increasingly unstable. In time the flow acquires an entirely different character: it becomes turbulent. The crystal leaves a wake of vortexes and eddies, which cause it to spin as it falls.

The halos that result from spinning crystals are lateral arcs and tangent arcs. They can be divided readily into two categories: the arcs attending the 22-degree halo (the infralateral or Lowitz arc, the supralateral arc, the mesolateral arc and the circumscribed halo) and the ones that accompany the 46-degree halo (the infralateral arc, the supralateral arc and the upper and lower tangent arcs). There are actually nine arcs because at low solar elevations the circumscribed halo looks like an upper and lower two-tangent arc to the 22-degree halo. All of the arcs can be brightly colored because they are refractive phenomena.

Some halos are formed below the horizon. In order to see them one must look down into the ice crystals. Until airplane travel became common such halos could be seen only from high mountaintops and cliffs. Bright halos can sometimes be seen below the horizon because of reflection from horizontally oriented ice faces. A sub-sun is frequently observed by people in airplanes and is a sure sign that sub-halos are about. I once saw (but alas did not photograph) a splendid sub-halo complex over Canada, consisting of a sub-sun flanked by two sub-parhelia.

The sun is not the only source of light for halos. At night the moon often has a halo. In northern regions where ice fogs occur pillars can be seen standing over street lamps and runway lights. When snowflakes lie horizontally, pillars and sub-lights can appear below the headlights of an automobile.

Many halos, some common and some rare, remain a puzzle. Anthelic arcs appear frequently, but attempts to explain them have not fully succeeded. They have been observed in so many forms that more than one crystal may be involved. A number of other halos are likewise not satisfactorily explained. Plainly much remains to be learned.

Research on halos is inching its way into the 20th century. So far, however, no work has been done beyond classical optics, and little progress has been made in the areas of polarization and diffraction. What is needed is a large number of observations, on which new theoretical work can be based.

This situation offers an opportunity to the interested skywatcher, who can do valuable research with modest equipment: a camera, a notebook and a sharp eye. Calibrated photographic observations are badly needed. The observer should record carefully the radius of any halo, the angular shapes and extensions of the arcs, the color characteris-

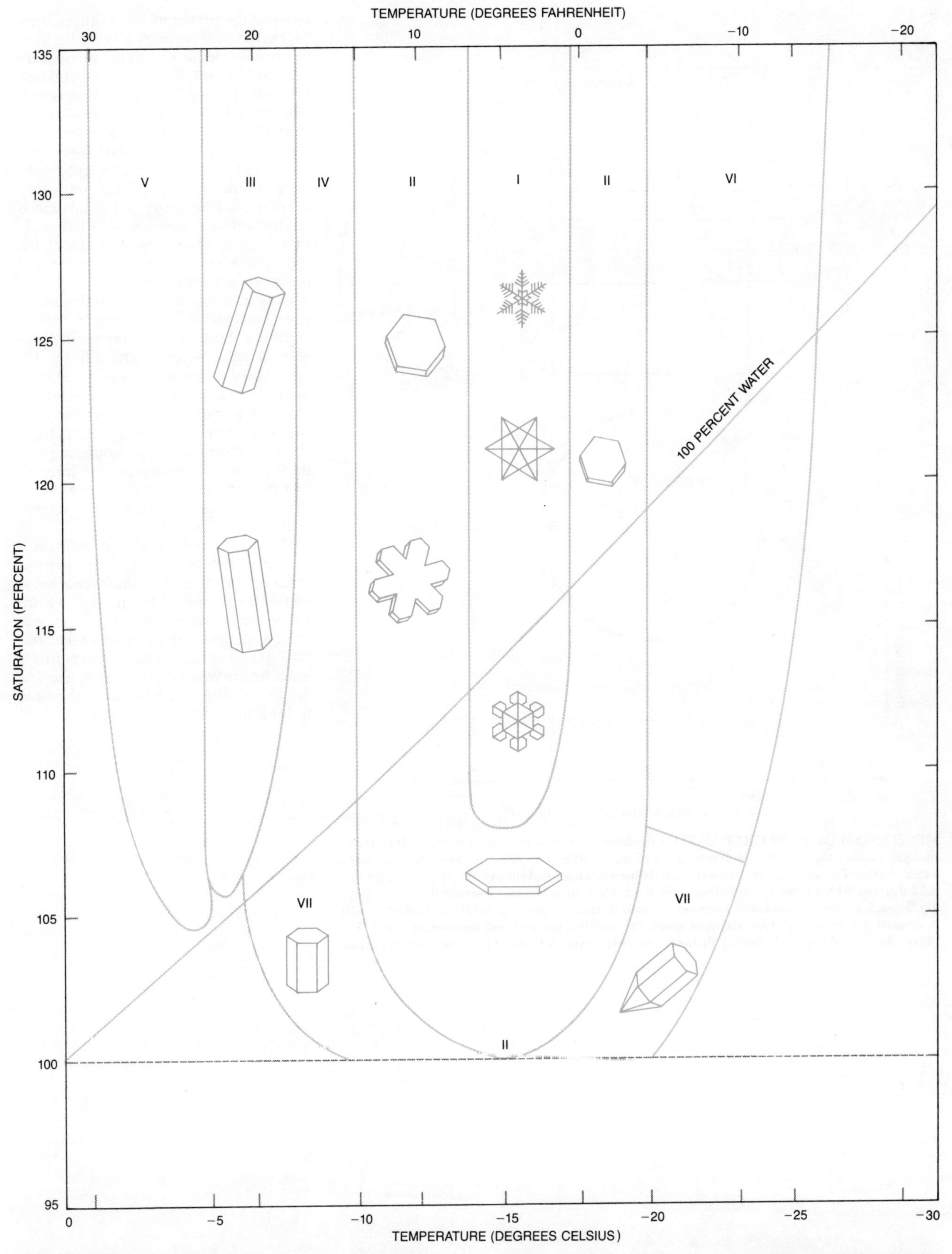

FORM OF CRYSTAL depends on the air temperature and the degree of water saturation of the ice. When the saturation is more than 108 percent, the temperature alone determines the type of crystal. The Roman numerals delineate regions with different forms, which can best be followed by reading the Celsius temperature scale from left to right. Irregular forms are not shown. The forms are irregular needles at minus 3 degrees (*V*), regular needles at minus 7 degrees (*III*), cups or scrolls at minus 9 degrees (*IV*), plates at minus 12 degrees (*II*), snowflakes at minus 15 degrees (*I*), plates at minus 17 degrees (*II*) and irregular plates at minus 23 degrees (*VI*). When the saturation is below 108 percent, only plates and columns grow. At saturations above 140 percent crystals grow so fast that they accumulate rime.

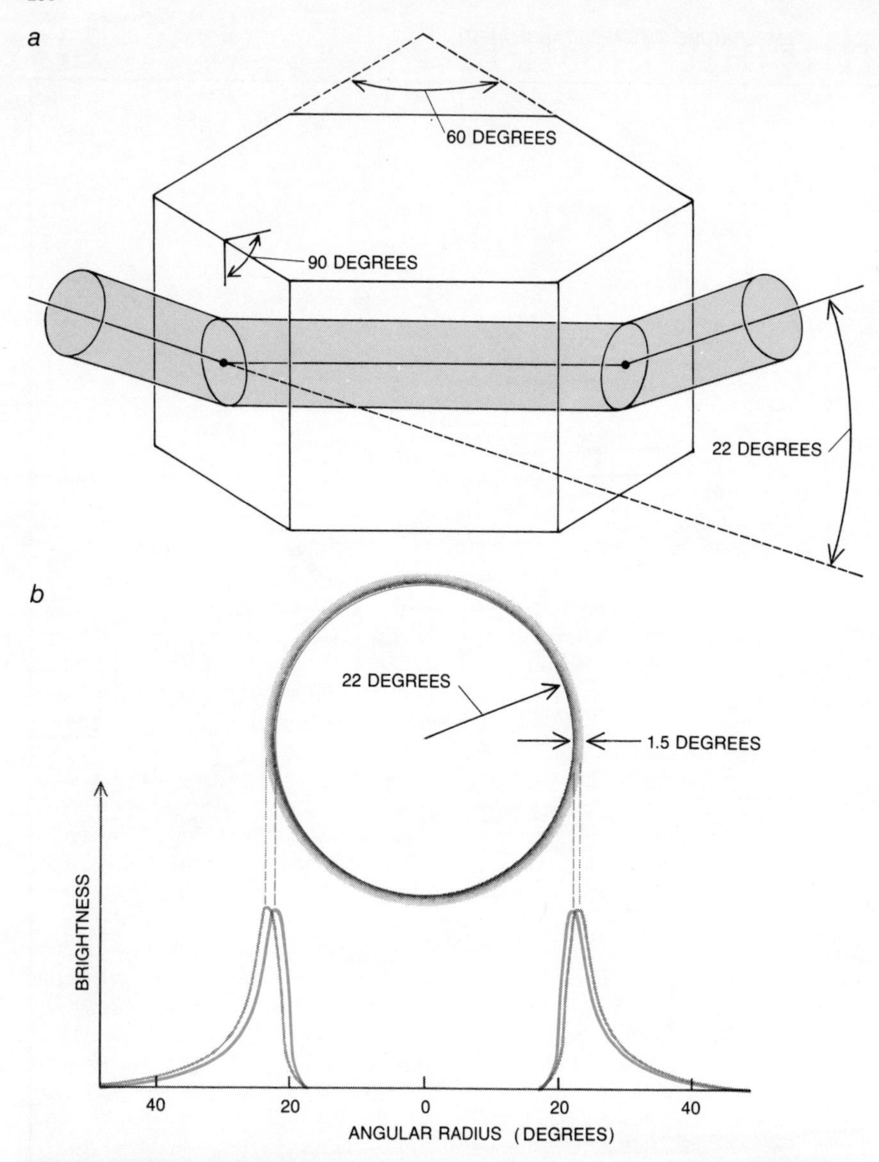

MECHANISM OF 22-DEGREE HALO is depicted. The halo is formed by the refraction of sunlight passing through the 60-degree faces of ice crystals. The effect is shown here (*a*) for a single crystal. The average deviation (the angle between the incident ray and the emergent ray) is 22 degrees, which is the mean radius of the halo. The halo is circular because the light is passing through billions of randomly oriented crystals. Because of dispersion (*b*) the 22-degree halo is actually a continuum of overlapping halos, the smallest one red and the largest violet. The inner edge is reddish and sharply defined; the outer edge is bluish-violet and much fuzzier.

tics and the period of observation. Photographs made through a Polaroid filter at several orientations to the vertical are most important. The filter's orientation for each photograph should be noted. The meteorological conditions at the time of the observations should be recorded, along with the date, the time, the altitude of the sun and the geographic location. Neither a large monetary grant nor a laboratory full of advanced equipment can compete with hundreds of energetic observers who are at the right place at the right time to see and report on a halo complex.

Only in a few fields of physics will a casual glance at the laboratory reveal anything about the experiments in progress. When a halo is sighted in the icy laboratory overhead, however, one immediately knows the temperature of the cloud; the state of the water; the size, shape and orientation of the ice crystals; the conditions of temperature and humidity in which the crystals are formed, and the subtle optics that are producing the halo. If several arcs are observed, even more is known.

Halos stir one's mind and soul, since they probe both the physical environment of the cloud and one's awareness and appreciation of the natural world. From the chaos of billions of pale, microscopic, angular crystals of ice nature spins a colorful fabric of expansive, graceful curves. All of them are accessible to the observer who takes the time to look up.

Noctilucent Clouds 15

by Robert K. Soberman
June 1963

These mysterious "night-shining clouds" that occur 50 miles up are seen occasionally in summer at high latitudes. In August 1962 rocket-borne collectors brought back samples of their substance

During the summer months in far northern and southern latitudes an alert observer will occasionally see tenuous clouds glowing faintly in the sky at twilight. Tinged a silvery blue, the clouds usually form billows resembling the waves of a ghostly ocean. Such clouds have been given the name noctilucent clouds, meaning "luminous night clouds." Although they must have aroused the wonder of the inhabitants of high latitudes for centuries, it is only recently that the real mysteries they present have come to be appreciated.

In 1896 the German astronomer O. Jesse measured the altitude of a display of noctilucent clouds over northern Germany from the parallax apparent in simultaneous photographs made with cameras 22 miles apart. He arrived at the startling figure of 250,000 feet (50 miles). Measurements made since have invariably come within a few thousand feet of this altitude. This explains the luminosity of the clouds: it is the reflection of sunlight reaching them from far below the horizon of the observer. The noctilucent clouds are the highest clouds of the planet. They ride in the sky above more than 99.9 per cent of the atmosphere and 40 miles higher than the realm of the weather.

At such high altitude what substance and what process can form clouds? Various investigators have argued that noctilucent clouds consist of tiny crystals of pure ice; that they are clouds of dust of terrestrial or meteoritic origin; that they are clouds of dust particles covered with ice. The third hypothesis has attracted the most support, but the issue could be settled only by examination of an actual sample of a noctilucent cloud.

Last August a rocket-borne collector trapped some of the substance of one of these clouds over Sweden and brought it safely back to earth. Analyses now being conducted in several laboratories in Sweden and the U.S., including my own in the Air Force Cambridge Research Laboratories, seem to have set aside the first two hypotheses without, so far, completely confirming the third. The evidence that the clouds contain ice-covered particles of meteoritic dust has now opened a series of new questions.

It is not at all clear, for example, why the clouds are seen only at high latitudes and almost exclusively in summer. On the one hand, conditions may be more favorable for seeing them at these latitudes in this season and they may be more widely distributed but unseen in the sky over the rest of the planet; on the other, the formation of the clouds may actually be confined, for reasons yet to be discovered, to the regions of the sky where they are seen and to the summer season. Why they always occur within a few thousand feet of an altitude of 50 miles presents still another question. The noctilucent clouds also invite investigation because the speed with which they move across the sky, sometimes exceeding 400 miles an hour, provides a measure of the wind velocity in the extremely thin atmosphere at such high altitude. Finally, in the mechanism that drives the winds there may be an explanation for the undulations of the clouds, which can extend dozens of miles from crest to crest and several miles in depth, sometimes standing still and sometimes even traveling in the opposite direction to the motion of the cloud as a whole.

Noctilucent clouds almost always make their appearance in the same well-defined region of the sky between the twilight arch at the horizon and the darkening vault of the night sky overhead. This is because they scatter the sunlight best in the forward direction and the optimum illumination and contrast exists when the sun is between six and 12 degrees below the horizon [*see upper illustration on page 135*]. A large part of the sky behind and partly above the observer is then in the shadow of the earth; there the sky is dark and stars are plainly visible. From the horizon to about 10 degrees above it, however, the dense lower atmosphere and the haze of dust and moisture in it catch and scatter the sunlight, outshining stars and noctilucent clouds. It is therefore in the region 10 to 20 degrees above the horizon, between the twilight arch and the night sky, that the clouds usually appear.

The clouds visible to the observer may be part of a system covering much larger regions of the sky. Although occasional bright displays have been seen even up to the zenith in the clear atmosphere of northern Scandinavia, the clouds that are overhead at one place are usually visible only to observers elsewhere who can see them at a lower angle. Noctilucent clouds are so thin that bright stars shine through them. Binoculars or a telescope can be used to exclude the light of the twilight arch from the eyes and increase the apparent contrast in the wave structure of the clouds. For some unknown psychological reason it is also helpful to observe the clouds inverted by a mirror, prism or lens or simply by bending over or lying down to get the inverted view. The illusion of depth and distance and the brightness of the clouds can be enhanced even in a photograph if one looks at the photograph upside down; the reader can verify this with the cloud photographs illustrating this article, allowing 30 seconds or so for the illusion to take effect.

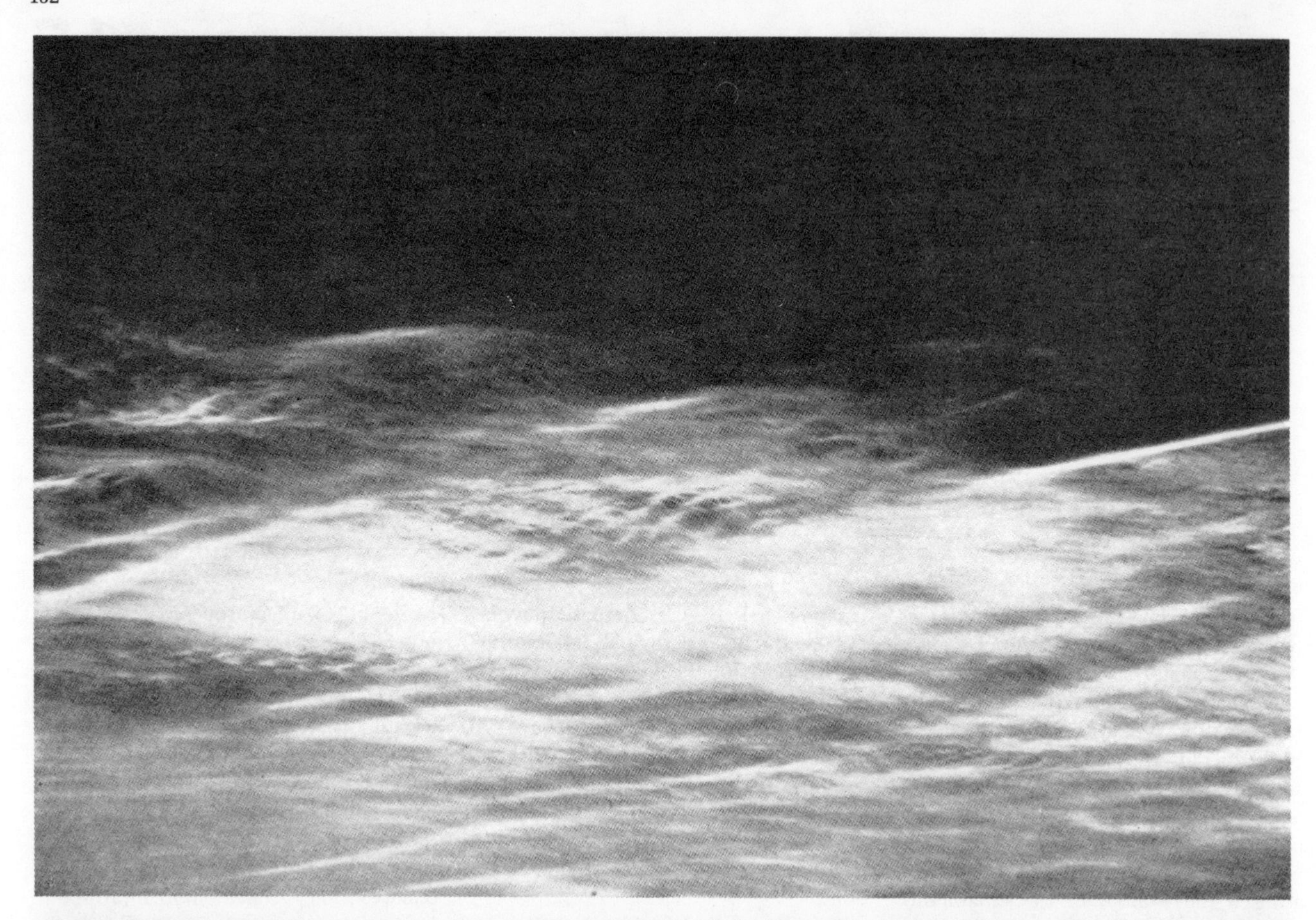

NOCTILUCENT CLOUDS were photographed near Östersund, Sweden, on the night of August 10–11, 1958, by Georg Witt. The upper photograph was made at 11:45 p.m., the lower at 12:10 a.m. The display moved higher in the sky and the clouds changed considerably between exposures. The waves average about 15 miles from crest to crest.

From the record of observations one might conclude that noctilucent clouds are rare and that they appear mostly during the summer and only between the latitudes of about 45 and 80 degrees in the Northern Hemisphere. The record becomes suspect when one notes that most sightings have been recorded in northwestern Europe and the U.S.S.R. above 58 degrees. According to Benson Fogle of the University of Alaska, only two sightings from Canada and one from Alaska had been recorded before 1962. Obviously the paucity of observations in the high latitudes of North America and the Southern Hemisphere must be more closely related to the lower density of population and hence to the smaller number of potential observers in these regions than to the physics of the upper atmosphere. At the suggestion of Sydney Chapman of the University of Alaska observers kept a cloud watch at College, Alaska (near Fairbanks) last August. The record for North America was immediately doubled by the recording of three new displays. The stir of interest in the subject has since brought a report from the U.S. Weather Bureau at Anchorage, Alaska, to the effect that noctilucent clouds are observed there at least once a year and that the bureau's record for 1962 showed a sighting in May and another in August. Arnold Hanson of the University of Washington has recently told of seeing noctilucent clouds on September 13, 1961, from a floating ice island, Arliss II, off the northern coast of Alaska at a latitude of 76 degrees, the second northernmost sighting ever reported. The sparse record of the Southern Hemisphere has been fleshed out by a report from the U.S. oceanographic vessel *Eltanin* of a sighting last December over the South Atlantic east of Tierra del Fuego.

The question still remains whether or not the clouds as well as the observations are restricted to the high latitudes during the summer. Without doubt the chances for sighting the clouds are best in those regions at that time because the sun does not dip far below the horizon before starting to rise again. Twilight and ideal seeing conditions can therefore last for hours [*see lower illustration on page 135*]. On the other hand, since the twilight centers around midnight at these latitudes, low temperatures and frost may deter all but the most dedicated cloud-hunters. The clouds have been sought at lower latitudes but without success. It is true that the twilight is shorter; it lasts for only 20 minutes at 30 degrees from the Equator. In addition, an untrained person can easily mistake noctilucent clouds for the familiar cirrus and cirro-stratus clouds. Offsetting these considerations is the fact that the observing time comes at a more reasonable hour. There are also many more qualified observers living in these latitudes, some of whom should have seen noctilucent clouds if any had occurred.

Two of the three hypotheses about the nature of the clouds offer some explanation for the apparent localization of the phenomenon in time and space. According to the first hypothesis, noctilucent clouds are true ice clouds formed by the spontaneous condensation of water vapor. This can occur when the moisture content of the air becomes so great in relation to the temperature that the air is supersaturated (a relative humidity of more than 100 per cent), a common condition in the atmosphere from sea level to 30,000 or 40,000 feet. Higher in the stratosphere, up to about 175,000 feet, the relatively high temperature and low moisture keep atmospheric moisture in the vapor state. At still higher altitudes the temperature drops until it reaches a second minimum at about 250,000 feet [*see illustration on opposite page*] in a region known as the mesopause. Here, given sufficiently high humidity along with sufficiently low temperature, condensation of the atmospheric moisture could occur. Rocket studies over Fort Churchill in northern Canada have established that the mesopause temperature at that latitude reaches its lowest value, about 160 degrees below zero Fahrenheit, during the summer months. This may explain why the clouds appear only in summer, and only in the mesopause.

The second hypothesis, which holds that noctilucent clouds consist of nonvolatile solid particles, has drawn some support from weak correlations between meteoritic showers and observations of the clouds. An unusually brilliant display over Russia, Sweden, Denmark, Germany and England appeared hours after the great Tunguska meteorite fell in Siberia in 1908. The lowest-latitude sighting, at 45 degrees, was reported on this occasion. Meteorites typically vaporize below 300,000 feet. Many are seen

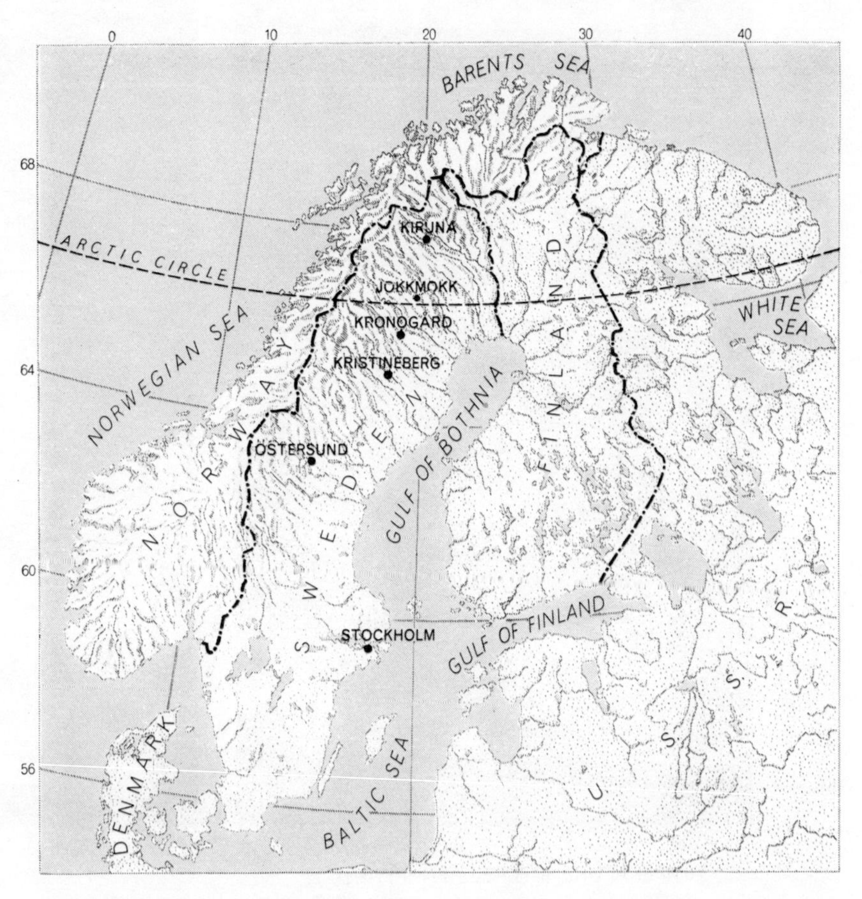

SITE OF ROCKET EXPERIMENT was Kronogård, a farm near the Arctic Circle. Observers at Kristineberg could see noctilucent clouds that were directly overhead at Kronogård, although invisible there. Airplane pilots also kept watch between Stockholm and Kiruna.

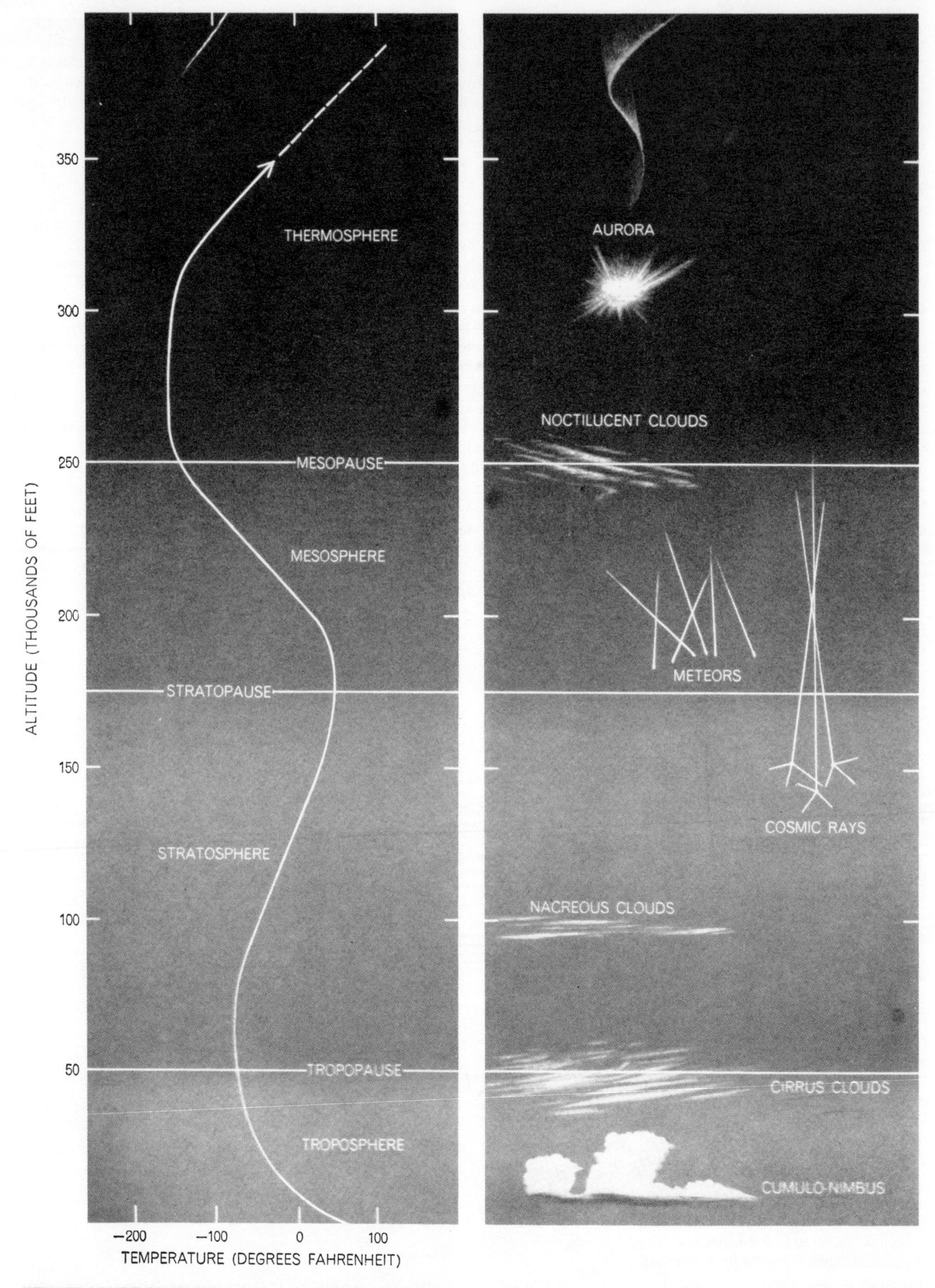

TEMPERATURE CHANGES define various regions of the atmosphere (*left*). Troposphere contains the weather clouds. At the tropopause the temperature ceases to drop. It starts to decline again at the stratopause, and at mesopause it reaches a second minimum. Noctilucent clouds occur only in the region of the mesopause. The lower nacreous clouds are also rare and poorly understood.

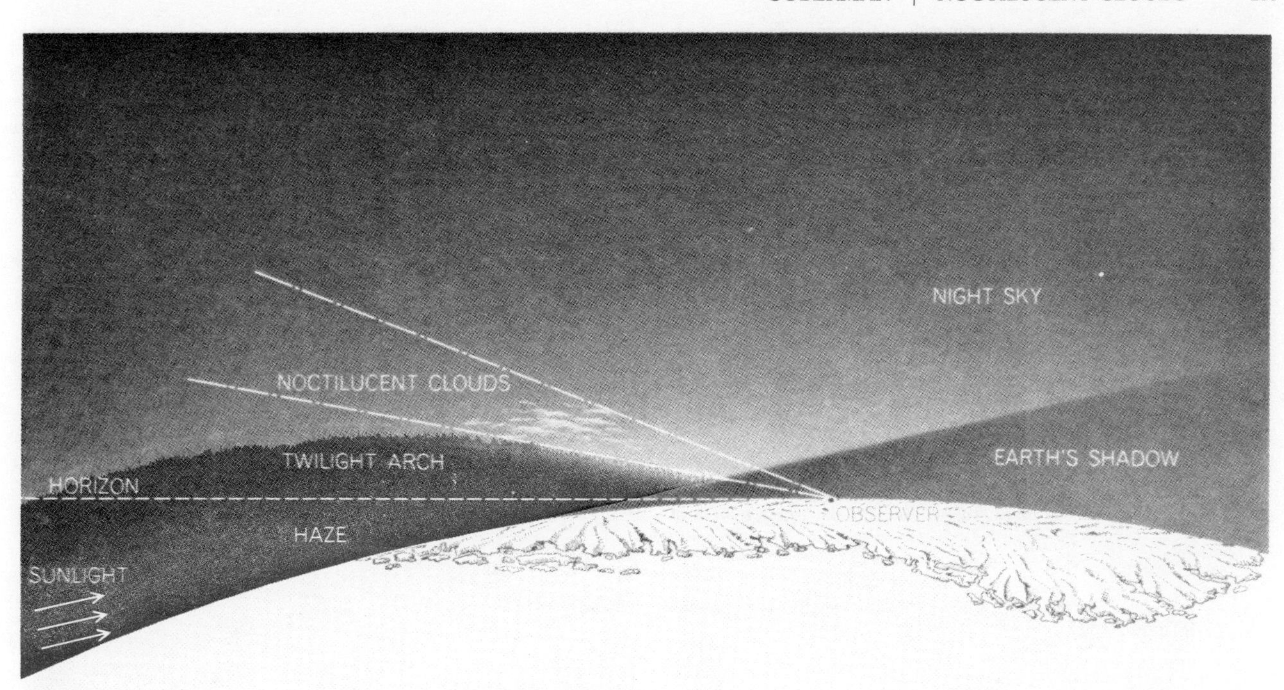

TWILIGHT CONDITIONS associated with observation of noctilucent clouds are illustrated in this diagram. Sunlight comes from below horizon; observer is in semidarkness. Haze layer absorbs sunlight and scatters red wavelengths toward observer, producing twilight arch (*color*) that blots out faint sources of light. The noctilucent clouds can be seen only above the arch. Occasional bright cloud displays are visible directly above the observer. Conditions shown here last for hours in summer in far north and south.

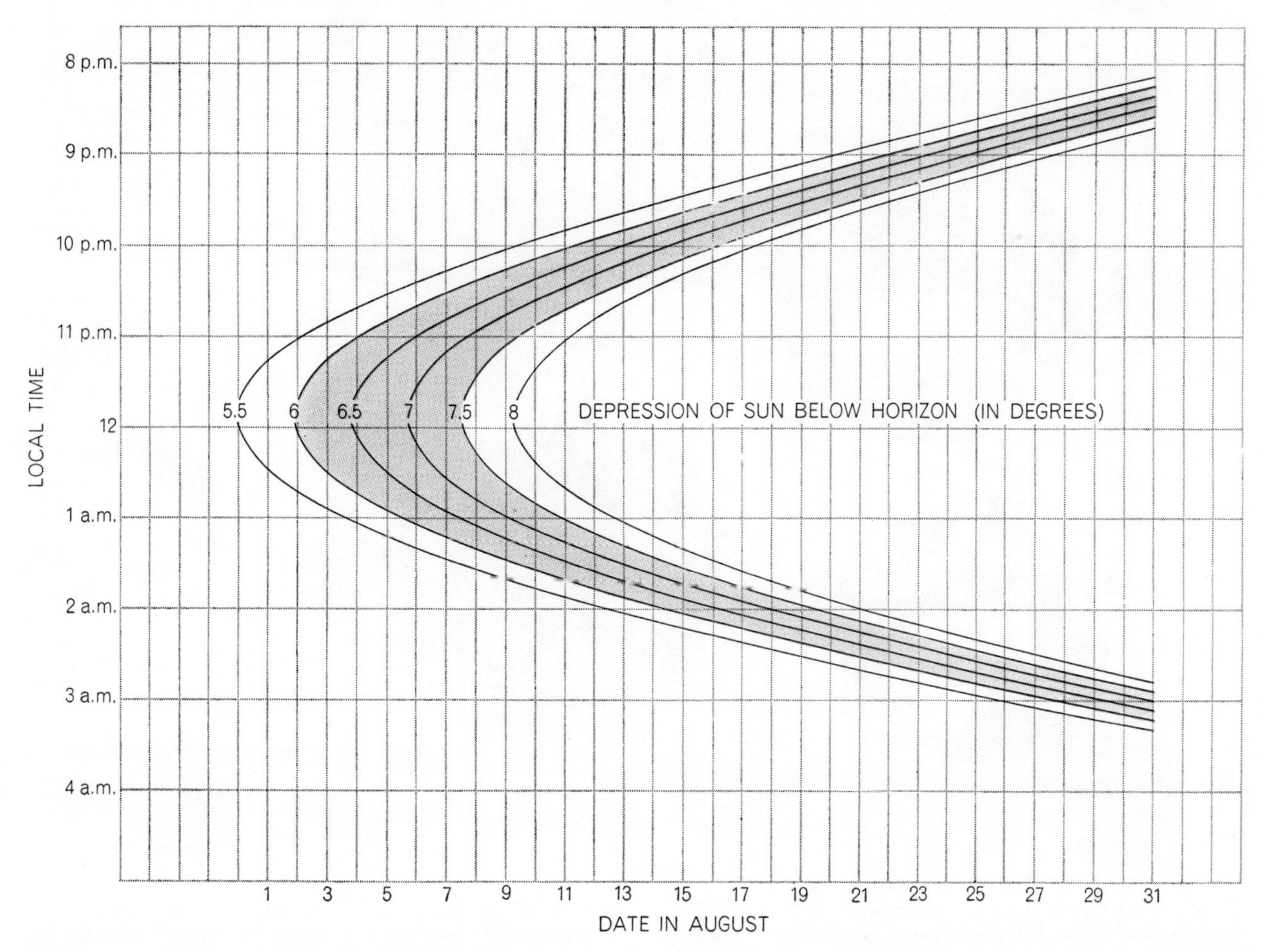

OPTIMUM OBSERVING TIME (*color*) during August at Kristineberg can be found on this chart devised by Witt. Clouds seen from Kristineberg at normal angle are directly overhead at Kronogård. Continuous observing time is longest early in the month.

PARTICLE COLLECTOR carried by rocket into noctilucent cloud consisted of (*from left*) can, two spacers, four coated surfaces to catch particles, spoked hold-down plate and a cover for can. Alternate spokes of hold-down plate crossed but did not touch center of each surface, enabling particles from air to settle there but keeping upper-atmosphere particles away from this area. This arrangement provided one of the controls that allowed the investigators to differentiate between matter from the lower and the upper atmosphere.

INNER NOSE CONE of rocket (*upper right*) had two oval holes to expose cans (*upper left*) for collecting particles from mesopause region of atmosphere. Slot in cone was port for nuclear-emulsion pack (*lower right*) of cosmic ray experiment. Interior cone (*lower left*) rotated 90 degrees at 275,000-foot altitude to close ports, and rod holding the can covers also turned 90 degrees to seal the cans. The nuclear emulsion was also turned away from slot. All the parts are arranged here in the "open" position in relation to each other.

to break up in their flight through the atmosphere. In addition, particles too small to leave a visible meteor trail—micrometeorites—constantly rain on the earth. They can be recovered from the air and from deep-sea sediments [see "Cosmic Spherules and Meteoritic Dust," by Hans Pettersson; SCIENTIFIC AMERICAN, February, 1960]. The idea that meteoritic dust alone makes up the clouds, however, leaves a number of aspects unexplained. Since the meteoritic dust particles are a permanent feature of the upper atmosphere and presumably distributed evenly around the world, there is no immediately apparent reason why the clouds should make such infrequent appearances, during the summer only, at high latitudes and at the 250,000-foot altitude.

Only a few investigators have suggested that the clouds could be composed of solid particles from the earth. Measurements of the intensity and polarization of the light from the clouds have indicated that the mean diameter of the cloud particles ranges from .1 to .4 micron. (A micron is a thousandth of a millimeter.) There is no evidence for vertical winds strong enough to lift particles of this size to 250,000 feet. Without the aid of cataclysms such as volcanic eruptions (rarer than noctilucent clouds) or the explosion of nuclear bombs (unknown before 1945) such particles could not reach the required altitude.

The idea that clouds consist of crystals of ice condensed around nonvolatile solid particles agrees with what is known about the formation of ordinary clouds in the troposphere. Solid particles of microscopic size form condensation nuclei that promote the translation of the water in the atmosphere from the vapor to the liquid and solid state. In line with this idea it could be shown that the condensation nuclei of the noctilucent clouds may have a terrestrial origin; they are presumably small enough to be raised by the same circulation mechanism that brings the water-vapor molecules to high altitude. Dust of meteoritic origin, however, could supply the condensation nuclei equally well. In either case the ice coating would cause the particles to "grow" to the postulated diameter of .1 to .4 micron, at which time they would become visible as clouds. The rarity and the peculiar distribution of the clouds could be explained in terms of the special meteorological conditions necessary to wring the water vapor out of the thin upper atmosphere. Some investigators have even suggested that variation in the delicate balance of condensation and evaporation at slightly different heights causes the wave formations so characteristic of the clouds.

In 1960 the Meteor Physics Branch under my direction at the Air Force Cambridge Research Laboratories started looking into the possibility of direct rocket investigation of noctilucent clouds. We were then working with groups at Union College in Schenectady, N.Y., and Northeastern University in Boston on a project employing rockets to sample meteoritic dust in the upper atmosphere. In June, 1961, the first successful flight of the "Venus Flytrap" rocket brought back particles of this material from 102 miles above its launching site at White Sands, N.M. We decided that direct sampling of noctilucent clouds offered an equally interesting challenge.

Since the project would entail waiting at the launching pad with a "hot" rocket until the clouds appeared, we needed a location where noctilucent

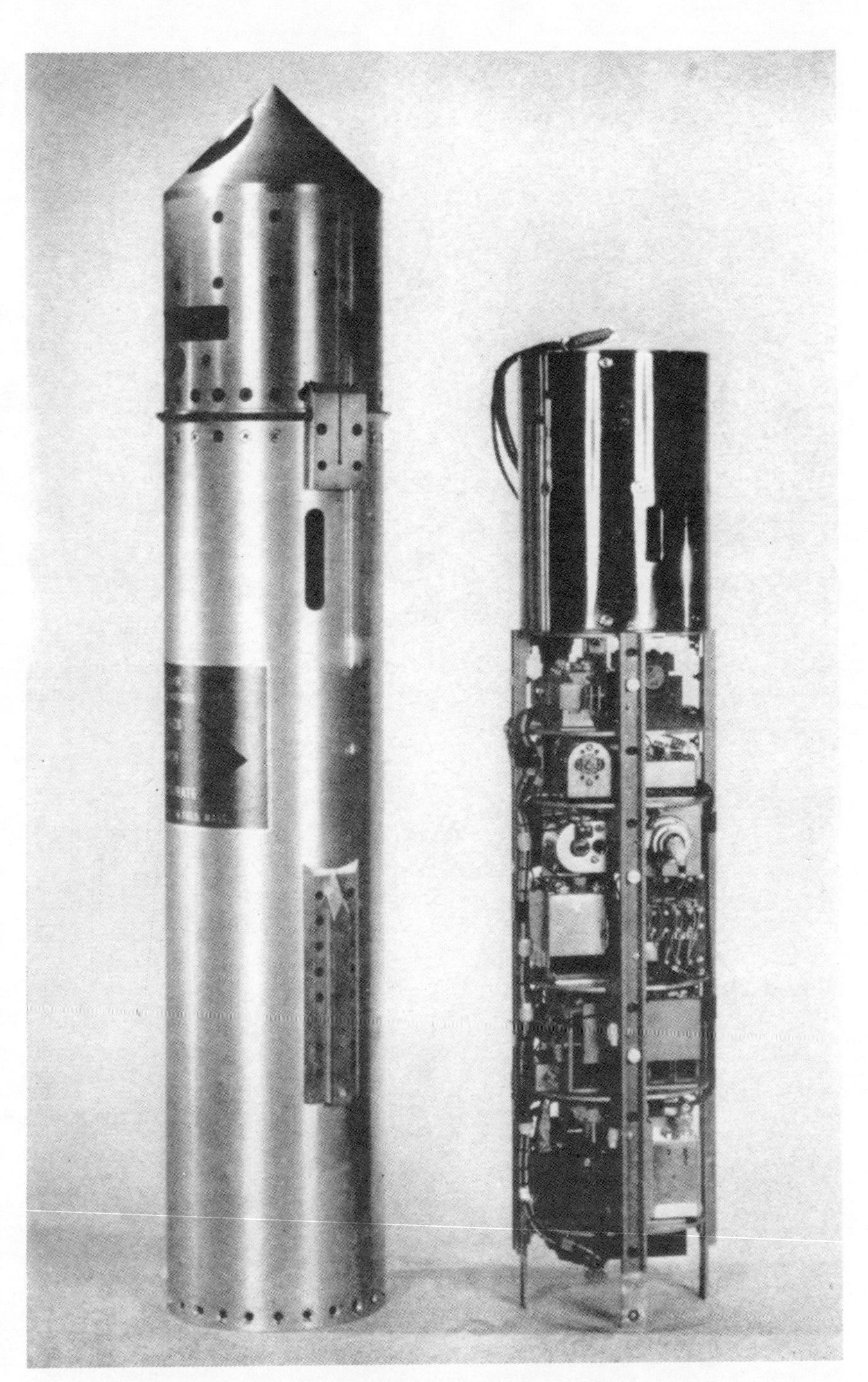

ROCKET PAYLOAD PACKAGE included electronic controls, radio transmitters and other equipment seen here. The nose cone, empty here, carried experiments shown on next page.

NIKE-CAJUN ROCKET is shown being prepared for launching in August 1962 at Kronogård. Four of these rockets were flown and two nose cones descended successfully from mesopause.

MOMENT OF LAUNCHING of the second rocket took place on August 11 at 2:40 in the morning. The two photographs on this page were supplied by the Swedish Space Committee.

clouds had been reported with reasonable frequency. We decided on northern Sweden, which has regions of flat terrain that would make possible overland firing of the rocket and subsequent recovery of the payload. We learned from Bert R. J. Bolin, director of the Institute of Meteorology at the University of Stockholm, that some of the workers at the institute had for several years been conducting an extensive program of ground-based observations of noctilucent clouds (known in Swedish as *nattlysende moln*). With the ready co-operation of the Swedish workers and with equipment supplied by the U.S. National Aeronautics and Space Administration, we set out to take the first samplings of noctilucent cloud particles in the summer of 1962. My principal colleagues on the project were Georg Witt of the University of Stockholm, Curtis L. Hemenway of Union College and William Nordberg of NASA.

Our scheme was to sample the particles from an observed cloud and also to sample the atmosphere at the same altitude at a time when no cloud was observed. For this purpose NASA supplied us with four Nike-Cajun solid-fuel rockets, capable of carrying a 90-pound payload to an altitude of 75 miles. Into the rocket we packed not only our own cloud-sampling gear but also the instrumentation for two other experiments. These were joint undertakings of the Kiruna Geophysical Observatory and the University of Lund, both in Sweden, and the Cosmic Ray Group at the Air Force Cambridge Research Laboratories. The payload in addition had to include the power supply and electronics for control functions and for the transmission of information to the ground. The package also included dive brakes, a parachute and a radio beacon to bring the payload safely to earth and direct us to the landing spot.

The cloud-sampling gear, carried in the nose of the rocket, consisted of two cans for collecting particles or evidence of particles. To expose the specially prepared collecting surfaces in the cans, the payload ejected its outer nose cone at 225,000 feet. At 275,000 feet the remaining inner nose cone rotated a quarter turn to close the collecting ports and place covers on the cans. The closing of the cans sealed in a sample of the high vacuum of this altitude as well.

At each point we took extreme precautions to prevent contamination of the collecting surface by terrestrial particles, which could have ruined the ex-

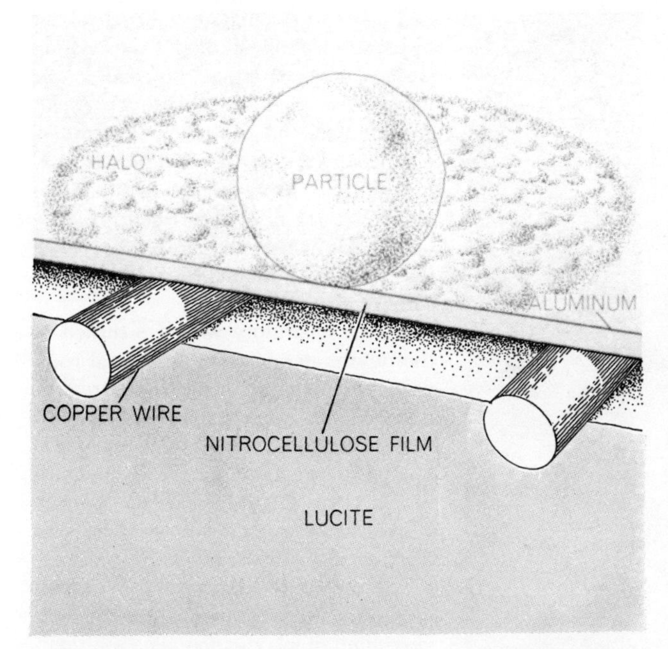

ONE NITROCELLULOSE SURFACE had coating of aluminum, as diagramed here. Another was coated with fuchsin dye. The nitrocellulose film rested on copper-wire mesh. Drawing shows only the arrangement, not the relative scale of the various components.

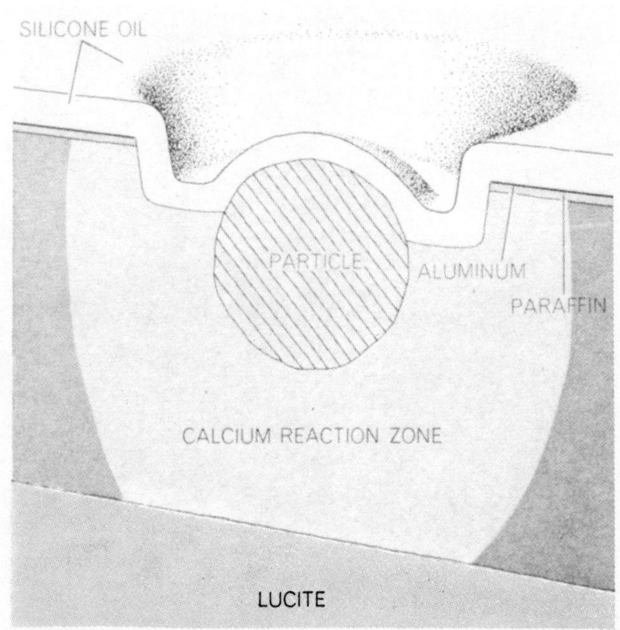

CALCIUM COLLECTING SURFACE was expected to react chemically with any ice surrounding particles. The silicone oil sealed the hole against atmospheric moisture after penetration by particles. The various parts of this diagram are also not drawn to scale.

periment. The surfaces were prepared and loaded into the cans in a dust-free laboratory. The payload tip was cleaned and the cans were placed inside in a portable dust-free chamber. Before launching we kept the cone in a plastic bag through which filtered air was circulated under positive pressure. Atmospheric friction on the ascent completed the rocket "scrubdown." We knew that, in spite of such precautions, we could not exclude all terrestrial contamination; therefore we built several controls into the experiment to enable us to recognize the contaminants. For example, we mounted a shield over a small area of each of the collecting surfaces, close to but not touching the surface. In the lower atmosphere particles of the size we expected to collect in the mesopause move by Brownian movement and essentially "float" in the air, so that any surface over which air passes will collect some of the particles. At the extremely low pressure of the upper atmosphere, particles of this size travel in ballistic trajectories and cannot strike a shielded surface. Hence we could discount as contaminants any particles found on the unshielded areas that looked like those on the shielded areas.

We prepared four different kinds of collecting surface. One was designed to collect solid particles that did not evaporate; it consisted of a nitrocellulose film with a thin coating of aluminum, resting on a copper-wire grid [*see illustration at left above*]. To record the presence of ice, which would have melted and evaporated before we could recover the nose cone, we coated another such film with a fuchsin dye that would show spots where particles or parts of them had melted. We made one surface out of the soft metal indium in the hope that it would show the impact craters of particles, whether or not they were still present. (Unfortunately the craters, if any, are so small we cannot as yet pick them out from the roughness of the surface attributable simply to the grain structure of indium.) The fourth collecting surface consisted of calcium coated with paraffin, aluminum and silicone oil. This surface too was designed to detect the presence of ice particles or ice-coated particles. The particles would have sufficient momentum to penetrate the protective coatings, and the water would react with the calcium to produce calcium hydroxide. Any solid particle or solid core of a particle would be sealed in the calcium by the oil, which would protect the calcium from reaction with water vapor in the atmosphere.

The launching site, prepared under the direction of Lars Rey and the Swedish Space Committee, was an old farm called Kronogård near the Arctic Circle. All the equipment was set up and made ready early in August, and we began our nightly vigil. Experienced observers searched the sky from ground stations throughout Scandinavia while Swedish aircraft carried trained men above the low clouds to watch for noctilucent clouds. On their nightly run from Stockholm to Kiruna commercial airline pilots were to report by radio if they saw what they thought were clouds of this type. Trained observers stationed at Kristineberg, about 90 miles south of our station, provided particularly valuable information. They looked for clouds 20 degrees or so above their northern horizon; these clouds would be directly over our heads at Kronogård. We probably would not be able to see them, although we might see extensions of the same display above our northern horizon. After a week of expectancy, shared by all of Sweden through the newspapers, we fired the first rocket. Before the month was over we had fired all four. Two of the flights were successful. We recovered cans from one flight into a noctilucent cloud and from one when no clouds were visible.

The rocket that penetrated the cloud collected some 10 million particles per square inch of collecting surface. This was between 100 and 1,000 times more than the number of particles collected by the rocket that traversed the same upper regions of the atmosphere on a cloudless night. From the solid particles or nuclei of particles [*see illustrations on following page*], we have learned that part, at least, of the cloud substance is solid and not volatile. About 20 per cent of the particles, however, are surrounded by a halo, or ring, on the collecting surface, indicating that these particles

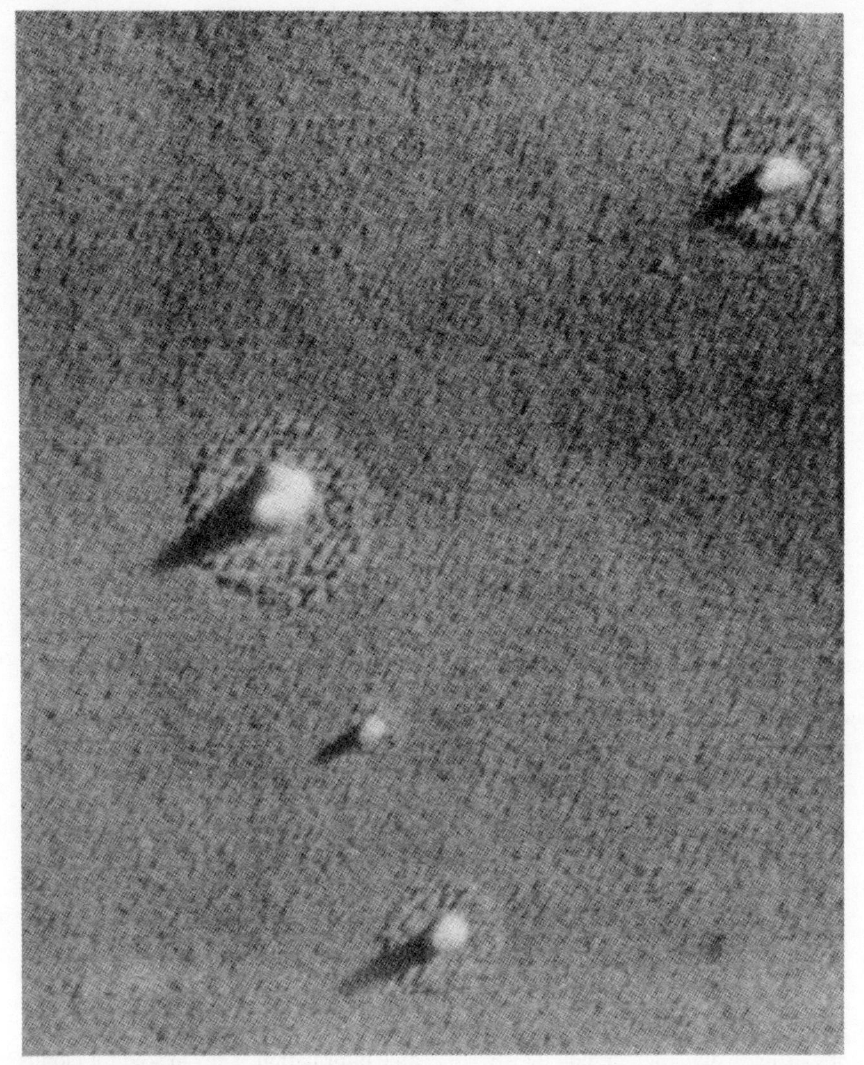

NOCTILUCENT CLOUD PARTICLES were captured on an aluminum-nitrocellulose surface during rocket flight. Particles are nonvolatile, but many are surrounded by halos presumably made by volatile material. The particles are enlarged about 30,000 diameters in this electron micrograph, which was made at Air Force Cambridge Research Laboratories.

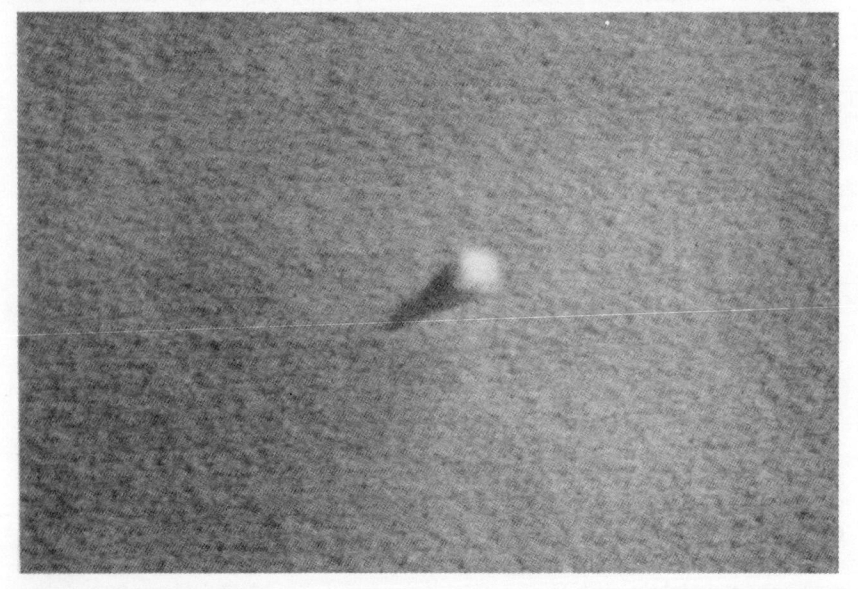

MESOPAUSE PARTICLES such as this one, collected during rocket flight into sky empty of noctilucent clouds, produced no halo on aluminum or dye surfaces. Collecting surfaces from cloudless mesopause contained 100 to 1,000 fewer particles than those flown into cloud.

had a coating of some substance that later evaporated. From studies of the calcium surface it appears that the coating was indeed ice. Significantly, the particles collected by the control rocket from a cloudless sky show no evidence of having had a coating of ice. Further detailed studies of the particles are now being conducted under at least seven different electron microscopes on both sides of the Atlantic.

In our laboratory we have turned to the determination of the chemical composition of the particles, an extremely difficult task because the total mass of the particles collected comes to only a ten-millionth of a gram. For chemical analysis we have to employ neutron activation in a reactor or bombardment with electrons to obtain an X-ray spectrum. These techniques are sensitive only to certain elements and at best yield limited information. The solid particles do seem to contain nickel, an element quite rare in terrestrial particles but common in those of meteoritic origin. The size of the solid particles, which mostly range from .05 to .5 micron in diameter, also indicates an extraterrestrial origin; this is in agreement with the size of the particles as estimated from optical studies and is considerably larger than the maximum size of terrestrial particles that could be carried upward in the atmosphere.

It would appear, therefore, that the noctilucent clouds consist of extraterrestrial dust particles coated with ice. This finding as to the substance of the clouds by no means explains their origin. The meteoritic particles are supposed to be uniformly distributed in the upper atmosphere, and there is no good reason why they should be distributed in any other way. Yet our flight into a cloudless sky yielded a particle count like that from the Venus Flytrap rocket over New Mexico, or between 100 and 1,000 times fewer particles than the flight into a cloud. No one has yet proposed an explanation for such a large local fluctuation in the concentration of extraterrestrial particles in the high atmosphere. It is also an open question whether the higher concentration of these particles, serving as condensation nuclei, initiates the formation of the clouds or whether the condensation of the clouds is more purely a matter of local temperature, air pressure and vapor pressure. We hope the rocket flights now being planned for 1964 will begin to answer these and the other questions that surround the rare and beautiful spectacle of the noctilucent clouds.

The Aurora

16

by Syun-Ichi Akasofu
December 1965

The information gathered by rockets and artificial satellites has contributed to a new physical description of the aurora in which the earth's magnetosphere acts like a gigantic cathode-ray tube.

It is almost impossible to capture in photographs or describe in words the unearthly beauty of the aurora as it shimmers and flames in the polar night sky. Familiar to almost everyone from pictures and descriptions but only occasionally visible where most people live, the phenomenon has long lacked a satisfactory explanation. Now, within the past few years, ground-based observations have been combined with information acquired by rockets and artificial satellites to produce a physical description of the aurora that relates it to the large-scale interaction of, on the one hand, the magnetic fields that surround the earth in space and, on the other, the high-velocity "wind" of electrically charged particles streaming from the sun. According to this view the magnetosphere of the earth acts like a gigantic cathode-ray tube that marshals charged particles into beams and focuses them on the earth's polar regions. The aurora is a shifting pattern of images displayed on the fluorescent screen provided by the atmosphere.

In more technical terms the aurora is a fluorescent luminosity produced by the interaction of atoms or molecules in the upper atmosphere and energetic charged particles entering the atmosphere from space. The incoming particles, guided by the lines of force in the earth's magnetic field, are electrons and protons. The atoms and molecules are chiefly those of oxygen and nitrogen. When they are struck by incoming particles, they are stripped of one or more electrons (ionized) or raised to a higher energy state (excited); when they return to their original condition, by acquiring electrons or by losing energy, they emit radiation of a characteristic wavelength. Thus the spectrum of the aurora can provide detailed information about the atoms and molecules present in the upper atmosphere.

To the eye most auroras are green or blue-green, with occasional patches and fringes of pink and red. Excited oxygen atoms account for both green and red light, at the respective wavelengths of 5,577 angstrom units and 6,300 angstrom units. Ionized nitrogen molecules emit intense light, particularly violet and blue light in a group of spectral bands between 3,914 and 4,700 angstroms. Excited nitrogen molecules account for a series of emission bands that are particularly intense in the deep red part of the spectrum between 6,500 and 6,800 angstroms [*see top illustration on page 143*]. The oxygen radiation at 5,577 angstroms and the nitrogen radiation at about 3,900 angstroms originate predominantly at an altitude of about 110 kilometers (70 miles). The 6,300-angstrom radiation of oxygen originates chiefly between 200 and 400 kilometers.

At an altitude of 100 kilometers there is often no dearth of oxygen atoms excited to the level at which they could emit red light at 6,300 angstroms, but such spontaneous emission does not occur until about 200 seconds after excitation has taken place. In this period the probability is large that an excited oxygen atom will lose part of its energy in a collision with another atom or molecule. On the other hand, spontaneous emission at 5,577 angstroms (green light) takes place in about .7 second; hence green radiation predominates over red at low altitudes. Higher up in the atmosphere collisions are infrequent enough so that the 6,300-angstrom emission of excited oxygen atoms has time to take place. At such altitudes, however, the density of oxygen is so low that the red radiation is faint unless the flux of incoming particles is high enough to excite a large fraction of all the oxygen atoms present.

Another weak source of red light is the emission of radiation by excited hydrogen atoms, which enter the atmosphere originally as protons (hydrogen nuclei). Along the way the protons pick up electrons to form hydrogen atoms. When these atoms are first created, they are in an excited state and identify themselves as they decay to lower levels by emitting the familiar Balmer series of spectral lines.

The excited and ionized states that supply most of the visible light of the aurora are produced by beams of incoming electrons that have energies of less than 10,000 volts, or less than half the energy of the electrons in the beam of a television picture tube. The energies of auroral electron beams have been measured by precisely coordinating ground-based measurements, which record luminosity profiles, and rocket or satellite measurements, which supply information on the interactions taking place in the upper atmosphere. Among those who have made important contributions to such studies are C. E. McIlwain of the University of California at San Diego, B. J. O'Brien of Rice University and a cooperating group made up of investigators at the Lockheed Aircraft Company and our laboratory at the University of Alaska.

Auroral luminosity takes two basic forms: ribbons and cloudlike patches. A vigorous auroral display generally evolves from the former to the latter, but many auroras disappear without ever breaking up into patches. A ribbon display has a vertical dimension of a few hundred kilometers and an east-west dimension of at least a few thousand kilometers. The ribbon itself is

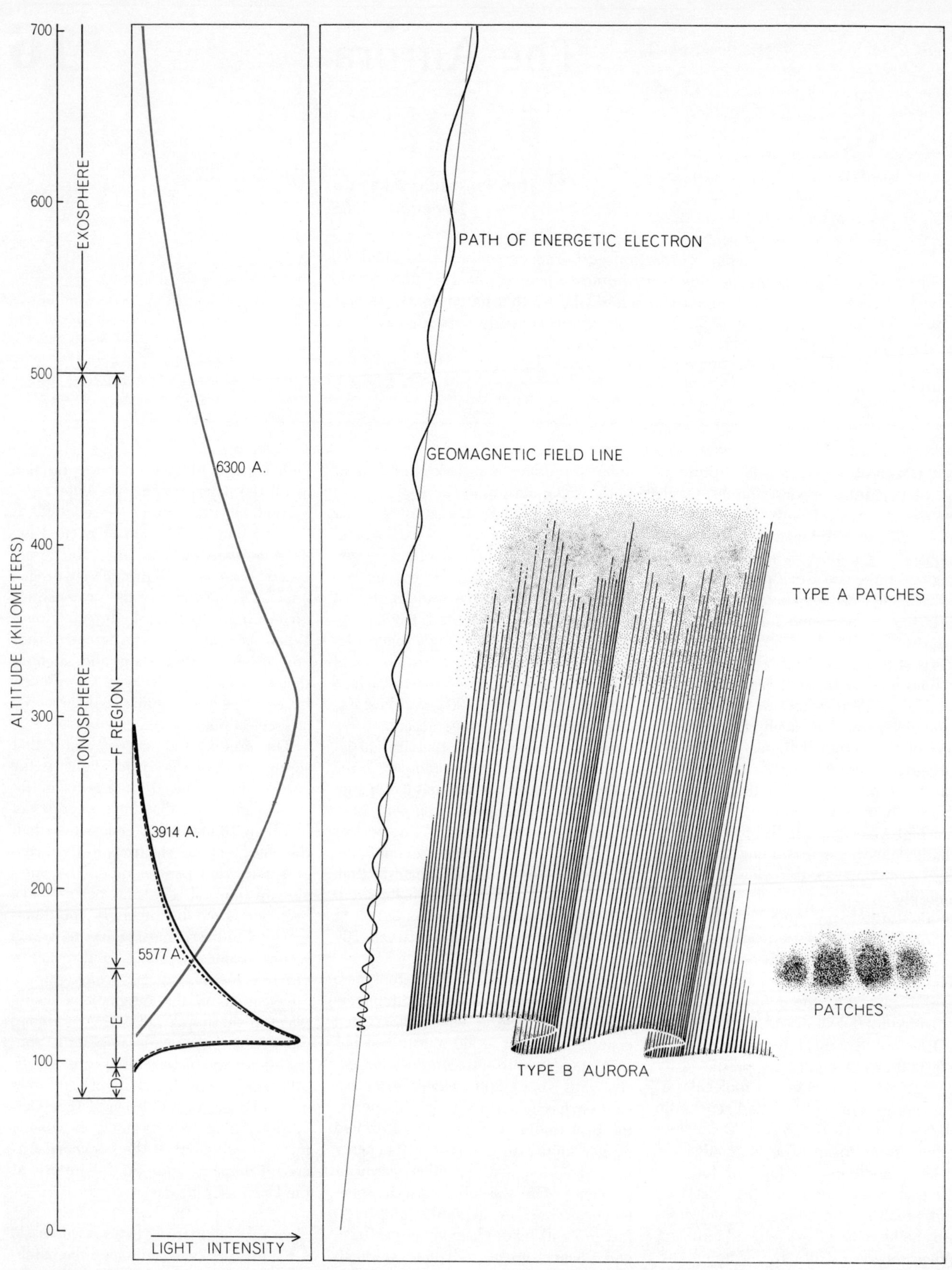

COMMON TYPES OF AURORA are shown with their average heights and characteristic radiations. The normal ribbon-like type, represented here by a "rayed band," has a vertical dimension of a few hundred kilometers and an east-west dimension of at least a few thousand kilometers. The ribbon itself is only a few hundred meters thick. An intensely active rayed band often develops a pink glow at the bottom; this is described as the Type *B* aurora. During the most intense activation the ribbon-like form collapses and is succeeded by cloudlike patches; these usually occur after midnight. When oxygen atoms become sufficiently excited to emit light at 6,300 angstrom units, rosy patches of Type *A* aurora appear at an altitude of 300 to 400 kilometers. A single energetic electron is shown gyrating down a geomagnetic-field line. Light is emitted as a result of its collisions with resident particles in upper atmosphere.

only a few hundred meters thick, which suggests that it is produced by a sheetlike electron beam that comes from the magnetosphere into the polar upper atmosphere. The ribbon form usually appears in multiple tiers—like stage curtains hanging one behind the other—stretching across the entire sky.

Students of the aurora have developed a series of descriptive terms to identify various subcategories of the ribbon form [*see bottom illustration at right*]. When the ribbon is in its simplest and quietest configuration, it is known as a homogeneous arc; at such times it has a fairly smooth luminosity, brightest at the bottom and fading into the night sky at the top. As the ribbon becomes slightly more active it develops fine folds a few kilometers in width, with the result that the aurora seems to be composed of aligned columns, or rays, of light; this is called a rayed arc. With more intense activation the folds spread to a few tens of kilometers in width. When the larger folds are superposed on the more delicate ones, the ribbon is called a rayed band. If the activation continues to increase, the rayed band develops a beautiful pink glow at the bottom of the folded ribbon; this is often described as a Type *B* aurora. Finally, if the activation rises still further, the folds or loops grow to a truly grand scale, with widths of a few hundred kilometers. As soon as the activation ceases, however, the folds tend to disappear and the ribbon resumes its homogeneous form. This suggests that the homogeneous form represents the fundamental structure of the aurora and that the folds and convolutions are indeed evidence of increased activation.

During the most intense activation the ribbon form collapses and is succeeded by the cloudlike patches; these appear most commonly after midnight. A comparison of reports from observers widely separated around the Arctic Circle leaves no doubt that the various active forms at different places and different local times are closely related to one another. For example, when one observer sees unactivated homogeneous arcs in the evening sky, other observers will report that auroras have been fairly quiet all around the Pole. On the other hand, when the quiet arcs become activated during the evening to form rayed arcs and rayed bands, observers watching the morning sky elsewhere will see previously quiet arcs break up into cloudlike patches.

I have spoken so far only of the most common types of aurora, but several

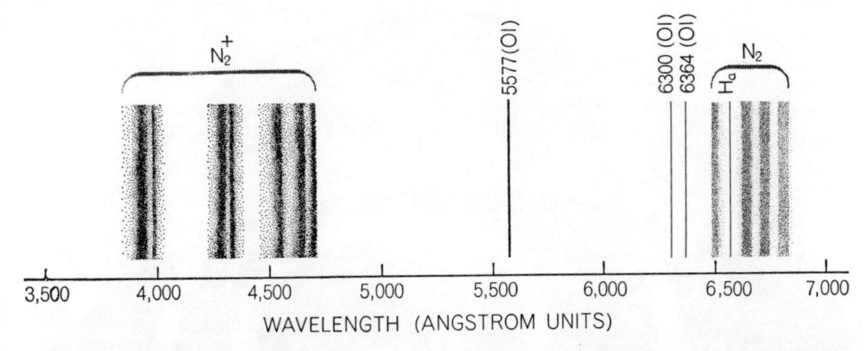

SPECTRUM OF THE AURORA can provide detailed information about the kind of particles present in the upper atmosphere and their normal energy states. Excited oxygen atoms account for both green and red light at the specific wavelengths of 5,577, 6,300 and 6,364 angstroms. Ionized nitrogen molecules supply violet and blue light in a group of spectral bands between 3,914 and 4,700 angstroms. Excited nitrogen molecules supply a series of emission bands in the deep red part of the spectrum between 6,500 and 6,800 angstroms.

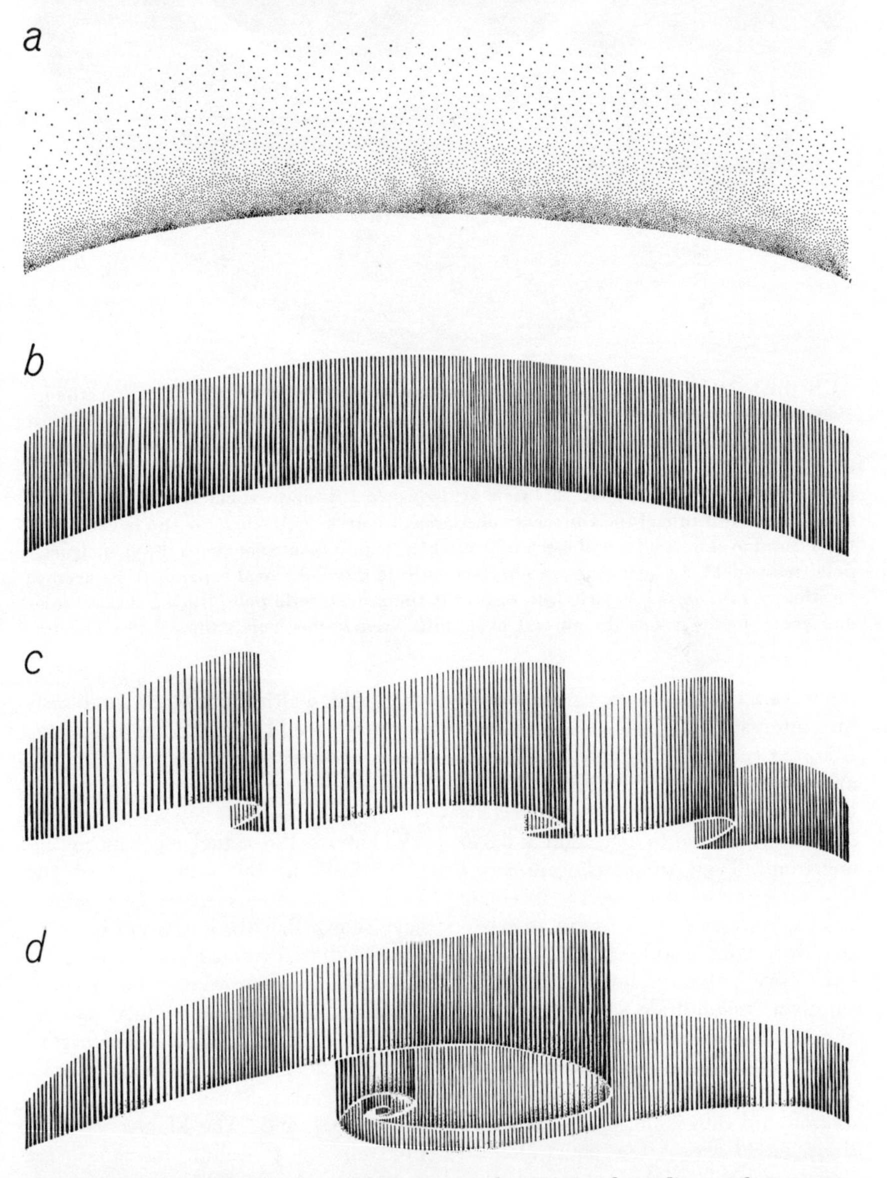

RIBBON-LIKE FORM of the aurora has various subcategories, depending on the intensity of the activation. When the ribbon is in its simplest and quietest configuration, it is known as a homogeneous arc (*a*). As the ribbon becomes slightly more active it develops fine folds and is called a rayed arc (*b*). With more intense activation larger folds are superposed on the more delicate ones and the ribbon comes to be known as a rayed band (*c*). An extremely active Type *B* aurora develops a pink glow at the bottom of the folded ribbon (*d*).

AURORAL ZONE, which is defined as a narrow band centered on the line of maximum average annual frequency of auroral visibility (*broken white line*), has only a statistical significance. At any given hour or minute slightly activated auroras tend to appear, on the average, along an oval zone that coincides with the nearly circular auroral zone only at the observer's local midnight. (In this view of the world it is approximately midnight at the Geophysical Institute of the University of Alaska at Fairbanks.) When the sun is very calm, the auroral oval is smaller and essentially circular, with its center at about the geomagnetic pole (*white dot*). As soon as the sun becomes a little active the oval expands to its average location and becomes eccentric with respect to the geomagnetic pole. During intense solar and geomagnetic storms the auroral oval shifts even farther south toward the Equator.

other varieties are seen, some only at rare intervals. One of these is the Type *A* aurora, a spectacular rosy variety that appears at an altitude of 300 to 400 kilometers when oxygen atoms become sufficiently excited to emit light at 6,300 angstroms. Type *A* auroras occur only a few times in a dozen years. Incoming protons frequently give rise to an extensive but faint band of whitish-green luminosity. During intense magnetic storms a "midlatitude red arc" appears at latitudes considerably south of New York City; it consists of a wide but not very bright band of 6,300-angstrom radiation and thus is not easily seen with the unaided eye. A few hours after an intense solar flare the entire polar region is bombarded by energetic protons expelled from the sun, producing the whitish-green "polar-cap glow." In addition to all these natural varieties of aurora, high-altitude nuclear-bomb tests have produced brilliant crimson auroras. None of these special types of aurora need concern us further.

What are the actual mechanisms involved in the production of the aurora? Instruments carried by satellites have shown that the earth and its magnetic field are confined in a huge cavity carved in the solar wind. The cavity is the magnetosphere; it contains all the belts of particles trapped in the earth's magnetic field [see "The Solar Wind," by E. N. Parker, SCIENTIFIC AMERICAN, April, 1964, and "The Magnetosphere" by Laurence J. Cahill, Jr., SCIENTIFIC AMERICAN, March, 1965].

Furthermore, the magnetosphere has a long cylindrical tail that is carried downstream by the solar wind; this was first suggested by J. H. Piddington of the Commonwealth Scientific and Industrial Research Organization in Australia. Norman Ness of the Goddard Space Flight Center has recently confirmed the existence of this comet-like tail with satellite measurements. Within the cylindrical tail the lines of force in the earth's magnetic field are bunched together like a bundle of spears. The lines of force above the plane of the magnetosphere's equator are directed toward the sun; those below the plane are directed away from the sun. The equatorial plane therefore constitutes a neutral sheet. The secret of the aurora seems to be hidden in the tail of the magnetosphere.

The role of the magnetosphere's tail can be visualized by comparing the magnetosphere to a cathode-ray tube; this analogy was suggested to me by C. T. Elvey, formerly director of the Geophysical Institute of the University of Alaska. In a cathode-ray tube electrons are emitted by a heated filament and accelerated toward an anode that is perforated with a small hole. Some of the electrons pass through the hole and form a pencil-like beam that is deflected, on its passage to the face of the tube, by electric fields between two pairs of plates or, in some tubes, by magnetic fields set up by coils. The electron beam strikes a fluorescent material on the tube face, producing a luminous image. The luminous display on the screen thus supplies evidence of changes in both electric and magnetic fields along the path of the electron beam.

In much the same way the shifting patterns of the aurora over the entire polar night sky supply evidence of changes in the magnetic and electric fields along the path of electrons streaming toward the earth. In some obscure fashion the tail of the magnetosphere accelerates and collimates electrons in the magnetosphere into ribbon-like beams that impinge sharply on the upper atmosphere. The task ahead is to identify the precise mechanisms that play the role of the electron gun, the anode, the electric plates and the magnetic coils.

Clues to these mechanisms can be found in changes in the magnetic field on the earth's surface and in space, as well as in the auroral display itself. Let us, therefore, "watch" the auroral display as it appears on the "screen" of the entire polar night sky. If one travels northward from the border between the U.S. and Canada, one will see the aurora with increasing frequency. The increase does not continue all the way to the North Pole; the frequency reaches a maximum over the southern part of Hudson Bay. Excellent maps have been prepared that show auroral "isochasms": the lines of equal average annual frequency of visible auroras. The auroral zone is defined as a narrow band centered on the line of the maximum isochasm. There is, of course, an auroral zone in the Southern Hemisphere as well as in the Northern. The center of the northern auroral zone is not, as one might think, the magnetic dip pole near Resolute Bay in Canada (73.5 degrees north latitude and 100 degrees west longitude), but what is known as the dipole, or geomagnetic pole, at the northwestern tip of Greenland (78.5 degrees north and 69 degrees west).

The auroral zone, however, has only a statistical significance. At any given hour or minute auroras tend to appear, on the average, in an oval zone that coincides with the nearly circular auroral zone only at the observer's local midnight [*see illustration on opposite page*]. Elsewhere the oval zone falls inside the auroral zone. The oval zone, if it were viewed from a point above the geomagnetic pole, would appear as an oval glow roughly fixed in space above the geomagnetic pole. The earth turns below the oval pattern once a day, and the locus of the midnight portion of the oval traces out a circle that coincides with the auroral zone. The auroral oval is a new concept that has evolved gradually as the result of cooperation among workers in Australia, Canada, Denmark, Finland, Norway, Sweden, the U.S.S.R., the United Kingdom and the U.S.

From investigations of the outer belt of particles trapped in the earth's magnetic field (particularly studies by James A. Van Allen and L. Frank of the State University of Iowa), it seems likely that the oval belt of the aurora lies immediately poleward of the curve of intersection between the ionosphere (which begins about 80 kilometers above the earth's surface) and the shell of trapped particles that forms the outer boundary of the inner magnetosphere [*see illustration below*]. This region is populated with electrons whose energies range upward from about 40,000 volts. Such electrons are produced in the tail of the magnetosphere and flow along the boundary of the inner magnetosphere, creating a more or less steady glow where they intersect with the upper atmosphere. The oval has the shape it

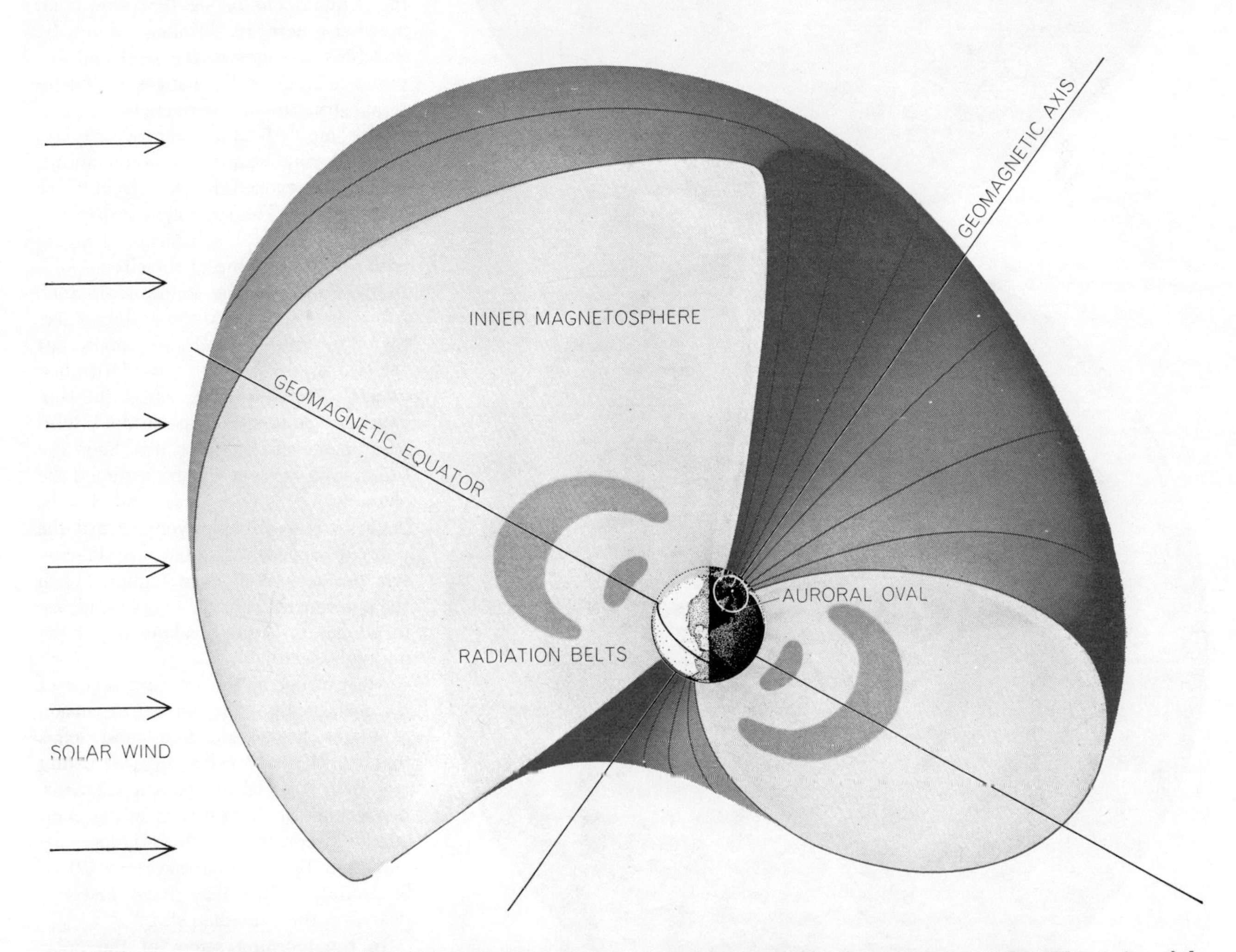

THE MAGNETOSPHERE is a vast cavity in the solar wind that contains the earth's magnetic field; its role in the production of the aurora is to act like a gigantic cathode-ray tube that focuses electron beams, as well as proton beams, toward the earth in the vicinity of the magnetic poles. The oval belt of the aurora coincides roughly with the intersection curve between the ionosphere, which begins about 80 kilometers above the earth's surface, and the outer boundary of the inner magnetosphere (*color*). This region of the magnetosphere is populated with electrons whose energy ranges from about 40,000 volts upward; the extended tail of the magnetosphere, whose magnetic-field lines terminate inside the auroral ovals, is not shown here. The auroral oval has the shape it does because the inner magnetosphere is highly asymmetric with respect to the solar wind, and hence to the earth's noon-midnight axis.

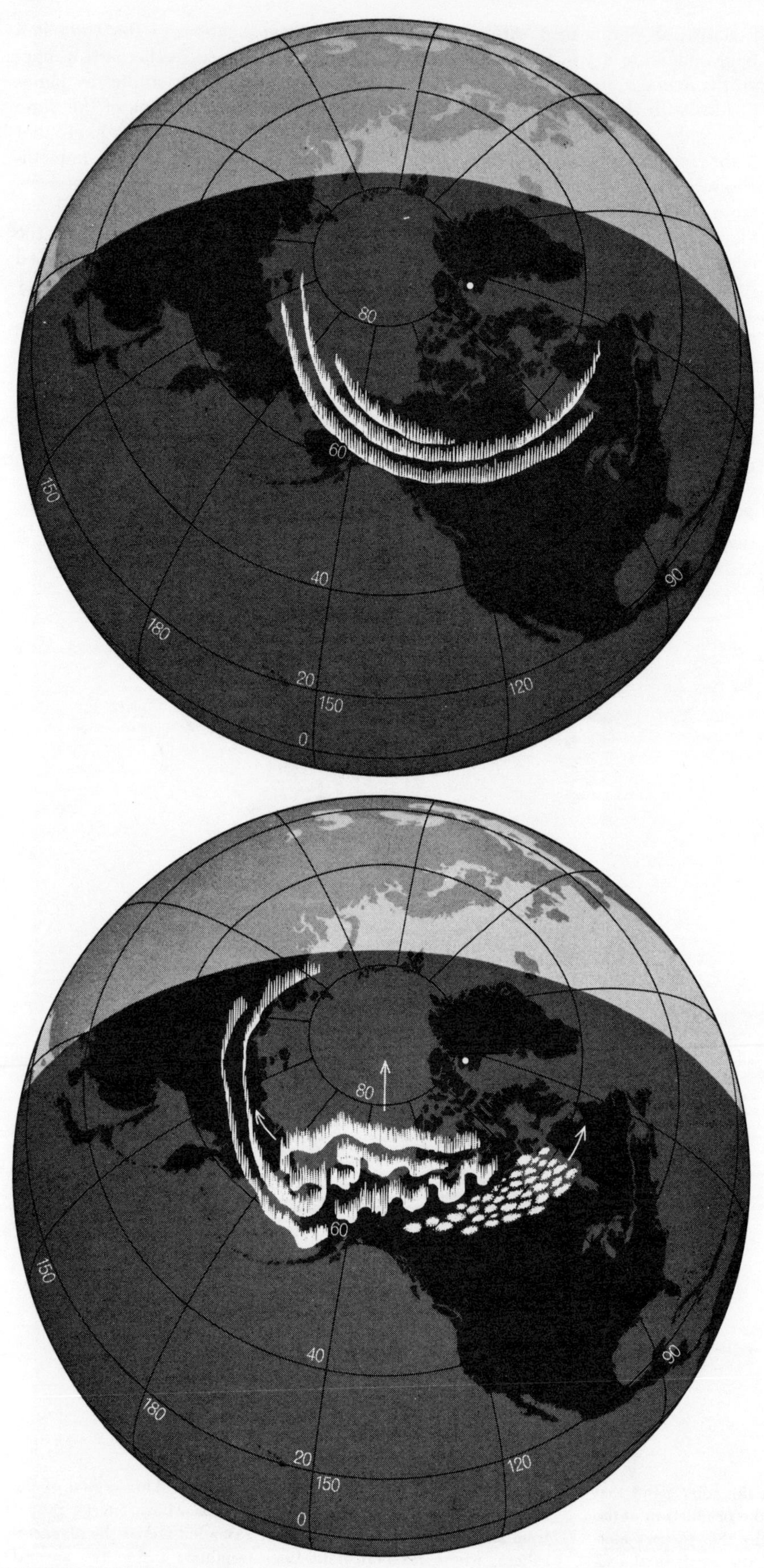

AURORAL SUBSTORM, the most spectacular auroral display, occurs during major magnetic storms and coincides with a rapid buildup of circumpolar ring currents. Such substorms originate in the midnight sector of the auroral oval and usually last for two or three hours. These two views show the quiet homogeneous arcs of the aurora before the substorm (*top*) and the complex display of many auroral types at the substorm's height (*bottom*).

does because the magnetosphere is highly asymmetric with respect to the solar wind and hence to the noon-midnight axis on the earth's surface.

The size of the oval changes greatly with solar activity. When the sun is extremely quiet, the midnight portion of the oval recedes toward the geomagnetic pole and the oval becomes almost circular and faint. The oval occupies its typical position when the sun is in a state of average activity. After a major solar storm the oval expands rapidly toward the Equator, the amount of expansion being roughly proportional to the volume of electric current flowing in the tail of the magnetosphere and also to the intensity of a gigantic ring current that grows within the magnetosphere and flows westward around the earth. The expansion of the oval toward the Equator indicates that the inner magnetosphere is shrinking. Thus by watching changes in the oval one can visualize large-scale changes in the internal structure of the magnetosphere.

The most dynamic auroral displays occur during major magnetic storms, which are evoked in turn by intense solar activity. The most violent displays, known as auroral substorms, coincide with a rapid buildup of the circumpolar ring currents in the ionosphere. Such substorms originate in the midnight sector of the auroral oval and usually last for two or three hours [*see illustration at left*]. The first indication of the substorm is a sudden increase in the brightness of one of the quiet arcs. Soon the brightened auroras, having assumed the character of rayed arcs and bands, begin to spread explosively toward the poles at speeds of about five kilometers per second or even higher. When the poleward expansion is very rapid, the rayed bands begin breaking up in the midnight sector.

Meanwhile, in the evening sector of the polar region, the auroral expansion generates large-scale folds and loops that travel westward along preexisting arcs with a speed of about a kilometer per second. Such westward surges commonly sweep across the Alaskan sky and then the Siberian sky only 20 or 30 minutes after they have first appeared in the Canadian sky.

In the morning sector of the polar region the auroral arcs or bands that lie in the northern part of the oval also expand explosively, but those that lie in the southern part of the oval often disintegrate into cloudlike patches. Both the bright bands and the patches drift rapidly eastward. As the substorm subsides, the scattered auroral frag-

BRILLIANT COLORS displayed by several different types of aurora are captured in these photographs made by the author in the neighborhood of the University of Alaska at Fairbanks. The photograph at top left is of a quiet and diffuse auroral arc; at top right is a similar arc with a strong enhancement of the pink light emitted by excited nitrogen molecules in the earth's upper atmosphere. The remaining four photographs are of a ribbon-like auroral type known as an active rayed band. The photographs were all made with a 35-millimeter camera, using a high-speed color film and an $f/1.2$ lens. The exposure time was between one and five seconds.

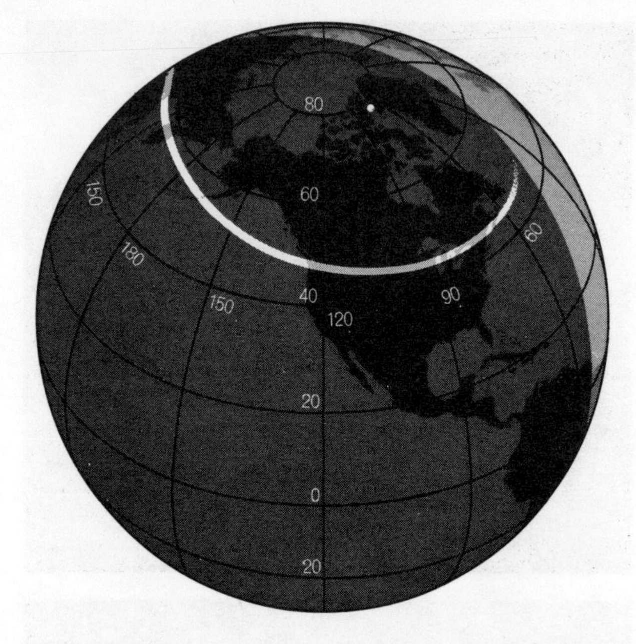

MOST INTFNSE AURORAL SUBSTORM of recent years occurred during the great magnetic storm of February 11, 1958, which resulted from an intense solar flare on February 9. By 1020 hours (universal time) on February 11 at least three large substorms were observed. At the height of the magnetic storm (*left*) the auroral oval was driven so far south that it formed a line connecting Redmond, Ore., Vermillion, S.D., Williams Bay, Wis., Ithaca, N.Y., and Hanover, N.H. Curiously no bright auroras could be seen north of this line. Then a fourth substorm began and within 30 minutes the active auroras had spread northward until they covered a band some 2,000 kilometers wide (*right*). An intense shower of X rays was recorded at the height of the fourth substorm.

ments converge slowly and reassemble into quiet homogeneous arcs, tracing out the auroral oval as it was before the substorm began.

Perhaps the most intense auroral substorm of recent years occurred during the great magnetic storm of February 11, 1958, which resulted from an intense solar flare on February 9. About 29 hours after the flare the magnetosphere was enclosed in a violent wind of charged particles. Within a few hours ring currents began building up within the magnetosphere and reached maximum intensity between 1000 and 1100 hours (universal time) on February 11. During this buildup period at least three large auroral substorms occurred. At the height of the magnetic storm the auroral oval was driven so far south that it formed a line connecting Redmond,

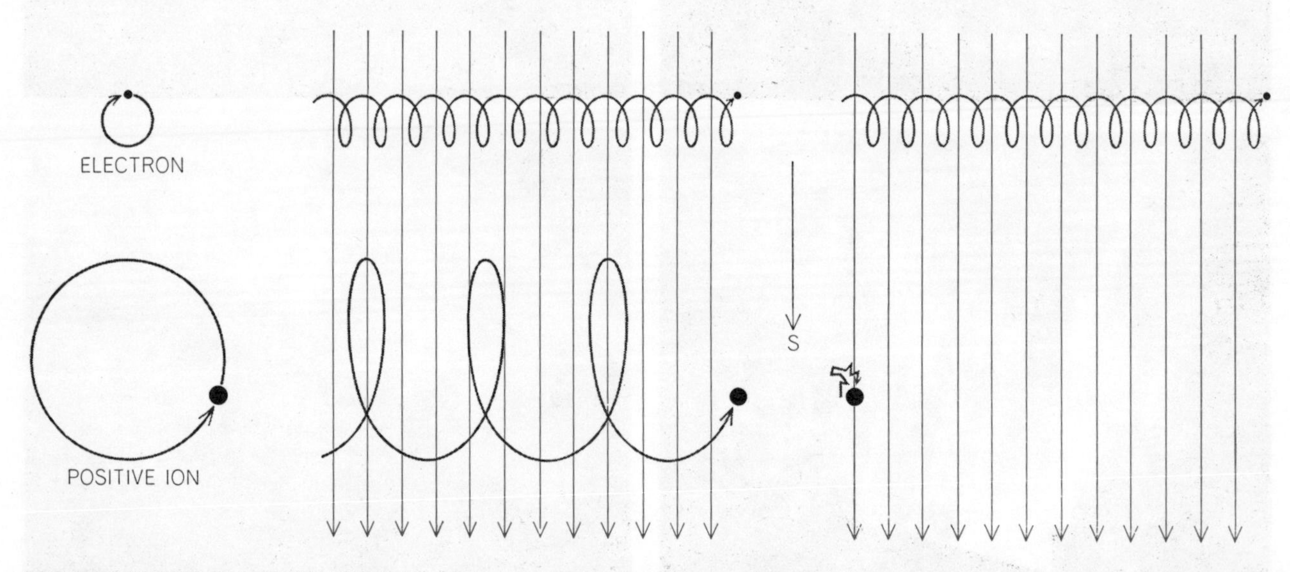

ELECTRONS AND POSITIVE IONS in the earth's ionosphere respond differently to the onset of an auroral substorm, depending on their altitude. Before such a substorm (*left*) electrons are forced to rotate clockwise, whereas positive ions are forced to rotate counterclockwise, around the geomagnetic-field lines, which in this view must be imagined as entering the page at right angles from above. The events that underlie the substorm also give rise to an electric field (*colored arrows*) that points southward, causing both electrons and positive ions to be deflected eastward. At levels above the *E* region of the ionosphere (*middle*) the density of gas particles is so slight that electrons and positive ions have about equal mobility; even though both may travel violently eastward together there is no net displacement of electric charge and hence no flow of current. In the *E* region (*right*), however, the positive ions, being larger, collide more frequently with other particles than the electrons do and thus tend to be stopped, whereas the electrons continue to drift eastward. This separation of electric charges gives rise to the westward-flowing polar electrojet (*see illustration on next page*).

POLAR ELECTROJET (*black arrows*), **an intense electric current that flows westward around the auroral oval, occurs only during an auroral substorm. Part of the electrojet leaks away from the auroral oval and spreads southward over the whole ionosphere** (*white arrows*).

Ore., Vermillion, S.D., Williams Bay, Wis., Ithaca, N.Y., and Hanover, N.H. Curiously no bright auroras could be seen north of this line. Then the fourth substorm began, and within 30 minutes the active auroras had spread northward until they covered a band some 2,000 kilometers wide [*see top illustration on preceding page*]. In anticipation of the magnetic storm J. R. Winckler and his co-workers at the University of Minnesota sent a balloon aloft over Minneapolis with radiation-recording instruments. At the height of the fourth substorm the instruments recorded an intense shower of X rays generated when energetic auroral electrons collided with atoms and molecules in the upper atmosphere.

X-ray production is just one of several interesting phenomena associated with the auroral substorm. For example, during such a storm intense electron beams penetrate deeper into the atmosphere than usual and greatly increase the ionization of the lower parts of the ionosphere known as the *E* and *D* regions. The *E* region reflects short radio waves and makes long-distance radio communication possible. When abnormal ionization occurs in the *D* region, communication is often disrupted. On the other hand, heavy ionization in that region tends to scatter shorter radio waves: the microwaves that ordinarily pass through the ionosphere. When such waves are scattered, television viewers sometimes receive a channel that originates in a distant city. It has also been observed that the atmosphere is heated and expands upward during the auroral substorm, causing satellites to decelerate at a slightly faster rate.

The large-scale activity of the magnetosphere during the substorm has at least one other major effect. When the magnetosphere interacts with the *E* region of the ionosphere, it generates the "polar electrojet," an intense electric current that flows westward around the auroral oval [*see illustration above*]. To understand the origin of this current, let us imagine that we are looking down on the polar ionosphere, where we find that the lines of force in the earth's magnetic field are penetrating the ionosphere from above. Under the influence of this field electrons moving through the upper ionosphere are forced to travel in a clockwise direction, whereas positive ions are forced to travel in a counterclockwise one [*see bottom illustration on preceding page*].

When energetic charged particles flow out from the tail of the magnetosphere toward the earth, they develop (in the Northern Hemisphere) an electric field that points southward. Such a field, being directed at right angles to the magnetic field, causes both electrons and positive ions to be deflected eastward. At levels above the *E* region the density of gas particles is so low that electrons and positive ions have about equal mobility; even though both may flow strongly eastward together, there is no significant net displacement of electric charge and hence no flow of current. In the *E* region, however, the positive ions, being larger, collide more frequently with other particles than the electrons do and thus tend to be stopped, whereas the electrons continue to drift eastward. This separation of electric charges gives rise to the westward polar electrojet. (According to convention, current flow is opposite to electron flow.) It has been found that part of the electrojet leaks away from the auroral oval and spreads southward over the whole ionosphere.

The study of the auroral substorm owes much to photographic records made by an all-sky camera, the first models of which were designed by Carl W. Gartlein of Cornell University. This instrument consists of a camera mounted above a convex mirror that reflects the whole sky from horizon to horizon. The all-sky camera was first used extensively during the International Geophysical Year (1956–1957), when 67 nations participated in a worldwide geophysical program. Since then all-sky photographs from as many as 115 stations in the Arctic and Antarctic have been collected and studied by our laboratory at the University of Alaska.

Our task has been to analyze successive photographs of the auroral substorm taken simultaneously at various polar stations and to infer minute-to-minute changes in the magnetosphere. These analyses and inferences are then correlated whenever possible with measurements made simultaneously by instrumented satellites. Out of this coordinated study has emerged the account of the aurora I have presented in this article.

The Airglow

by Robert A. Young
March 1966

A faint but constant light is emitted by a layer of excited atoms and molecules in the earth's upper atmosphere. The events that give rise to this radiation have now been reproduced in the laboratory

The night sky is illuminated by a number of different sources, located both outside the atmosphere and within it. Among the major contributors to this collective illumination are the moon, the stars and the lights of cities. Yet if it were possible in some way to eliminate all the light from these familiar sources, the sky would still not be completely black. In the latitudes of the Temperate Zones a faint multicolored glow would remain, almost uniform in brightness over the entire sky. (In arctic and antarctic latitudes the glow is often overpowered by the light of the aurora.)

The general term for this phenomenon is the airglow; depending on the time of day it is observed, it is also called the nightglow, the twilightglow and the dayglow. Because the airglow is most difficult to observe during the day, studies of the phenomenon have usually concentrated on the nightglow, with the expectation that the processes that give rise to the nightglow are similar to those that give rise to some of the dayglow and the twilightglow (both of which are now under intensive study). Since some quite different processes may be initiated by the direct action of sunlight on the earth's atmosphere, however, considerable differences between the nightglow and the dayglow are to be expected.

It has long been known that the airglow originates within the atmosphere. The first spectrogram of the airglow, obtained early in this century, was clearly distinguishable from spectrograms of stars or the aurora. Until 1927, however, none of the radiation from the airglow was associated with a specific atomic or molecular event. Then a distinct green line at a wavelength of 5,577 angstrom units in an airglow spectrogram was identified as originating with excited oxygen atoms. Later it was found that most of the airglow radiation in the blue region of the spectrum was associated with molecular oxygen (O_2), and that radiation in the infrared region was associated with the hydroxyl radical (OH). Airglow spectrograms have also revealed the existence of sodium in the upper atmosphere. So far, however, no emission from nitrogen, the most abundant gas in the atmosphere, has been reliably identified in such spectrograms.

For the past few years my colleagues and I in the department of atmospheric sciences at the Stanford Research Institute have been engaged in studying the processes that excite the airglow. We have concentrated particularly on the green spectral line of atomic oxygen at 5,577 angstroms. A large part of our research effort has gone into reproducing this and other airglow radiations in laboratory apparatus [*see illustration on opposite page*].

Before proceeding to a discussion of our experiments, it may be useful to review briefly how excited atoms and molecules emit light. As is often said, the electrons that surround the nuclei in atoms and molecules are in a very general way analogous to planets traveling in orbit around the sun. When an electron is farther from its "sun" than it is ordinarily, the atom or molecule is electronically excited. When the electron returns to its lowest energy level, or smallest orbit, its excess energy is emitted in the form of an electromagnetic wave. In the case of atoms this emission is at a specific wavelength and accounts for a single bright line in a spectrogram. The atomic nuclei in molecules, on the other hand, both revolve around a common center of mass and vibrate back and forth; for each jump of an electron back to a smaller orbit the emission is fanned out into bands by the revolution of the nuclei and the bands are displaced by the vibration of the nuclei.

Most of the atomic lines and molecular bands in the spectrum of the airglow are seldom observed in the laboratory. One reason is that these lines and bands arise from changes in the internal structure of atoms and molecules that seldom happen; in the language of quantum mechanics such changes are said to be "forbidden." Time is an important factor in determining if a forbidden change will occur, and on the atomic scale events that can preempt the forbidden change are plentiful and occur with great rapidity. For example, excited atoms and molecules generally lose their excess energy by radiation approximately a hundred-millionth of a second after being excited. Molecules vibrate some 10 trillion times per second, and collisions between atoms and molecules begin and end in a trillionth of a second. It would appear, therefore, that when forbidden emissions of energy are observed that require a full second (as in the case of the green line of atomic oxygen), 10 seconds or 100 seconds (as in the case of other forbidden emissions of atomic and molecular oxygen), these events must be occurring in extreme slow motion in the atomic world.

ARTIFICIAL AIRGLOW was produced in the author's laboratory at the Stanford Research Institute. The predominantly green color inside the bottom pipe arises from the de-excitation of nitrogen dioxide molecules (NO_2) flowing through the system . . . A schematic drawing of the entire apparatus appears on the next two pages.

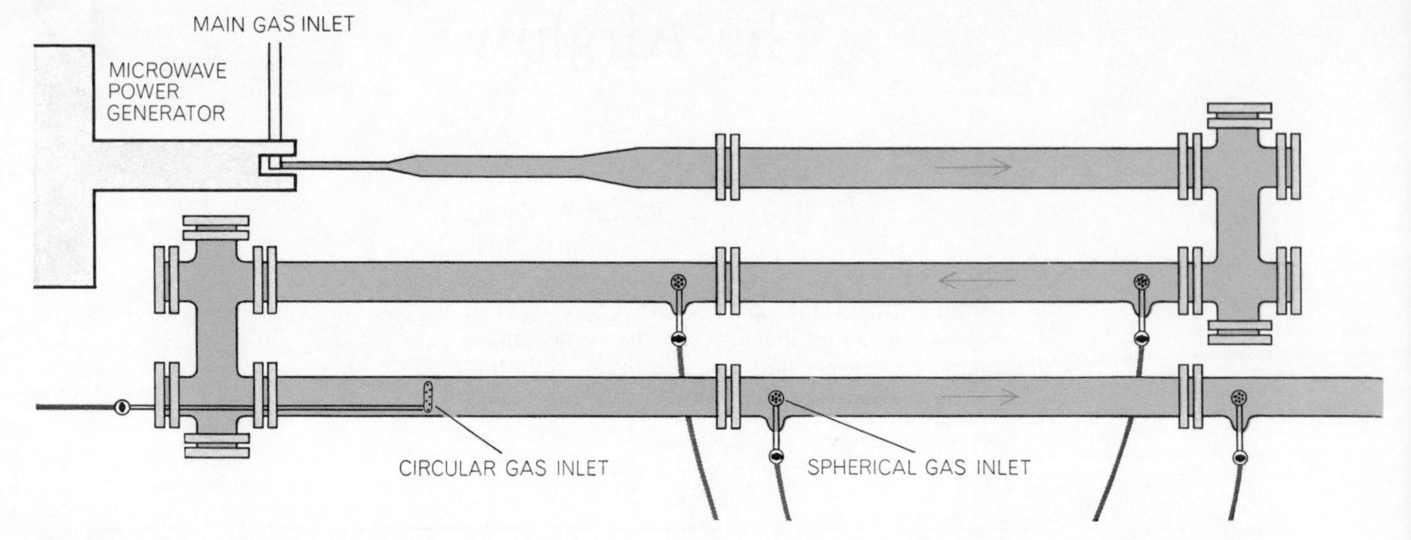

EXPERIMENTAL SYSTEM in which the author and his colleagues studied the excitation of atomic oxygen is depicted schematically on these two pages. Molecular nitrogen gas is admitted to the system at top left, where a microwave discharge dissociates some of the gas into atomic nitrogen. After the gas travels through the two upper sections of the glass pipe, nitric oxide is added to the system through the gas inlets located along the bottom section of pipe. The nitric oxide reacts with the nitrogen atoms to form oxygen atoms and more molecular nitrogen. The oxygen atoms in turn interact to produce excited oxygen atoms, which are de-excited both by emitting light and by colliding with other, nonexcited atoms. (These processes are studied in the large spherical bulb at the far right, where measurements of light intensity are made.) In order to separate these two effects another gas (called a quenching gas) that does not react with nonexcited atomic oxygen but does remove the excess energy of excited atomic oxygen is introduced to the system in measured amounts until it removes excited oxygen atoms as quickly as they are removed by collisions with nonexcited atomic oxygen plus radiation. The amount of quenching gas needed to reduce the light intensity by half verified the author's analysis that attributed the major loss of excited oxygen atoms in the experimental apparatus to reactions with nonexcited oxygen atoms.

Several forbidden emissions of oxygen have been recorded in the spectrum of the airglow. If one were to try to reproduce such emissions in a laboratory experiment at, say, a hundredth of atmospheric pressure at sea level, each excited atom and molecule would collide at least a million times per second with other atoms and molecules; thus there would be a great many opportunities for the excess energy to escape in the collisions before being radiated. Under such circumstances emission from the excited particles is said to be quenched. It is clear that at least part of the reason for the peculiar spectrum of the airglow is the low collision rate of atoms and molecules in the upper atmosphere, where the air pressure is on the order of a millionth of the pressure at sea level.

The best estimate of the location of the airglow in the atmosphere has been made with the aid of rocket-borne light detectors. According to these studies the green light from atomic oxygen and the blue light from molecular oxygen is produced in a rather narrow layer some six miles thick approximately 60 miles above the surface of the earth. Infrared radiation from hydroxyl radicals comes from a somewhat lower level, and red light from atomic oxygen comes from a much higher one [*see bottom illustration on page 155*]. It seems reasonable to expect some change in the condition of the atmosphere in the vicinity of these emission layers.

A suggestion as to what the change might be was made in 1931 by the British physicist Sydney Chapman, who was engaged at the time in analyzing the effect of the sun's radiation on the composition of the earth's atmosphere. Chapman's theory of how ozone (O_3) was formed in the atmosphere led him to surmise that some molecular oxygen would be dissociated into individual oxygen atoms at altitudes above 55 miles by the action of ultraviolet radiation from the sun. Molecular nitrogen (N_2) would be similarly affected only at much greater altitudes. Chapman further suggested that some of the oxygen atoms produced by ultraviolet radiation during the day would interact at night to produce the nightglow.

Since the nightglow persisted undiminished through the entire night, only a small fraction of the available oxygen atoms could be used up in this process, and a more or less constant

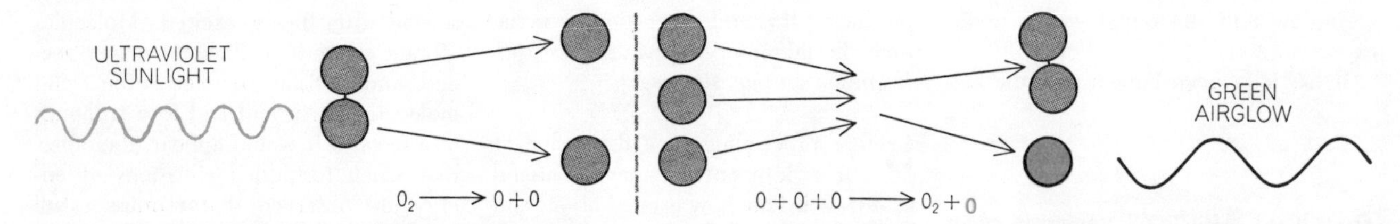

RADIATIVE MECHANISM responsible for the green spectral line at 5,577 angstroms was suggested by the British physicist Sydney Chapman in 1931. According to Chapman's hypothesis, some molecular oxygen (O_2) would be dissociated into individual oxygen atoms (O) at altitudes greater than 55 miles by the action of ultraviolet light from the sun. Some of the atoms produced in this way during the day would react in threes at night to produce molecular oxygen plus excited atomic oxygen (*color*). In about a second the excited oxygen atoms would emit their excess energy in the form of an electromagnetic radiation with a wavelength of 5,577 angstroms.

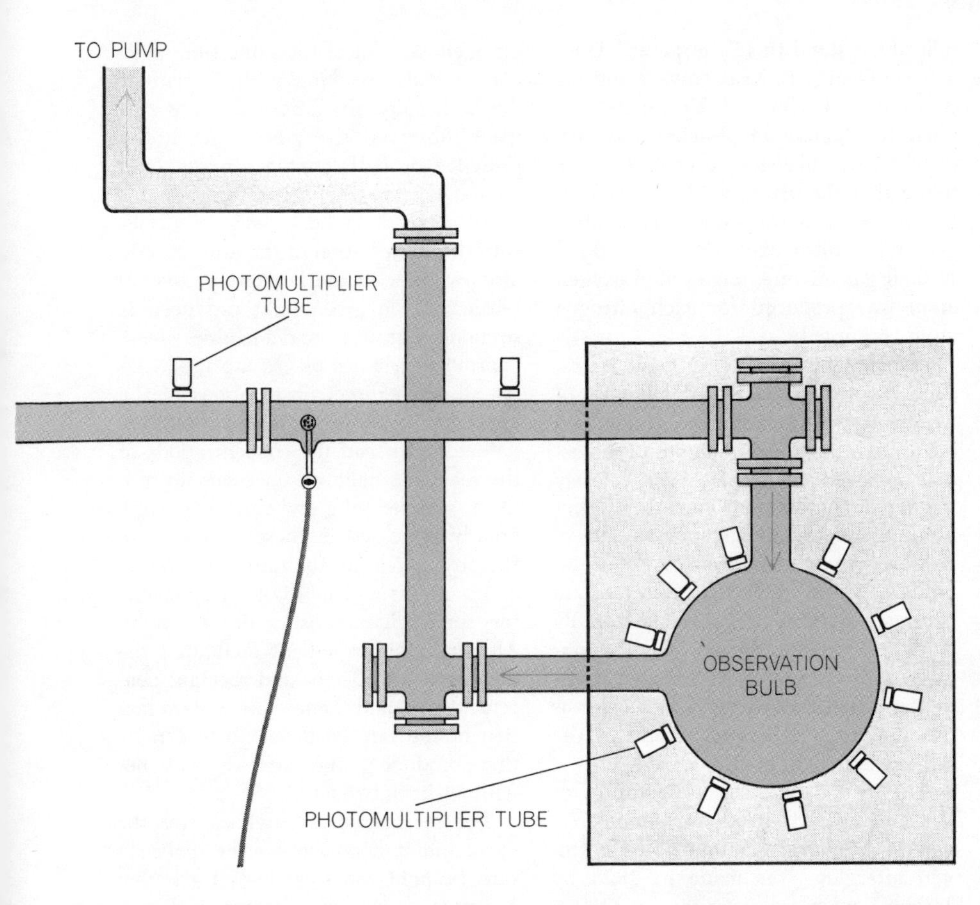

layer of atomic oxygen would be required. At the time Chapman formulated his hypothesis only the green line at 5,577 angstroms had been positively identified in the airglow spectrogram as originating from atomic oxygen; other lines and bands were observed, but their origin was unknown. Chapman proposed that the green emission from atomic oxygen was produced by the interaction of three oxygen atoms to produce an oxygen molecule plus an excited oxygen atom [*see bottom illustration on opposite page*].

Emission bands produced by excited oxygen molecules were later identified in the airglow spectrum. It was at first supposed that these bands were produced when pairs of oxygen atoms combined (with the aid of a stabilizing collision of each pair with a third atom or a molecule) to form excited oxygen molecules. Our laboratory experiments have confirmed this mechanism in the case of the ultraviolet bands produced by excited molecular oxygen, but it appears that some other mechanism involving ozone molecules is responsible for oxygen bands in the infrared. The intense infrared emission band produced by vibrationally excited hydroxyl radicals, first identified in the airglow spectrum in 1950, can also be achieved in the laboratory (and presumably in the airglow as well) by the reaction of a hydrogen atom (H) with an ozone molecule to form an excited hydroxyl radical plus an oxygen molecule. The process whereby atomic oxygen is excited to produce red emission lines at 6,300 and 6,364 angstroms is apparently quite complicated, and several mechanisms have been proposed.

Although in terms of the total light emitted by the airglow the green light of atomic oxygen at 5,577 angstroms is negligible, this single emission line in principle contains as much information about conditions in the airglow layer as any of the brighter emission lines or bands. Whether or not this potential information is received depends on the detectability of the emission. Two factors other than brightness determine this detectability: first, how distinct the emission is from other lines or bands, and second, how sensitive the light detector is. The green line of atomic oxygen at 5,577 angstroms is conspicuously isolated from other airglow emission lines, and modern photoelectric detectors are particularly efficient in the green region of the spectrum. Together these two factors more than compensate for the comparative weakness of the atomic-oxygen emission at this wavelength.

The same two factors also account for the fact that the green spectral line of excited atomic oxygen has been more fully studied than any other aspect of the airglow. The relative ease of observing this particular emission line of atomic oxygen in the airglow holds true for artificial airglows produced in the laboratory, and much of our present work is devoted to exploiting these advantages.

If the major fuel that is "burned" to power the airglow is atomic oxygen, as Chapman suggested, the largest quantum, or unit, of energy available for excitation by a single reaction is equal to the energy that would be given up if these atoms were to form molecules. The molecules most likely to be formed when atomic oxygen reacts with other constituents of the upper atmosphere are molecular oxygen and ozone. The energy given up by either of these reactions is less than five electron volts, which is quite small compared with the energy, say, of the charged particles that produce the aurora or excite radiation in laboratory discharges. This small quantum of energy available to excite the constituents of the upper atmosphere is another major factor in determining the unique spectral properties of the airglow.

There is still only indirect evidence of the existence of a layer of atomic oxygen in the upper atmosphere. Refined measurements of the intensity of the sun's ultraviolet radiation, the rate at which molecular oxygen can absorb this radiation, and the mean density and temperature of the upper atmosphere all strongly suggest, however, that an average of a trillion oxygen atoms per cubic centimeter are present in the airglow emission layer at an altitude of about 60 miles. Laboratory measurements indicate that comparatively few of these atoms are consumed during the night by chemical reactions. It follows that each day's accumulation of oxygen atoms must be removed by bulk transport to lower regions of the atmosphere. In sum, it appears very likely that the layer of atomic oxygen envisioned by Chapman does indeed exist.

In order to discuss in a quantitative way the chemiluminescent processes that may be responsible for the airglow it is necessary to know how the speed of a particular reaction is related to the concentrations of the reactants. The rate at which any two substances react to form other substances is proportional to their collision rate, and this rate in turn is proportional to the multiplication product of their concentrations. When more than two reactants must interact simultaneously, the rate of the reaction involves additional concentration factors. If an excited atom or molecule can lose its excess energy only by radiating a photon, or quantum of light, then the number of photons radiated per second must be equal to the number of atoms or molecules excited per second.

On this basis Chapman's mechanism for the production of the green spectral line of atomic oxygen implies that the intensity of the emission is proportional to the cube of the atomic-oxygen concentration. The "proportionality constant" in such a relation is called the rate coefficient; the speed at which the reaction takes place is found by multiplying the rate coefficient by the concentrations of the reactants. From estimates of the density of atomic oxygen in the airglow layer and from measurements of the rate of emission for the green line in airglow spectrograms, the rate coefficient can be computed. It is this rate coefficient, then, that must be measured in order to verify Chapman's excitation mechanism.

In attempting to validate Chapman's hypothesis experimentally, we have made use of a phenomenon first studied systematically by Lord Rayleigh in 1911. This phenomenon occurs after nitrogen at low pressure is subjected to an intense electrical discharge. When the discharge is stopped, the gas continues to glow—in some cases for hours. For the next 40 years many workers, including Rayleigh, attempted in vain to fully understand this "afterglow." Then in 1956 George B. Kistiakowsky and his colleagues at Harvard University applied the technique of mass spectroscopy to the nitrogen afterglow. They found that the major reactant responsible for the luminosity was atomic nitrogen. When nitric oxide (NO) was added to their gas mixture, one excited oxygen atom was produced for each nitrogen atom consumed.

Kistiakowsky's discovery made it possible to prepare measured amounts of atomic oxygen from atomic nitrogen and nitric oxide in an environment of molecular nitrogen to simulate very closely the actual condition of the earth's upper atmosphere. (Small amounts of molecular oxygen would improve the simulation.) Since molecular oxygen, in contrast to molecular nitrogen, rapidly de-excites atomic oxygen, some stratagem for producing excited oxygen atoms in the absence of oxygen molecules is essential to a laboratory study of the airglow emissions of atomic oxygen.

The first laboratory observation of the green spectral line of atomic oxygen in an oxygen-contaminated nitrogen afterglow was made in 1928 by Joseph Kaplan, who was then at Princeton University. In 1960 Charles A. Barth of the Jet Propulsion Laboratory of the California Institute of Technology and Kenneth C. Clark and I, then at the University of Washington, independently made the first measurements of the intensity of the green line in an afterglow produced by known amounts of oxygen atoms diluted with molecular nitrogen. At almost the same time William Schade, working with Kaplan at the University of California at Los Angeles, observed the green line in the presence of both atomic nitrogen and atomic oxygen.

Although it is fairly easy to duplicate the composition of the atmospheric airglow layer in the laboratory, the intensity of the green light produced is so faint that it is undetectable unless enormous volumes of gas are used. After all, even direct observations of the nightglow require extremely sensitive detectors. Instead the concentrations of the reactants believed necessary to produce selected airglow emissions were greatly increased in these experiments, thereby increasing the rate of excitation of the atoms and molecules responsible for some characteristic airglow light. Although this expedient facilitated the measurement of light and reactant concentrations—and hence the determination of the rate coefficient of the reaction producing the light—it was not without its drawbacks.

The most serious drawback was the concomitant increase in the collision rate, brought about partly by the higher concentrations of the reactants and partly by the requirement that diffusion to the boundaries of the experimental system be slow. As I have mentioned, the low collision rate of excited atoms and molecules in the upper atmosphere allows these systems to radiate light rather than exchanging their energy in collisions. It was not at all clear five years ago whether or not the same ab-

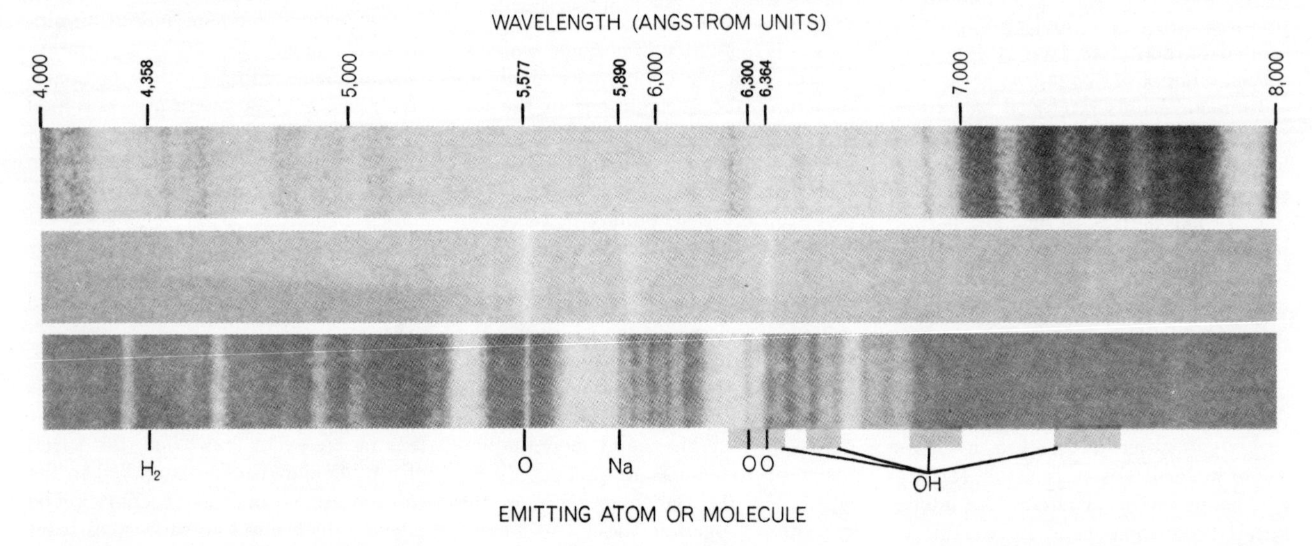

SPECTROGRAMS of the aurora (*top*), the airglow (*middle*) and an artificial airglow produced in the author's apparatus (*bottom*) are compared. The main difference between the aurora and the airglow spectrograms is the absence of molecular nitrogen (N_2) emission bands in the right half of the airglow spectrogram. The laboratory spectrogram represents an incomplete approach to simulated airglow, but it does contain the conspicuous green line of atomic oxygen (O) at 5,577 angstroms. It also contains molecular nitrogen bands, which are not present in the spectrogram of the complete simulation. Other atoms and molecules responsible for emission lines and bands in the airglow spectrogram are molecular hydrogen (H_2), sodium (Na) and the hydroxyl radical (OH).

sence of quenching would obtain in a laboratory experiment, and a complete study of quenching under these circumstances had to be undertaken. This in itself was a difficult project, and it was combined with the necessity of making absolute measurements of both very faint light fluxes and highly reactive atomic fragments. It is not surprising that more than 35 years elapsed before Chapman's suggested airglow-excitation mechanism could be verified.

In 1963 Robert Sharpless and I conducted a series of experiments at the Stanford Research Institute that measured both the intensity of the green line at 5,577 angstroms and the concentration of atomic oxygen produced by the reaction of atomic nitrogen with nitric oxide in a low-pressure, flowing-gas system. We found that the light intensity was proportional to the square of the atomic-oxygen concentration. This finding is in conflict with the prediction that the light intensity would be proportional to the cube of the atomic-oxygen concentration, which was made on the assumptions that the Chapman reaction produced excited oxygen atoms and that all the atoms radiated.

Clearly at least one of these assumptions is incorrect. This being the case, we must ask the question: What would be the relation between the light intensity and the concentration of atomic oxygen if the excited oxygen atoms that are responsible for the radiation we observe are indeed produced by the Chapman reaction but subsequently lose their energy by both emission and collision? It is necessary to bear in mind that the proportionality of light intensity to the concentration of excited atoms is fixed by the internal structure of the atoms and is not changed by other processes that may also de-excite these atoms.

If oxygen atoms are de-excited by radiation only, their rate of de-excitation equals the rate of emission, but if collisions also remove energy from excited atoms, their rate of de-excitation is larger than their rate of emission and consists of two components, one (radiation) independent of the environment and the other (collisions) linked to the density of the collision partners.

In a steady state (that is, one in which all the variables change slowly) the number of excited atoms produced per unit of time must equal the number lost per unit of time. The rate of loss is proportional to the number of atoms present only when one excited atom at a time is involved in the de-excitation

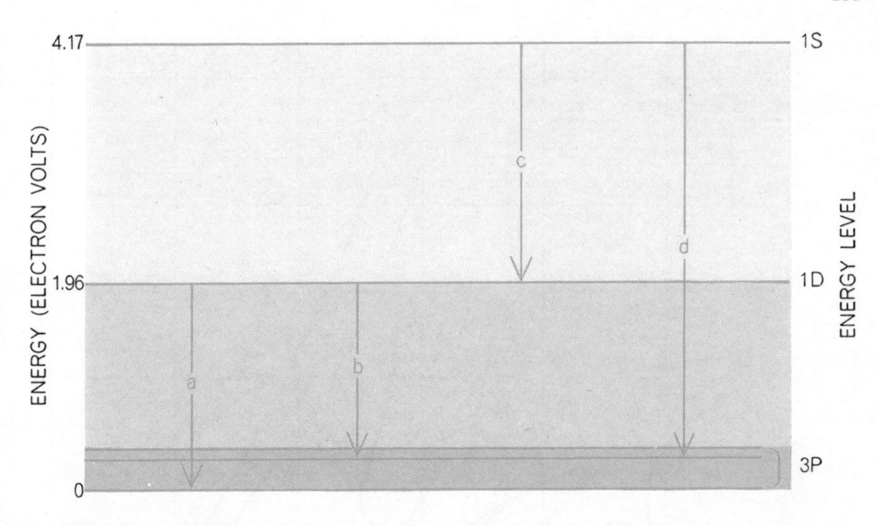

ELECTRON ENERGY LEVELS of atomic oxygen are shown in this diagram, together with four electron transitions between energy levels that account for characteristic emission lines in spectrograms of the airglow. Transitions *a* and *b* are responsible for red emission lines at 6,300 and 6,364 angstrom units respectively. Transition *c* produces the green spectral line at 5,577 angstroms, and transition *d* produces an ultraviolet line at 2,972 angstroms.

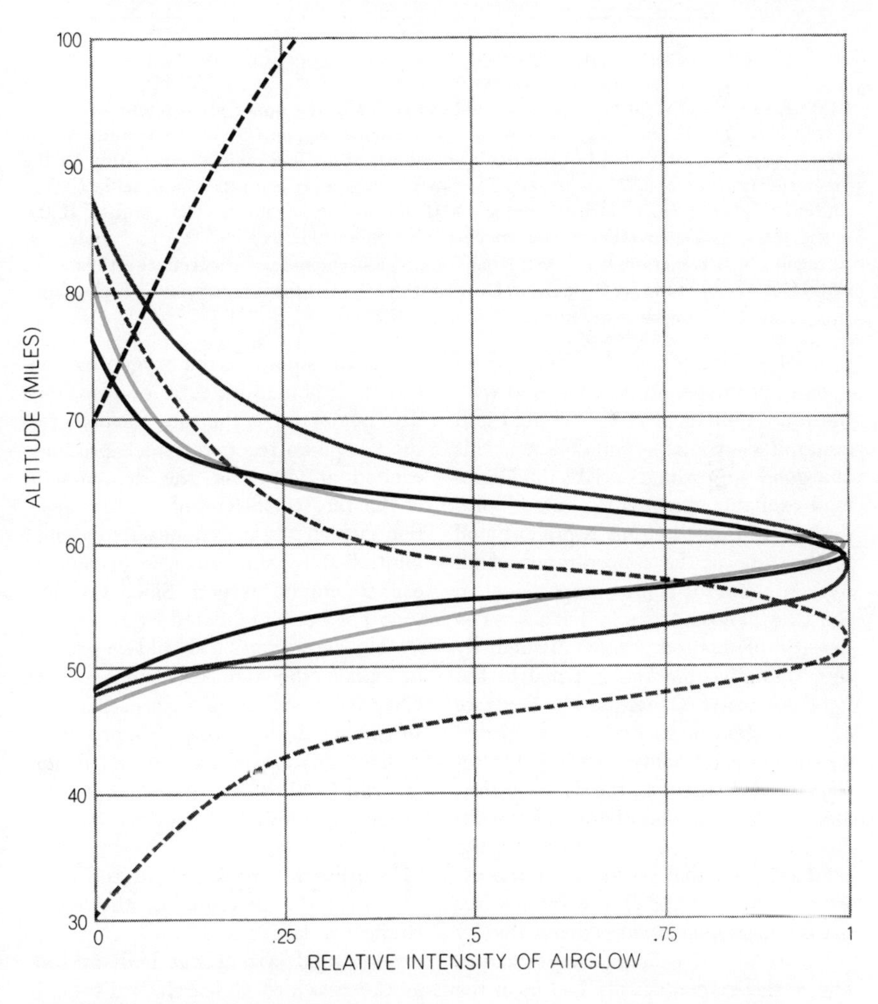

INTENSITY OF AIRGLOW produced by several different radiative mechanisms varies with altitude. Green light from atomic oxygen (*colored curve*) and violet and red light from molecular oxygen (*solid black curve and solid gray curve respectively*) is produced in a rather narrow layer some 60 miles above the surface of the earth. Infrared radiation from hydroxyl radicals (*broken gray curve*) comes from a somewhat lower level, and red light from atomic oxygen (*broken black curve*) comes from a much higher one. The data for the curves in this graph were obtained by D. M. Parker and his colleagues from rocket observations of the upper atmosphere conducted by the U.S. Naval Research Laboratory.

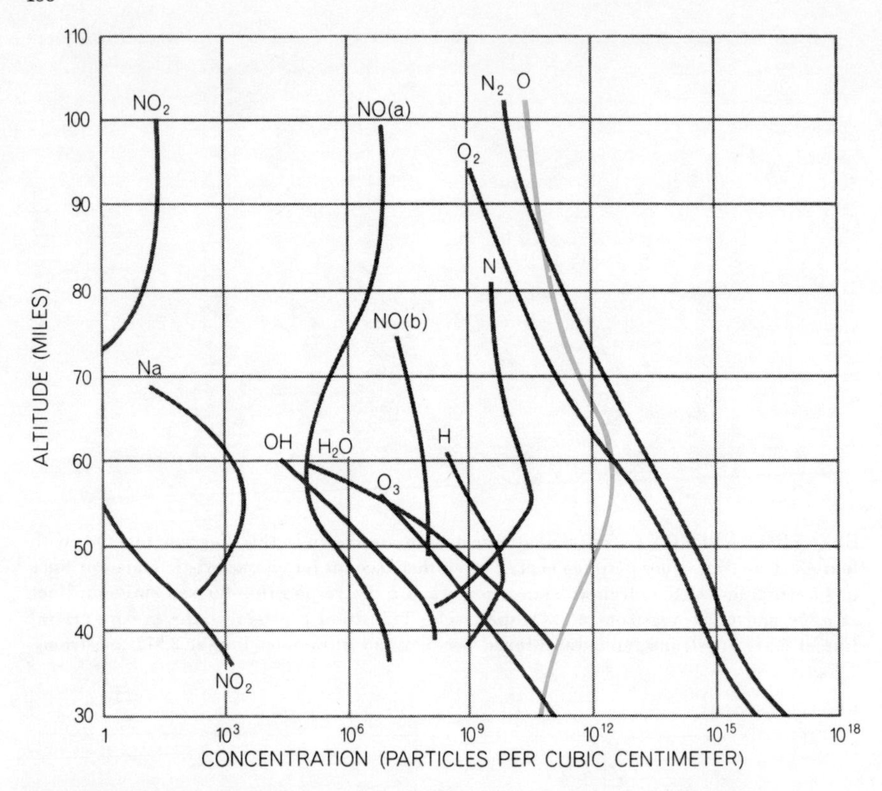

CONCENTRATIONS of various atoms and molecules in the upper atmosphere are represented here. The bulge in the colored curve for atomic oxygen (O) in the vicinity of the airglow layer was predicted in Chapman's hypothesis of the mechanism responsible for the green spectral line at 5,577 angstroms. The other curves represent nitrogen dioxide (NO_2), sodium (Na), two forms of nitric oxide (NO), the hydroxyl radical (OH), water (H_2O), ozone (O_3), hydrogen (H), atomic nitrogen (N), molecular oxygen (O_2) and molecular nitrogen (N_2). The graph was constructed from measurements and theoretical estimates by Charles A. Barth of the Jet Propulsion Laboratory of the California Institute of Technology.

process. (Because of the very small number of excited atoms compared with the total number of atoms—excited and nonexcited—available for collision, this is a good approximation.) In the Chapman excitation process the rate of production of excited atoms is proportional to the cube of the concentration of all oxygen atoms; it is also proportional to the rate of loss of excited atoms. The density of excited atoms divided by their radiative lifetime is equal to the light intensity. Consequently the rate of production of excited atoms divided by the proportionality constant that relates excited atoms to their rate of loss, and by the radiative lifetime of the excited state, will be proportional to the light intensity that results when the excited atoms radiate. If the rate of loss of excited oxygen atoms is dominated by collisions with nonexcited oxygen atoms, one of the oxygen-density factors in the production rate will be canceled by the oxygen-density factor in the loss rate, and the light intensity will be proportional to the square of the concentration of oxygen atoms.

Thus it appears that the proportionality constant that relates the light intensity to the square of the oxygen concentration is actually a ratio of constants. The numerator is the rate coefficient of the Chapman reaction (which produces excited atoms) and the denominator is the rate coefficient of another reaction (which produces nonexcited atoms) multiplied by the radiative lifetime of excited atomic oxygen. Since this lifetime has been calculated (it is approximately one second), it should be possible to obtain the rate coefficient of the Chapman reaction from our earlier measurements, if these are supplemented by a determination of the rate coefficient of collisional de-excitation of excited oxygen by nonexcited oxygen.

To make this measurement and to increase the precision of the earlier studies, a larger and much improved experimental system was built in 1964 at the Stanford Research Institute. In this apparatus Graham Black and I made extensive new observations of the quenching of excited atomic oxygen.

I shall briefly describe the operation of this new system and the findings it yielded. Since changes in the concentration of atomic oxygen affect both the production rate and the loss rate of excited oxygen atoms, it is difficult to separate these effects. If some gas could be found, however, that does not react with nonexcited atomic oxygen but does remove the excess energy of excited atomic oxygen, then this gas could be added to the mixture of excited and nonexcited oxygen until it removes excited oxygen atoms as quickly as they are removed by collisions with nonexcited atomic oxygen plus radiation. When this point is reached, the intensity of the emission from the excited oxygen atoms will be half what it was before the quenching gas was added.

We found that nitrous oxide (N_2O) and carbon dioxide (CO_2), which did not react with nonexcited oxygen atoms, were gases suitable for quenching the radiation from excited atomic oxygen. Since the rate at which these gases de-excite oxygen atoms is proportional to their concentration in the gas mixture, one would expect that the amount of quenching gas added to reduce the light intensity by half would be proportional to the density of nonexcited atomic oxygen. This was indeed true when the density of nonexcited atomic oxygen was large enough. The finding verified our analysis that attributed the major loss of excited oxygen atoms in the experimental apparatus to reactions with nonexcited oxygen atoms. In order to obtain the rate of the atomic-oxygen quenching reaction, however, the rate of the added-gas quenching reaction must be known.

If the concentration of atomic oxygen could be reduced sufficiently, radiation would be the predominant form of energy loss from excited oxygen atoms; if this radiation were detectable, the rate of excitation of oxygen atoms could be measured directly. As the concentration of atomic oxygen is decreased, however, the rate of excitation decreases much faster than the rate of quenching loss decreases, and as a result the light intensity is reduced to an undetectable level. If it were not for this complication, the rate of the Chapman reaction could be obtained directly.

These circumstances suggest an extrapolation procedure to obtain the rate coefficient for the quenching of excited oxygen atoms by the added gas. Once this quantity is found, the same procedure will also yield the rate coefficient for the de-excitation of oxygen atoms by nonexcited oxygen atoms. It is this latter parameter that is needed to convert our previously measured proportionality constant (relating light intensity to the square of the nonexcited-atomic-oxygen concentration) to the rate coefficient of the Chapman reaction. If the amount of quenching gas needed to reduce the light intensity by half is plotted on a graph against the density of atomic oxygen, a straight line results, since the densities are proportional to each other [*see illustration below*]. Because the excited oxygen atoms can radiate, however, this line does not pass through the point of origin but rather crosses the added-quenching-gas axis at a finite value. The amount of the quenching gas indicated by this crossing is then the amount needed to de-excite oxygen atoms at the same rate as that at which they radiate. Thus the rate coefficient of the added-gas quenching reaction is equal to the rate of light emission per excited atom divided by the density of the quenching gas. The slope of the line is the ratio of the added-gas quenching rate coefficient to the nonexcited atomic-oxygen quenching rate coefficient. By dividing the added-gas quenching rate coefficient by the slope of the line, the rate coefficient of quenching by nonexcited atomic oxygen can be obtained. Finally, the measured proportionality constant that relates light intensity to the square of the oxygen density is multiplied by the radiative lifetime of the excited oxygen atoms and by the rate coefficient of quenching for nonexcited atomic oxygen to yield the rate coefficient of the Chapman reaction.

Amazingly, everything happened in our apparatus just as we had anticipated. Several gases that react slowly with oxygen and nitrogen—for example nitrous oxide, carbon dioxide and molecular oxygen—were used to quench the radiation from the excited oxygen atoms. All these quenching gases indirectly gave the same value for the quenching rate coefficient for nonexcited oxygen atoms. When combined with either our original measurements relating light intensity to the square of the atomic-oxygen density or the measurements made in our new system, the rate coefficient of the Chapman reaction turned out to be the same as that deduced from the direct measurements of the nightglow. Chapman's hypothesis was finally confirmed: three nonexcited oxygen atoms can combine in the upper atmosphere to form an oxygen molecule and an excited oxygen atom, which can then emit light at 5,577 angstroms.

The airglow still has some secrets. In the very near future, however, the other chemical processes that power the airglow will undoubtedly be identified and their rate coefficients measured. The phenomenon of the airglow has become a tool to be exploited in other investigations by physicists, chemists and meteorologists.

For example, the mass transport of air to and from the airglow region of the atmosphere may be part of an important global circulation system. By using atomic oxygen as a tracer of air motion this circulation system could be studied. In the polar regions the sun is below the horizon for several months and therefore cannot replenish excited atomic oxygen lost by radiation and by massive air movements. A study of the concentration of excited oxygen atoms under these conditions may lead to a determination of the speed of the air mass of which they are a part. Fortunately new observational techniques now make it possible to observe the polar airglow in spite of the interference of auroral light.

The ultraviolet radiation from the sun that powers the airglow does not reach the earth's surface. Hence the airglow is a potentially useful indicator of changes in a component of sunlight that are not measurable directly by conventional ground-based techniques. A reanalysis of airglow data for the past 50 years may reveal significant fluctuations in the amount of ultraviolet radiation from the sun. Because of the slow response of the airglow to solar excitation, however, it is not certain that brief solar disturbances can be monitored in this way.

It is difficult to predict how our new knowledge of the airglow will be used in the future. But since the airglow mirrors the chemical state of the upper atmosphere it cannot fail to attract increasing attention as long as man insists on traversing this boundary between the earth and the rest of the universe.

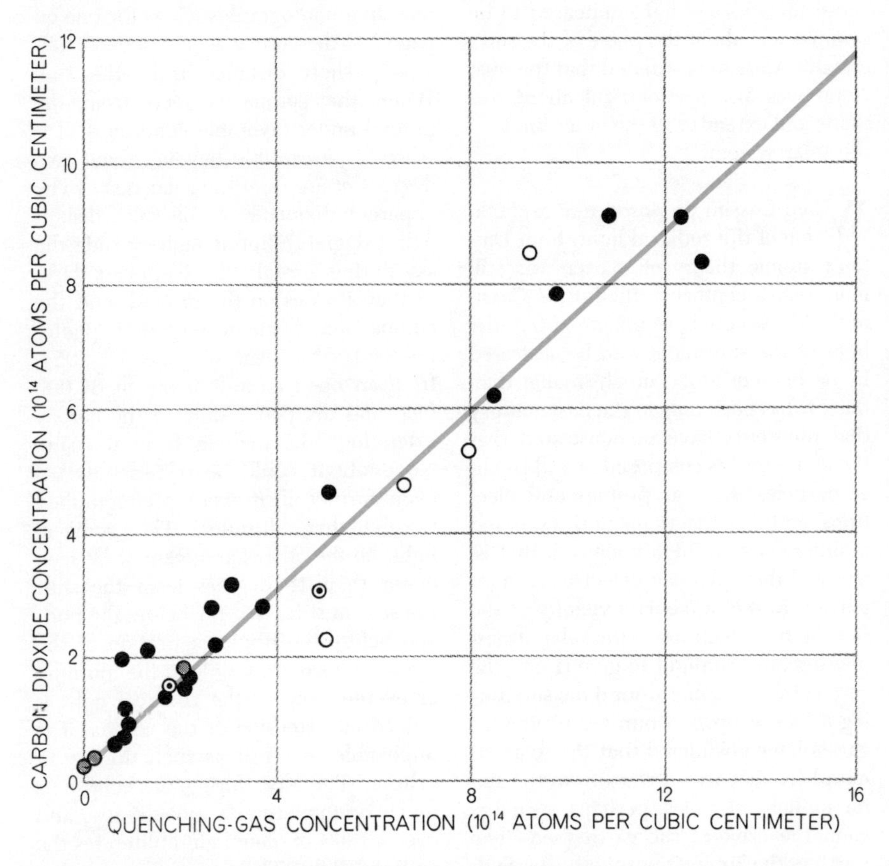

EXTRAPOLATION PROCEDURE is employed to find the amount of quenching gas needed to de-excite oxygen atoms at the same rate as that at which they radiate. If the amount of quenching gas needed to reduce the light intensity by half is plotted against the density of atomic oxygen, a straight line results, since their densities are proportional to each other. Because the excited oxygen atoms can radiate, however, this line does not pass through the point of origin but rather cuts the added-quenching-gas axis at a finite value. The amount of quenching gas indicated by this intercept is then the amount needed to de-excite oxygen atoms at the same rate as that at which they radiate. From the slope of the line the rate coefficient for de-excitation of atomic oxygen by nonexcited atomic oxygen can be obtained. The four different kinds of dots indicate readings for different mixtures of nitrogen and other gases in which the excited and nonexcited oxygen atoms were embedded.

18 The Zodiacal Light

by D. E. Blackwell
July 1960

On a clear, dark night in the tropics a luminous pyramid is seen in the sky after sunset and before sunrise. The phenomenon seems to be caused by the scattering of sunlight by interplanetary dust

About an hour after sunset on a clear, moonless night in the tropics a faint pyramid of light glows in the western sky. About as luminous as the Milky Way, the pyramid is broadest and brightest at the horizon and fades out toward the zenith. Because it stretches along the zodiac (the imaginary belt in the sky that contains the paths of the sun, the moon and the major planets), the glow is known as the zodiacal light.

As the earth turns on its axis, the glowing pyramid sinks beneath the horizon, the last and faintest vestige disappearing a few hours after sunset. Toward morning the spectacle is repeated in reverse order in the eastern sky. The faint apex appears a few hours before dawn; the entire pyramid attains its full splendor about an hour before sunrise. Thereafter the sky brightens, and the zodiacal light vanishes from sight. Away from the equatorial regions the zodiacal light can be seen occasionally, but at higher latitudes it is tilted at a sharp angle to the horizon, and so is quite difficult to detect except during certain seasons.

The origin of the light has been a subject of speculation for centuries. The classical explanation was worked out nearly 300 years ago by the French astronomer Jean-Dominique Cassini, who began a 10-year study of the phenomenon in 1683. Although some of his contemporaries argued that the zodiacal light was an atmospheric phenomenon, Cassini observed that its position in the sky was the same when it was viewed from different locations, and concluded that it must originate somewhere in space. He speculated that the light is caused by a disk-shaped cloud of interplanetary dust that reflects the light of the sun. Such a cloud would scatter sunlight in much the same way that a cloud of moths scatters the light from a street lamp at night: to an observer some distance away the lamp seems to be surrounded by a luminous haze. Because the zodiacal light appeared to be symmetrical about the plane of the sun's equator, Cassini postulated that the dust cloud was also symmetrical about this plane and extended to the outer limits of the solar system.

When Cassini proposed this explanation of the zodiacal light, John Dalton's atomic theory of matter was still more than a century in the future. Cassini could scarcely have imagined that the light of the sun might also be scattered by particles or atoms much smaller than dust. It has been only in the past century that physicists have demonstrated that the atom itself is composed of still smaller particles, such as protons and electrons, and that the atoms in the sun and in other stars are highly ionized, that is, many of their planetary electrons are removed. In the immediate vicinity of the sun the free electrons in the solar atmosphere scatter sunlight to give rise to the corona that is visible around the sun during a total eclipse. A number of investigators have postulated that the solar atmosphere reaches outward from the sun for millions of miles, perhaps even beyond the orbit of the earth [see "The Earth in the Sun's Atmosphere," by Sydney Chapman; SCIENTIFIC AMERICAN, October, 1959]. They have therefore been amending Cassini's explanation, contending that the zodiacal light is an extension of the solar corona and represents the scattering of sunlight by electrons rather than by dust.

Quite apart from the question of what kind of particles are involved, there are observers who hold, as some did in Cassini's time, that the zodiacal light originates in the earth's atmosphere. In deference to these workers it must be admitted that the evidence connecting the zodiacal light to the corona is indirect. In a photograph such as the one on page 162 the corona seems to end a relatively short distance from the sun. When the corona is seen from the ground under favorable conditions, it is scarcely perceptible beyond about two degrees of arc from the solar disk. (The apparent diameter of the solar disk is half a degree.) But at high altitude the sky during a total eclipse appears darker than it does on the ground, and the corona seems much larger. I photographed the eclipse of June 30, 1954, from an open aircraft flying at 30,000 feet, and my plates showed the corona extending 13.5 degrees from the sun. No doubt it could be traced outward even farther if it were photographed from higher altitudes. The zodiacal light, on the other hand, cannot be seen closer than 18 degrees from the sun: the sun must be that far below the horizon before the brightest portion of the pyramid becomes visible in the morning or evening sky. If the zodiacal light is indeed an extension of the corona, it is impossible to see it as such during an eclipse. The sky during an eclipse is much brighter than a moonlit sky, and even a trace of moonlight obliterates the zodiacal light. Thus there is a gap be-

ZODIACAL LIGHT was photographed by the author and M. F. Ingham from Chacaltaya in the Bolivian Andes, at an altitude of 17,100 feet. Stars caused the narrow streaks as the earth turned during the 10-minute exposure. Dark line in middle was caused by cross hair in telescope system.

tween the inner edge of the zodiacal light and the observed outer edge of the corona.

The gap can be seen in the diagram on page 163, which plots the brightness of the corona and the zodiacal light as a function of angular distance of the sun. The values for the zodiacal light are those obtained by Franklin E. Roach and his colleagues on Cactus Peak in California, at a latitude of 36 degrees North; the values for the inner corona are my aerial measurements from the eclipse of 1954. The two sets of measurements form two straight-line curves. Were it not for the gap in the region where no observations are available, the two curves would join quite smoothly to form a single one. They would hardly match so well if the corona and the zodiacal light were of different origin. This piece of indirect evidence therefore weighs against the possibility that the zodiacal light is an atmospheric effect.

Another item of evidence in favor of extraterrestrial origin is the position of the zodiacal light in the sky. When there are no conspicuous eruptions on the sun's surface, the glow is symmetrical about a line that is close to the ecliptic: the projection of the plane of the earth's orbit in the sky. To produce such a pattern the earth's atmosphere would have to be extended in some way in the plane of the earth's orbit. This is not an impossible idea; indeed, some Soviet investigators have supposed that the earth does possess such a gaseous tail.

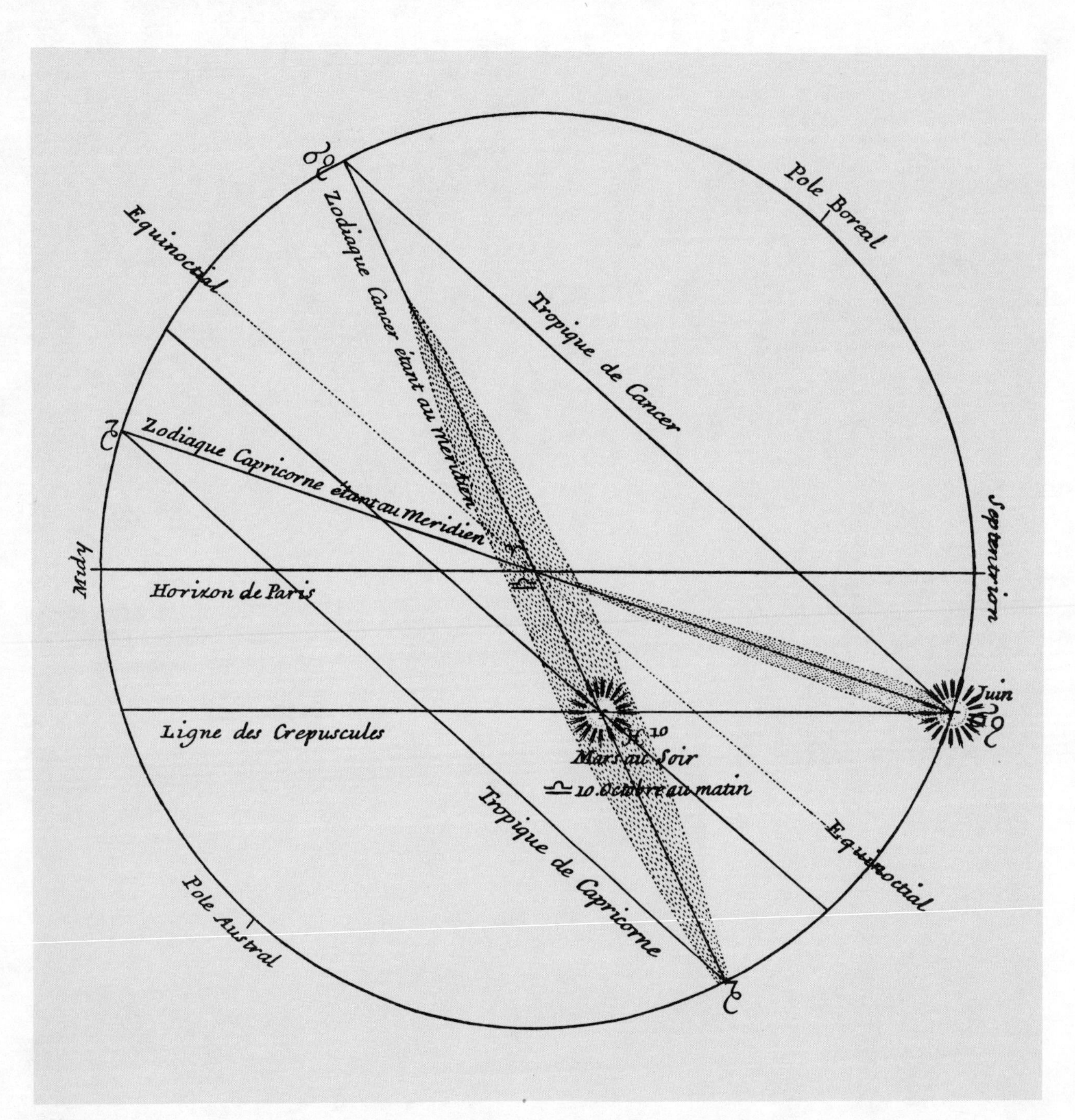

CLASSICAL VIEW of the zodiacal light was developed by the 17th-century French astronomer J. D. Cassini. His diagram of the zodiacal light is basically correct, although he was mistaken in his beliefs that the light was symmetrical about the plane of the solar equator and that the light was broader in March than in June. He argued that the light was due to a dusty nebula around the sun.

But the axis of symmetry of the zodiacal light does not lie precisely on the ecliptic. The pyramid of light is more nearly symmetrical about the plane of the orbit of Jupiter, which lies at an angle of 1.3 degrees to the ecliptic. In placing the axis of symmetry on the solar equator Cassini was about 7 degrees off, an error that is understandable in view of the fact that he made his observations in northern latitudes. But the orbital plane of Jupiter comports even better with his model of an interplanetary dust cloud. The gravitational forces accounting for the distribution of the cloud would logically find their plane of symmetry close to the orbital plane of Jupiter, the most massive planet. Thus it is safe to conclude that, although scattering by the atmosphere may play a small role in the zodiacal light, the major part of the light is scattered by matter in interplanetary space.

The question that remains is whether the scattering agent is dust, as Cassini proposed, or some other kind of particle. To make this choice it is necessary to consider the nature of the solar corona and to compare it to the zodiacal light. When the sun is photographed with a telescope, the edge of the solar disk appears to be perfectly sharp. Studies of the corona have shown, however, that the density of solar material does not change so abruptly as this would suggest, but diminishes gradually. The sharp difference in brightness is due to a change in the mechanism of radiation. Up to the edge of its clearly visible disk the sun radiates light simply because it is hot. Beyond the edge of the disk, in the region of the corona, the solar atmosphere is even hotter, but it is too thin to radiate much light; it can be seen during an eclipse chiefly because it scatters the light of the visible disk. The behavior of the flame of a gas stove illustrates the difference between the two regions. When the air supply is limited, the flame is luminous but opaque; when the air supply is increased, the flame becomes hotter and more transparent but less luminous. The analogy should not be carried too far, however, because conditions on the sun are quite different from those in stoves.

Karl Schwarzschild of the Potsdam Astrophysical Observatory first advanced (in 1905) the notion that free electrons in the atmosphere surrounding the sun scatter sunlight and thus give rise to the faint luminosity of the corona. He observed, in the first instance, that the color of the coronal light is almost iden-

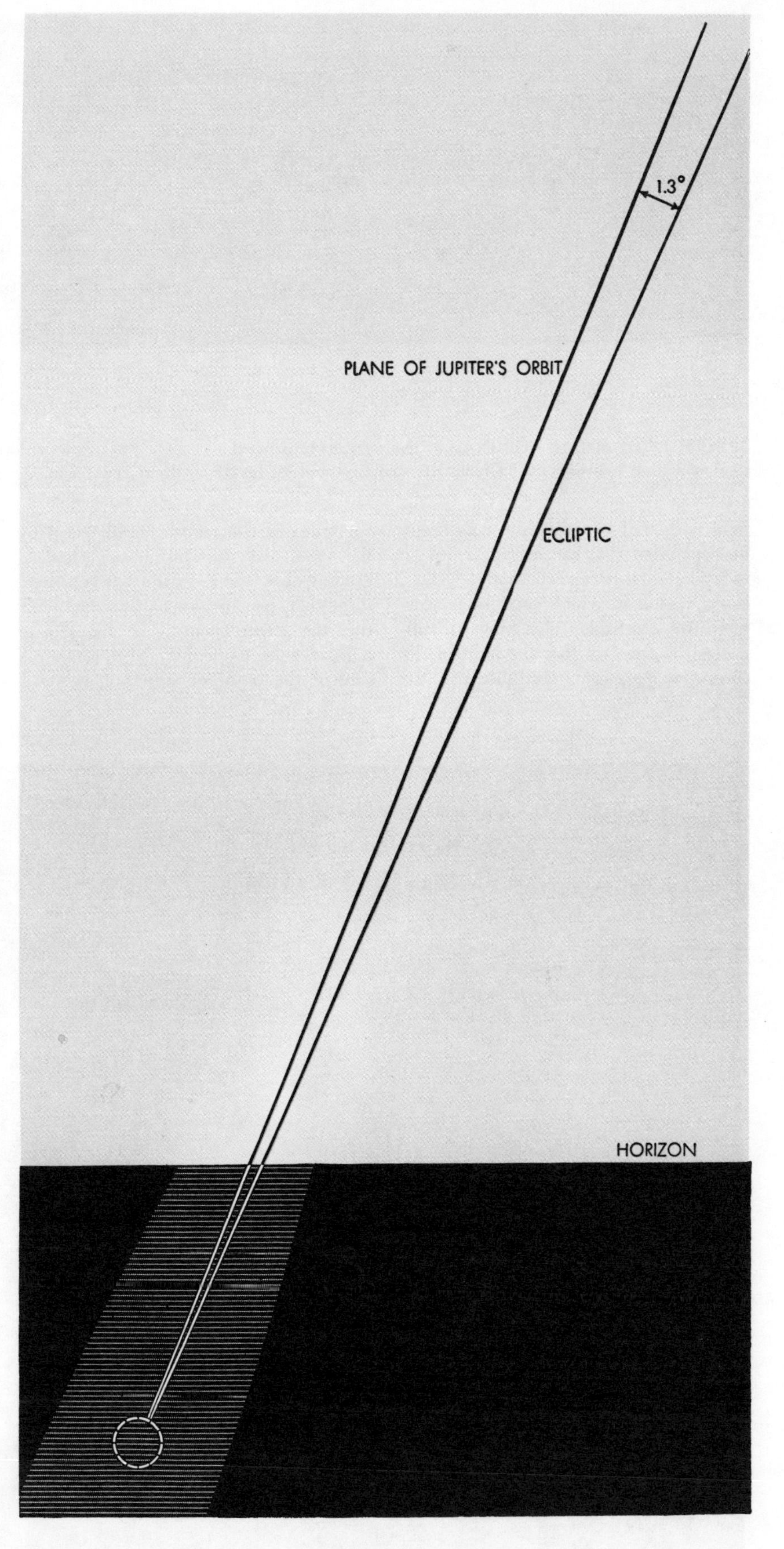

MODERN VIEW depicts the zodiacal light as the outer fringe of the solar corona, visible only when the sun (*broken circle*) has sunk beneath the horizon. The light is symmetrical about the plane of Jupiter's orbit, which lies at an angle of 1.3 degrees to the ecliptic.

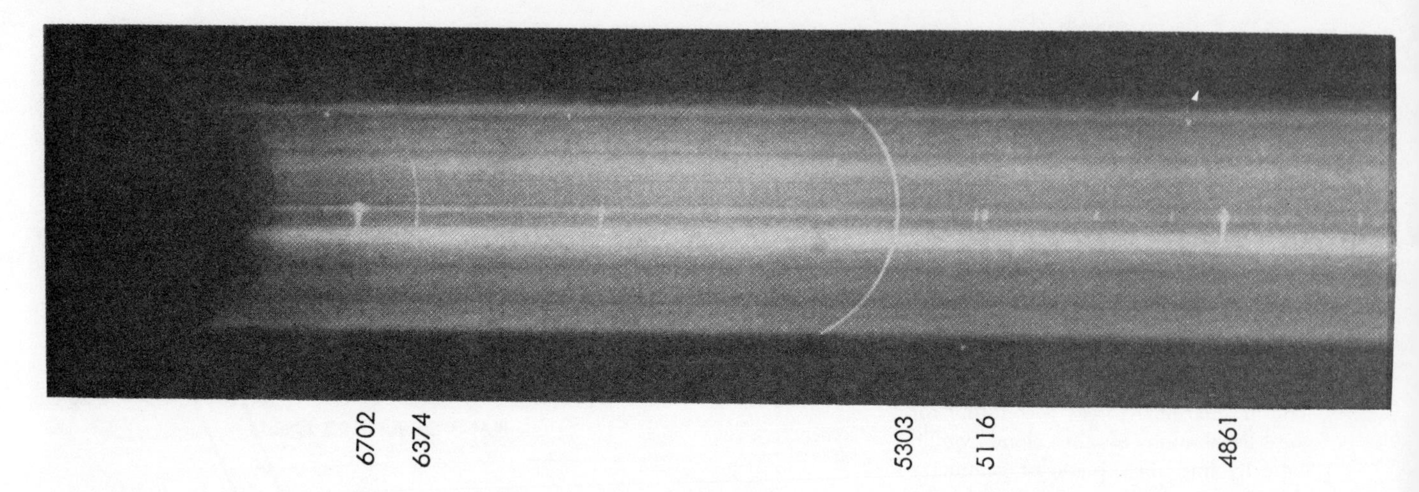

SPECTRUM OF SOLAR CORONA was photographed during the total eclipse of February 25, 1952, by Bernard Lyot and M. K. M. Aly. Spectrum is almost continuous, shading without interruption from red *(left)* through violet *(right)*. The labels indicate the

tical with that of sunlight; accordingly he concluded that the corona is not an independent source of thermal radiation, but a region in which sunlight is scattered by electrons. This view is supported by the fact that the light of the corona is polarized. And because the spectrum of the corona, unlike that of the solar disk, is continuous (that is, grading almost uninterruptedly from red through green to violet) he concluded that the electrons must be moving at high speeds, equivalent to a temperature of the order of a million degrees Fahrenheit. If the temperature were much lower, the spectrum of the scattered light would be interrupted by the same dark absorption lines—the Fraunhofer lines—that cross the spectrum of the solar disk. In a gas at a million degrees F., however, the velocity of the

SOLAR CORONA was also photographed during the eclipse of 1952. The brightness of the eclipse sky makes the corona appear to end a short distance from the sun. Photograph was made by H. Von Klüber during the Cambridge expedition to Khartoum.

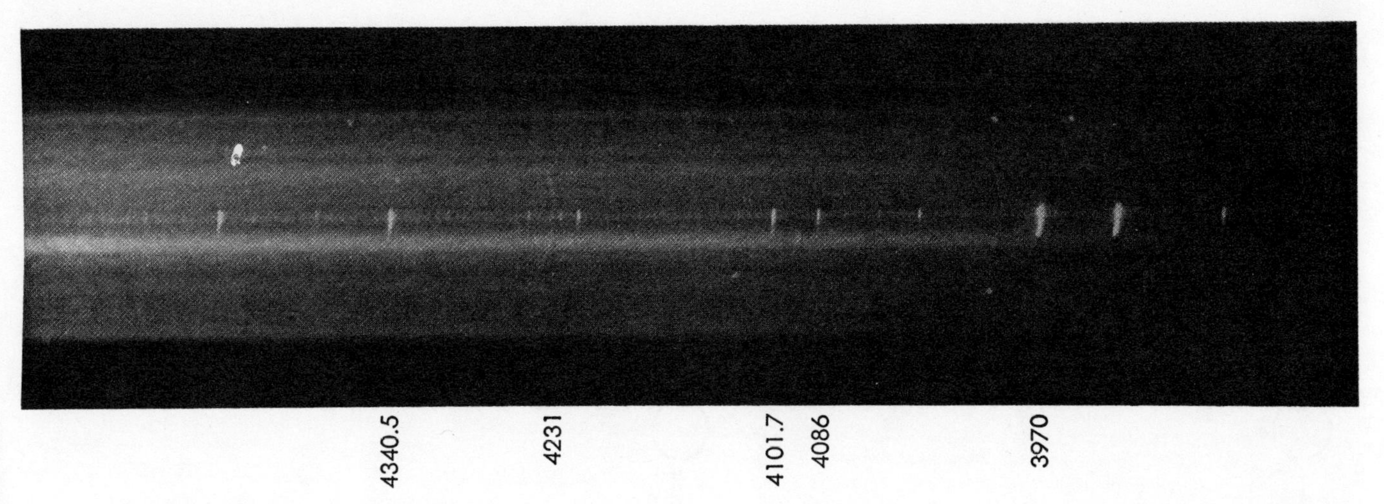

wavelength (in angstroms) of the bright spots and lines. The lines are bent, because the slit of the spectrograph is curved to fit the edge of the sun. Both the lines and the spots are produced by hydrogen and ionized heavier atoms in the atmosphere of the sun.

electrons is high enough to induce a Doppler shift in the light incident upon them. An electron traveling at a high velocity away from an observer shifts the wavelength of light toward the red end of the spectrum; conversely, an electron moving toward the observer shifts the wavelength toward the violet. The random motion of electrons in a gas at a million degrees would scatter light of a given wavelength over a band of wavelengths 250 angstrom units wide. Since most of the Fraunhofer lines in the solar spectrum are less than one angstrom wide, the scattering of light by electrons in a gas at this temperature would obliterate all but perhaps the broadest of them, producing a continuous spectrum. The extent to which the Fraunhofer lines are suppressed thus provides a measurement of the coronal temperature. Spec-

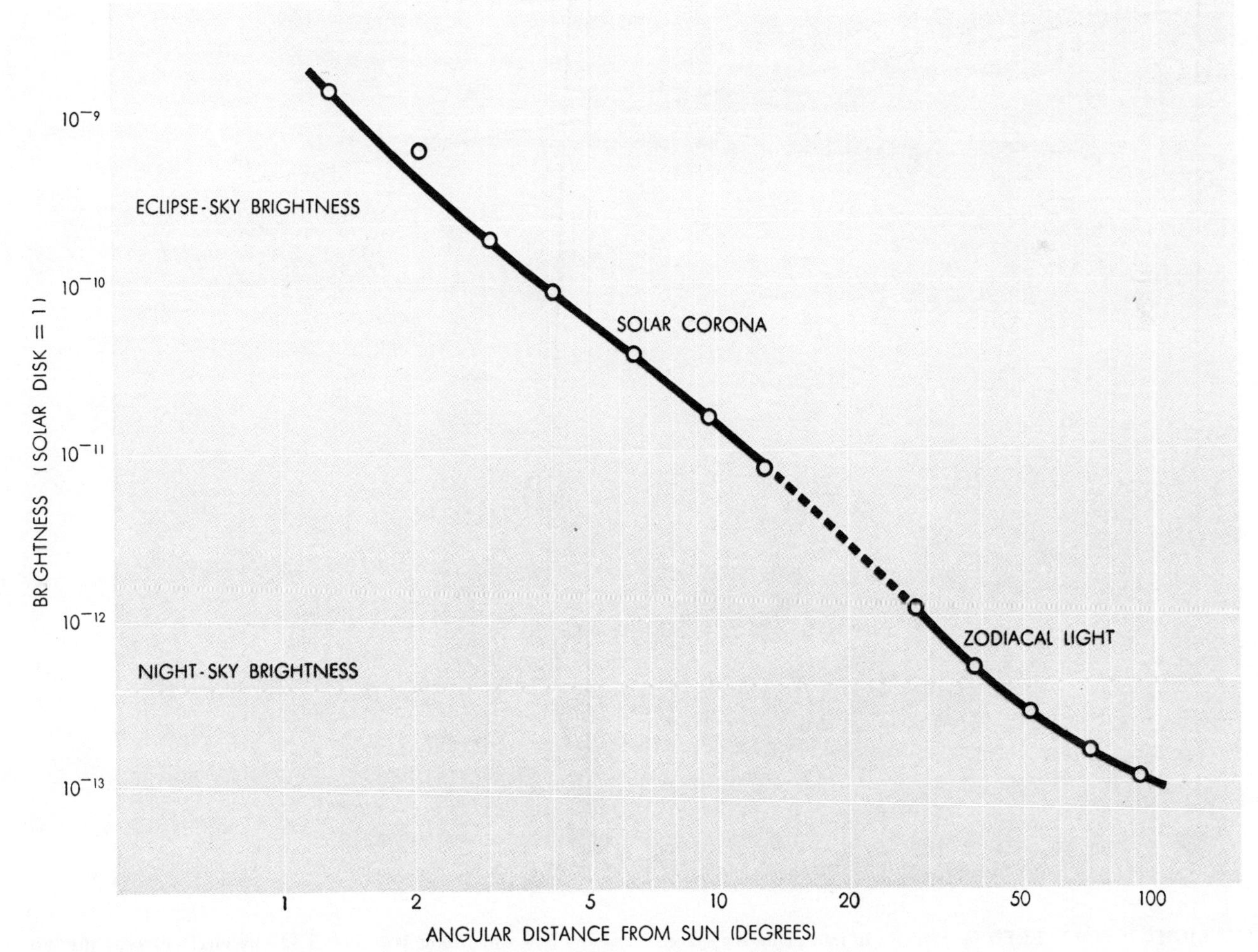

BRIGHTNESS CURVES of the solar corona and the zodiacal light almost join to form a single straight-line curve. Broken line indicates region where no observations are available. Shading indicates brightness of the sky at night and during total eclipse.

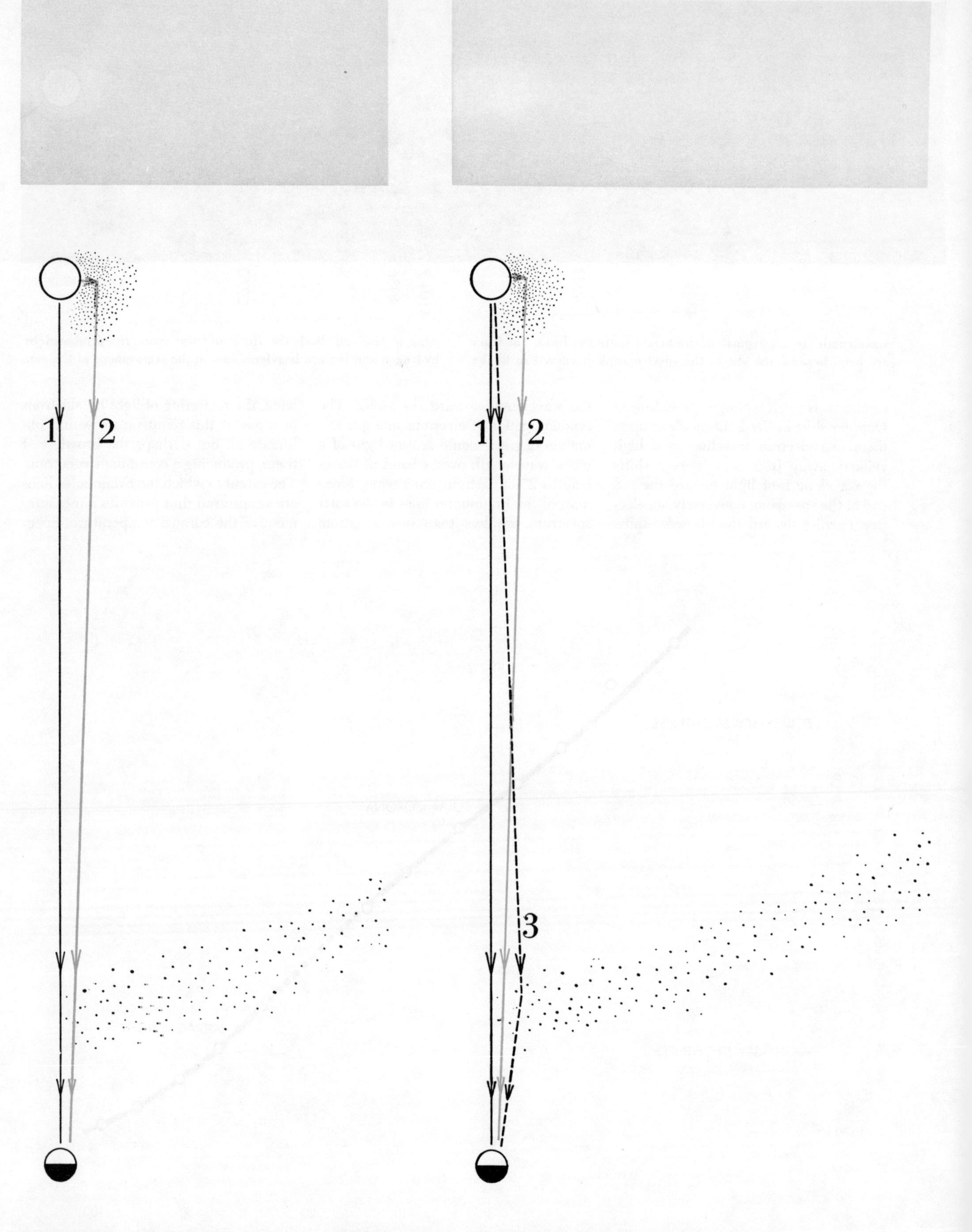

SUNLIGHT IS SCATTERED by particles in interplanetary space, and so reaches the earth via several optical paths (*arrows*). Most of the sunlight travels in a straight path (1). The diagram at left shows how some light is scattered (2) by electrons near the sun (*small dots*) to produce the inner corona (*top left*). The diagram at center shows how sunlight is diffracted (3) by dust particles

trograms of the inner corona, within about a quarter of a degree of the edge of the solar disk, show that the temperature there is about a million degrees F., which agrees with Schwarzschild's calculations.

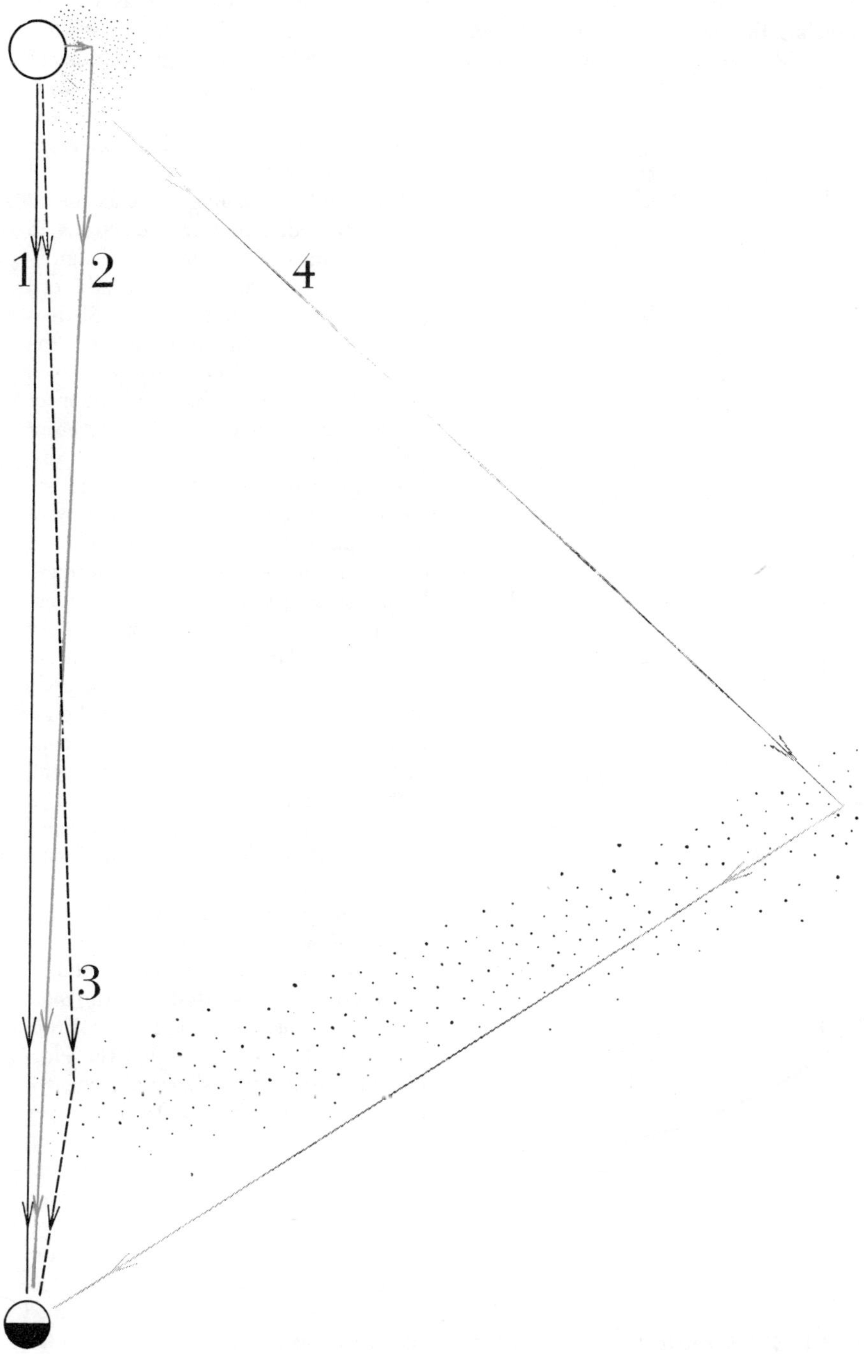

Such a high temperature makes it clear where the electrons in the gas come from: they are torn from atoms in high-speed collisions. Practically every hydrogen atom is stripped of its single electron, and some iron atoms lose as many as 13 electrons. These iron atoms can be detected spectroscopically because of their characteristic emission lines, but the stripped hydrogen atoms (protons), although much more abundant, cannot be detected, because naked atomic nuclei cannot radiate. Electrons thus account for most of the light of the inner corona. From the total brightness of the corona it is possible to calculate the coronal electron-density and so derive an estimate of the total density of the solar atmosphere. Within one degree of the edge of the solar disk the electron density comes to about a million electrons per cubic centimeter, but at greater distances the density falls off rapidly.

This decrease is reflected by a gradual change in the spectrum of the corona. Beyond a quarter of a degree from the solar disk, in the region sometimes called the middle corona, Fraunhofer lines begin to appear. Although the lines are weak, they are just as sharp as those in the solar spectrum. Their presence indicates that not all the light of the corona can be explained by electron scattering; at larger angular distances from the sun most of the corona appears to be due to scattering by larger, slow-moving particles—probably dust—in interplanetary space. This explanation was first suggested by Walter Grotrian of the Potsdam Astrophysical Observatory in 1934, but it immediately encountered a major difficulty. It was originally supposed that the dust involved in the middle corona is mixed with electrons, protons and ions such as compose the inner corona; yet the high temperatures encountered so close to the sun would instantly vaporize any dust particles.

In 1946 the difficulty was resolved almost simultaneously by two astrophysicists, C. W. Allen of the Commonwealth Observatory in Australia and H. C. van de Hulst of the Leiden Observatory, who suggested that the sunlight in this region is not scattered by reflection but by diffraction. The angle at which the light is diffracted makes the dust particles appear to be closer to the

1 2 4 3

(*large dots*) near the earth, thus increasing the size of the solar corona (*top center*). The diagram at right shows how sunlight is reflected (4) by dust particles farther from the earth to give rise to the outermost or zodiacal-light region of the corona (*top right*).

sun than they actually are [*see illustration on preceding two pages*]. Allen and van de Hulst postulated that the dust cloud forming the middle corona is situated far enough from the sun to permit the particles to remain intact. The middle corona thus arises in much the same way as does the halo that sometimes encircles the moon or the sun when they are seen through a thin layer of cloud. The only difference is that in one case the particles are in interplanetary space, while in the other they are in the earth's atmosphere.

By measuring the brightness of the middle corona it is possible to calculate the size of the dust particles and their distribution in space. On the assumption that the particles are the same size, their diameter turns out to be about one micron (.001 millimeter) and their density about one particle per cubic kilometer. A more refined calculation, which allows for particle sizes ranging from one to 300 microns, gives a somewhat greater number of particles per cubic kilometer. But the number is still so small that a space vehicle six feet in diameter traveling through the region would collide with perhaps one particle every 200,000 miles. Such thinly distributed particles produce a perceptible scattering effect only because of the great length of the optical path from the sun to the earth.

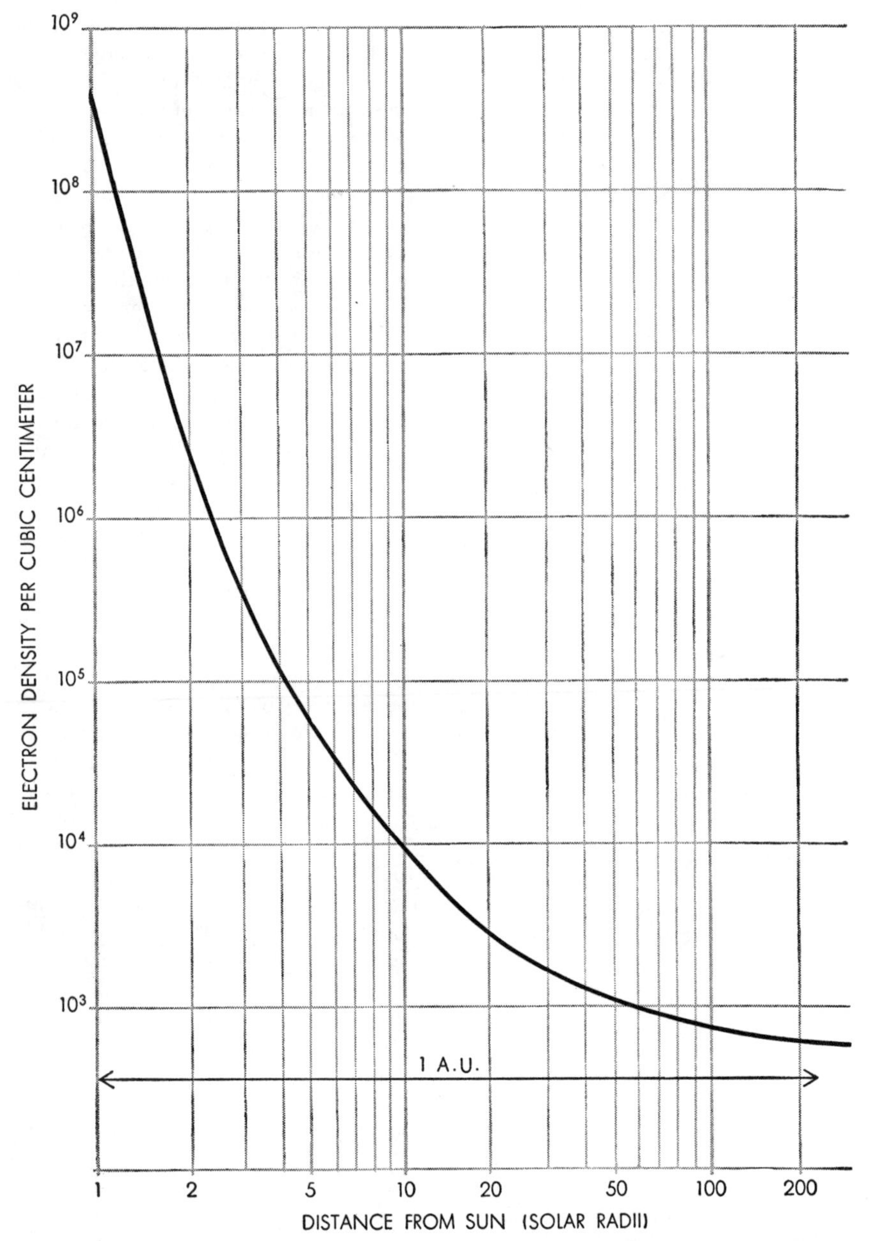

ELECTRON DENSITY of interplanetary space decreases with increasing distance from the sun. Arrow indicates one astronomical unit, the distance from earth to the sun. This curve is deduced from polarization of zodiacal light and from data on "whistlers." Author and M. F. Ingham found density at one astronomical unit lower than value shown here.

Thus free electrons close to the sun and interplanetary dust farther away from the sun both contribute to the scattering of sunlight in the corona. Moving outward to the zodiacal light, the question of what causes it may now be rephrased. Do electrons as well as dust particles contribute to the scattering of sunlight in this region of space? Or, to put the question still another way, can the study of the zodiacal light help determine how far the atmosphere of the sun reaches into space?

The most compelling evidence in favor of the idea that free electrons in interplanetary space contribute significantly to the zodiacal light was adduced by the German astronomer H. Siedentopf in 1953. From a station on the Jungfrau, 11,000 feet above sea level, he made a study of the polarization of the light and found that the degree of polarization increases with angular distance from the sun. According to his measurements, it reaches a maximum of about 22 per cent approximately 35 degrees from the sun, and then gradually decreases. In other words, if it were possible to view the zodiacal light through a Polaroid screen, the apparent brightness of the light would change over the ratio of 1.22 to 1 with the rotation of the screen. (It would change in the ratio of 1.5 to 1, or 50 per cent, if the experiment were performed in an artificial satellite, where the zodiacal light would not be diluted by night-sky radiation.)

Siedentopf believed that such a high degree of polarization could not be attributed to dust scattering alone, and he proposed electron scattering as the most likely mechanism to account for it. From his results he calculated that the electron density at the distance of the earth from the sun is roughly 600 electrons per cubic centimeter. This value is unexpectedly high, although an electron density of this order of magnitude had been predicted seven years earlier by Fred Hoyle, R. A. Lyttleton and Hermann Bondi of the University of Cambridge. Their "solar accretion" theory explained the corona as an effect incidental to the accretion of mass by the sun; the solar gravitational field was supposed to be drawing in matter from interstellar space. In line with this idea, the gas would gradually increase in density clos-

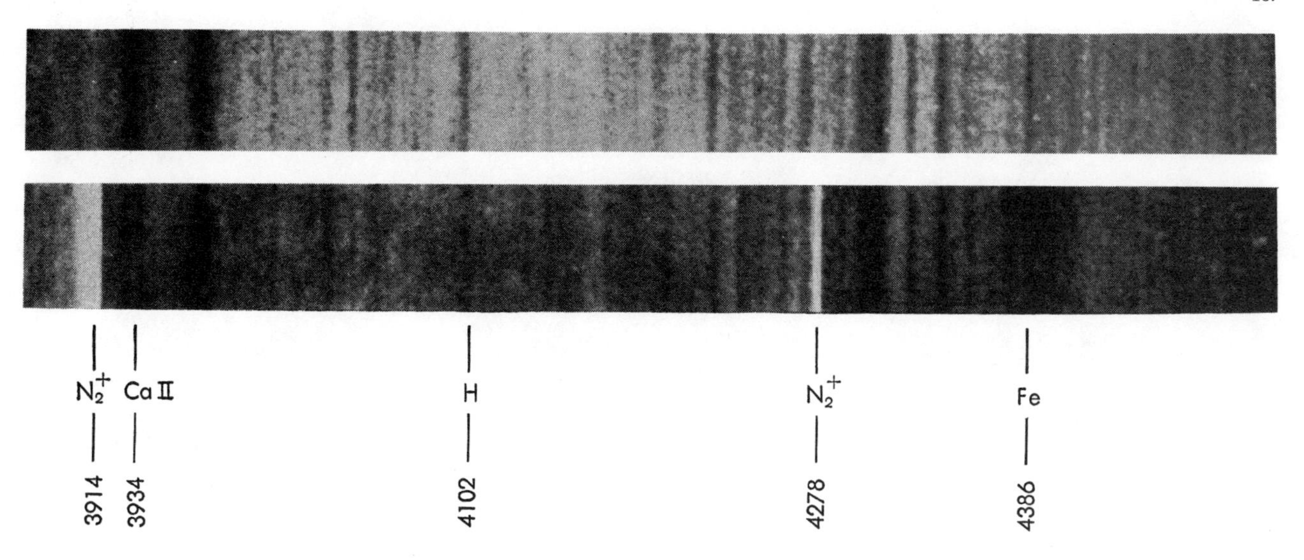

SPECTRUM OF ZODIACAL LIGHT (*bottom*) **resembles the absorption spectrum of the sun** (*top*) **rather than the continuous spectrum of the solar corona** (*shown at top of pages 162 and 163*). **The labels indicate the wavelengths of the spectral lines (in angstroms) and the elements that emit them: hydrogen (H), iron (Fe), ionized molecular nitrogen (N_2^+), and calcium with one electron removed (Ca II). The bright emission lines that appear in the spectrum of zodiacal light are due to night-sky radiation.**

er to the sun, and in the vicinity of the earth's orbit it would have a density of approximately 1,000 electrons per cubic centimeter.

There was soon an apparent confirmation of this value from still another unexpected source. In 1953 L. R. O. Storey of the Cavendish Laboratory Radio Group at the University of Cambridge was investigating a curious radio phenomenon called whistling atmospherics, or whistlers; he showed that these whistling noises originate in lightning discharges in the hemisphere of the earth opposite the hemisphere in which they are detected. The disturbance caused by the lightning stroke is propagated along a line of force in the earth's magnetic field that arches from hemisphere to hemisphere. In the course of propagation the original "white" radio noise of the stroke is sorted out according to frequency: the higher frequencies arrive first, and the noise is heard as a whistle of descending pitch [see "Whistlers," by L. R. O. Storey; SCIENTIFIC AMERICAN, January, 1956]. Storey found that the particular sorting of frequencies depends on the number of electrons along the path of the disturbance, and thus provides a measure of the electron density in the vicinity of the earth. As a result he was able to estimate that the electron density at a distance of 1.7 earth radii (about 7,000 miles) is about 400 electrons per cubic centimeter. If the electron density at this distance from the earth can be taken as representative of interplanetary space, then the agreement with Siedentopf's conclusions is excellent.

It is remarkable that the three lines of argument developed independently by Siedentopf, Storey, and Hoyle and his colleagues all indicate the presence of an interplanetary gas, and that all give approximately the same value (within an order of magnitude) for the electron density at the distance of the earth from the sun. But there are also powerful arguments against the existence of this gas. One of these is especially persuasive: Since the plane of symmetry of the zodiacal light is close to the plane of Jupiter's orbit, it follows that if the light is scattered by an electron gas, the gas must also lie in this plane. But this is difficult to believe, because the distribution of a gas composed of charged particles would be controlled by the magnetic fields that are known to exist in the solar system. These fields would distribute the particles in a sphere around the sun and not in the plane of Jupiter's orbit.

M. F. Ingham and I, working at the University of Cambridge, found such arguments so convincing that we decided to carry out a detailed study of the spectrum of the zodiacal light to determine whether any of it could be ascribed to electron scattering. Since the experiment required the best possible viewing conditions, we undertook an expedition to the cosmic-ray station of Chacaltaya in the Bolivian Andes. Here the air is so transparent that the zodiacal light can be traced to the zenith. Our method was similar to that used in the studies of the inner and middle corona. We reasoned that if the zodiacal light is caused by dust scattering alone, its spectrum would be identical with that of the solar disk, but if it is caused by electron scattering it would show a continuous spectrum. Previously Sydney Chapman had shown that even at the distance of the earth from the sun the temperature of an electron gas should be high enough to scatter light in a continuous spectrum. By comparing the spectrum of the solar disk with that of the zodiacal light, we hoped to settle the question of which mechanism is chiefly responsible for the zodiacal light.

With the help of the most sensitive photographic plates, a very fast spectrograph and exposure times up to seven hours we obtained several successful spectrograms. Since the zodiacal light can be observed for only an hour each evening, each spectrum required multiple exposures over the course of a week. The spectrograms of the zodiacal light show quite clearly the Fraunhofer lines of the solar spectrum [*see illustration above*].

Although the spectrograms are still being studied, we can already say that the electron density in the vicinity of the earth is far lower than has hitherto been supposed, and that electron scattering can make only a small contribution to the zodiacal light. It seems, therefore, that Cassini is right after all; the zodiacal light is due to dust scattering. By the same token the much-discussed interplanetary gas is too thin to be of significance.

BIBLIOGRAPHIES

I BASIC CONSIDERATIONS

1. Water

WATER ISN'T H_2O. A. M. Buswell in *Journal of the American Water Works Association,* Vol. 30, No. 9, pages 1433–1441; September, 1938.

2. Ice

THE ELECTRICAL PROPERTIES OF ICE. L. Onsager and M. Dupuis in *Electrolytes,* edited by B. Pesce. Pergamon Press, 1962.

LATTICE STATISTICS OF HYDROGEN-BONDED CRYSTALS: I, THE RESIDUAL ENTROPY OF ICE. J. F. Nagle in *The Journal of Mathematical Physics,* Vol. 7, No. 8, pages 1484–1491; August, 1966.

MECHANISM FOR SELF-DIFFUSION IN ICE. L. Onsager and L. K. Runnels in *Proceedings of the National Academy of Sciences,* Vol. 50, No. 2, pages 208–210; August, 1963.

RESIDUAL ENTROPY OF ICE. E. A. DiMarzio and F. H. Stillinger, Jr., in *The Journal of Chemical Physics,* Vol. 40, No. 6, pages 1577–1581; March, 1964.

SELF-DISSOCIATION AND PROTONIC CHARGE TRANSPORT IN WATER AND ICE. M. Eigen and L. De Maeyer in *Proceedings of the Royal Society,* Series A, Vol. 247, No. 1251, pages 505–533; October 21, 1958.

3. Snow Crystals

SNOW CRYSTALS: NATURAL AND ARTIFICAL. Ukichiro Nakaya. Harvard University Press, 1954.

SNOW CRYSTALS. W. A. Bentley and W. J. Humphreys. Dover Publications, Inc., 1962.

THE SIX-CORNERED SNOWFLAKE. Johannes Kepler. Oxford University Press, 1966.

4. The Growth of Snow Crystals

THE GROWTH OF ICE CRYSTALS FROM THE VAPOUR AND THE MELT. B. J. Mason in *Advances in Physics,* Vol. 7, No. 26, pages 235–253; April, 1958.

ICE-NUCLEATING PROPERTIES OF SOME NATURAL MINERAL DUSTS. B. J. Mason and J. Maybank in *Quarterly Journal of the Royal Meteorological Society,* Vo. 84, No. 361, pages 235–241; July, 1958.

THE INFLUENCE OF TEMPERATURE AND SUPERSATURATION OF ICE CRYSTALS GROWN FROM A VAPOR. J. Hallett and B. J. Mason in *Proceedings of the Royal Society,* Series A, Vol. 247, No. 1,251, pages 440–453; October 21, 1958.

THE PHYSICS OF CLOUDS. B. J. Mason. Oxford University Press, 1957.

THE SUPERCOOLING AND NUCLEATION OF WATER. B. J. Mason in *Advances in Physics,* Vol. 7, No. 26, pages 221–234; April, 1958.

5. Hailstones

THE FALLING BEHAVIOR OF HAILSTONES. Charles A. Knight and Nancy C. Knight in *Journal of the Atmospheric Sciences,* Vol. 27, No. 4, pages 672–681; July, 1970.

HAILSTONE EMBRYOS. Charles A. Knight and Nancy C. Knight in *Journal of the Atmospheric Sciences,* Vol. 27, No. 4, pages 659–666; July, 1970.

LOBE STRUCTURES OF HAILSTONES. Charles A. Knight and Nancy C. Knight in *Journal of the Atmospheric Sciences,* Vol. 27, No. 4, pages 667–671; July, 1970.

6. Fog

FOG. Joseph J. George in *Compendium of Meteorology,* edited by Thomas F. Malone. American Meteorological Society, 1951.

WEATHER ANALYSIS AND FORECASTING, VOL. II: WEATHER AND WEATHER SYSTEMS. Sverre Petterssen. McGraw-Hill Book Company, Inc., 1956.

CLOUD PHYSICS AND CLOUD SEEDING. Louis J. Battan. Anchor Books, Doubleday & Company, Inc., 1962.

THE UNCLEAN SKY: A METEOROLOGIST LOOKS AT AIR POLLUTION. Louis J. Battan. Anchor Books, Doubleday & Company, Inc., 1966.

AIR POLLUTION. 1968. R. S. Scorer. Pergamon Press, 1968.

II ATMOSPHERIC PHENOMENA

7. The Theory of the Rainbow

LIGHT SCATTERING BY SMALL PARTICLES. H. C. Van de Hulst. John Wiley & Sons, 1957.

LIGHT AND COLOUR IN THE OPEN AIR. M. Minnaert, translated by H. M. Kremer-Priest and revised by K. E. Brian Jay. G. Bell and Sons Ltd., 1959.

THE RAINBOW: FROM MYTH TO MATHEMATICS. Carl B. Boyer. Thomas Yoseloff, 1959.

INTRODUCTION TO METEOROLOGICAL OPTICS. R. A. R. Tricker. American Elsevier Publishing Co. Inc., 1970.

THEORY OF THE RAINBOW. V. Khare and H. M. Nussenzveig in *Physical Review Letters,* Vol. 33, No. 16, pages 976–980; October 14, 1974.

8. The Glory

A STUDY OF THE BACKSCATTER OF LIGHT FROM A SINGLE WATER DROPLET (dissertation). T. S. Fahlen. The University of New Mexico, 1967.

NUCLEAR GLORY SCATTERING. H. C. Bryant and N. Jarmie in *Annals of Physics,* Vol. 47, pages 127–140; 1968.

RAINBOWS AND GLORIES IN MOLECULAR SCATTERING. E. A. Mason, R. J. Munn and F. J. Smith in *Endeavor,* Vol. 30, page 91; May, 1971.

9. Mirages

METEOROLOGISCHE OPTIK. J. M. Perntner. Wilhelm Braumüller, 1902.

THEOLOGICAL OPTICS. Alistair B. Fraser in *Applied Optics,* Vol.R. 14, No. 4, pages A92–A93; April, 1975.

10. The Green Flash

FACTORS WHICH DETERMINE THE OCCURRENCE OF THE GREEN RAY. R. W. Wood in *Nature,* Vol. 121, No. 3,048, page 501; March 31, 1928.

FURTHER EXPERIMENTS IN ILLUSTRATION OF THE GREEN FLASH AT SUNSET. Lord Rayleigh in *The Proceedings of The Physical Society,* Vol. 46, Part 4, No. 255, pages 487–498; July 1, 1934.

THE GREEN FLASH AT SUNSET AND AT SUNRISE. T. S. Jacobsen in *Sky and Telescope,* Vol. XII, No. 9, pages 233–235; July, 1953.

THE GREEN FLASH, AND OTHER LOW SUN PHENOMENA. D. J. K. O'Connell, S. J. Interscience Publishers, 1958.

THE GREEN RAY. M. Minnaert in *Light and Colour in the Open Air,* pages 58–63. G. Bell and Sons Ltd., 1940.

11. The Mechanism of Lightning

PHYSICS OF THE AIR. W. J. Humphreys. McGraw-Hill, 1929.

THE MECHANISM OF THE ELECTRIC SPARK. L. B. Loeb and J. M. Meek. Stanford University Press, 1941.

FUNDAMENTAL PROCESSES OF ELECTRICAL DISCHARGE IN GASES. L. B. Loeb. John Wiley and Sons, 1939.

LIGHTNING. M. A. Uman. McGraw-Hill, 1969.

12. Ball Lightning

BALL LIGHTNING: A COLLECTION OF SOVIET RESEARCH IN ENGLISH TRANSLATION WITH AN ANNOTATED BIBLIOGRAPHY. Edited and compiled by Donald J. Ritchie. Consultants Bureau, 1961.

BALL LIGHTNING AS A PHYSICAL PHENOMENON. E. L. Hill in *Journal of Geophysical Research,* Vol. 65, No. 7, pages 1947–1952; July, 1960.

DER KUGELBLITZ. Walther Brand in *Probleme der Komischen Physik,* Vol. II/III; 1923.

THE FLIGHT OF THE THUNDERBOLTS. B. F. J. Schonland. Oxford University Press, 1950.

THE NATURE OF BALL LIGHTNING. P. L. Kapitsa in *Ball Lightning.* Edited and compiled by Donald J. Ritchie. Consultants Bureau, 1961.

THE NATURE OF BALL LIGHTNING. S. Singer. Plenum Press, 1971.

13. Thunder

LIGHTNING CHANNEL RECONSTRUCTION FROM THUNDER MEASUREMENTS. A. A. Few in *Journal of Geophysical Research,* Vol. 75, No. 36, pages 7517–7523; December 20, 1970.

THUNDER SIGNATURES. A. A. Few in *E ⊕ S Transactions of the American Geophysical Union,* Vol. 55, No. 5, pages 508–514; May, 1974.

HORIZONTAL LIGHTNING. Thomas L. Teer and A. A. Few in *Journal of Geophysical Research,* Vol. 79, No. 24, pages 3436–3441; August 20, 1974.

14. Atmospheric Halos

LIGHT AND COLOUR IN THE OPEN AIR. M. Minnaert, translated by H. M. Kremer-Priest. G. Bell and Sons, Ltd., 1959

FIELD GUIDE TO SNOW CRYSTALS. Edward R. LaChapelle. University of Washington Press, 1969.

INTRODUCTION TO METEOROLOGICAL OPTICS. R.A.R. Tricker. Elsevier North-Holland, 1971.

15. Noctilucent Clouds

COLLECTION AND ANALYSIS OF PARTICLES FROM THE MESOPAUSE. G. Witt, C. L. Hemenway and R. K. Soberman in *Proceedings of the Fourth International Space Science Symposium, Warsaw, Poland, June, 1963.* North Holland Publishing Co.

HEIGHT, STRUCTURE AND DISPLACEMENT OF NOCTILUCENT CLOUDS. Georg Witt in *Tellus,* Vol. 14, No. 1, pages 1–18; February, 1962.

NOCTILUCENT CLOUD OBSERVATIONS. Georg Witt in *Tellus,* Vol. 9, No. 3, pages 365–371; August, 1957.

NOCTILUCENT CLOUDS. F. H. Ludlam in *Tellus,* Vol. 9, No. 3, pages 341–364; August, 1957.

NOTE ON THE NATURE OF NOCTILUCENT CLOUDS. Eigil Hesstvedt in *Journal of Geophysical Research,* Vol. 66, No. 6, pages 1985–1987; June, 1961.

STUDIES OF MICROMETEORITES OBTAINED FROM A RECOVERABLE SOUNDING ROCKET. C. L. Hemenway and R. K. Soberman in *The Astronomical Journal,* Vol. 67, No. 5, pages 256–266; June, 1962.

OBSERVATIONS OF NOCTILUSCENT CLOUDS. C. I. Vill mann, Editor. NASA Tech. Translation T T F-546, 1969.

16. The Aurora

THE AURORA. S. Akasofu, S. Chapman and A. B. Meinel in *Handbuch der Physik: Vol. XLIX,* Springer-Verlag, Inc., in press.

AURORAL PHENOMENA. B. J. O'Brien in *Science,* Vol. 148, No. 3669, pages 449–460; April 23, 1965.

DYNAMIC MORPHOLOGY OF AURORAS. Syun-Ichi Akasofu in *Space Science Reviews,* Vol. 4, No. 4, pages 498–540; June, 1965.

POLAR AURORAS. V. I. Krasovskij in *Space Science Reviews,* Vol. 3, No. 2, pages 232–274; 1964.

17. The Airglow

ACTIVE NITROGEN AT HIGH PRESSURE. J. F. Noxon, thesis. Physics Department, Harvard University, 1957.

PHYSICS OF THE AURORA AND AIRGLOW. Joseph W. Chamberlain. Academic Press, 1961.

REACTIONS OF OXYGEN ATOMS. Frederick Kaufman in *Progress in Reaction Kinetics: Volume I.* Edited by G. Porter. Pergamon Press, 1961.

THE SPECTRUM OF ACTIVE NITROGEN. K. P. Bayes, thesis. Chemistry Department, Harvard University, 1959.

18. The Zodiacal Light

OBSERVATIONS FROM AN AIRCRAFT OF THE ZODIACAL LIGHT AT SMALL ELONGATIONS. D. E. Blackwell in *Monthly Notices of the Royal Astronomical Society,* Vol. 116, No. 4, pages 365–379; 1956.

OBSERVATIONS OF ZODIACAL LIGHT. F. E. Roach, Helen B. Pettit, E. Tandberg-Hanssen and Dorothy N. Davis in *The Astrophysical Journal,* Vol. 119, No. 1, pages 253–273; January, 1954.

THE SPECTRUM OF THE CORONA AT THE ECLIPSE OF 1940 OCTOBER 1. C. W. Allen in *Monthly Notices of the Royal Astronomical Society,* Vol. 106, No. 2, pages 137–153; 1946.

ZODIACAL LIGHT IN THE SOLAR CORONA. H. C. van de Hulst in *The Astrophysical Journal,* Vol. 105, No. 3, pages 471–488; May, 1947.

THE ZODIACAL LIGHT AND INTERPLANETARY MEDIUM. J. L. Weinberg, Editor. NASA SP-150, 1967.

INDEX